ARCHITECTURE, ENERGY, MATTER

STUDIO AS BOOK
NO. 01

Studio as Book is a new series of yearly publications that tender the extraordinary creative work undertaken in the Department of Architecture's design studios – in detail. The series includes undergraduate and graduate level work, and is intended to sit alongside the Open Exhibition and catalogue. Each book in the series covers the work of a single design studio over the course of at least two years. Its objectives are:

- To record, archive, and present the pedagogical programme and creative student outputs of a design studio
- To position the work of a design studio within a broader intellectual, scientific or aesthetic field
- To advance the design driven research being undertaken in the Department's design studios
- To provide a reference for future iterations and variations of a design studio

Reducing the creative output of a multi-year design studio to a single volume, using a pre-designed book template is no easy undertaking, and it is necessarily selective. At the same time, it provides a consistent, sure platform for the wide range of approaches to the discipline of teaching architectural design which characterise the department.

Each *Studio as Book* has been peer-reviewed on the basis of a proposal submitted by the studio's tutors to an editorial committee. In addition to studio briefs and student work, each book includes content that draws out the studio's research and pedagogical agenda. The format that this takes varies from book to book – reflective essays by tutors or past students, interviews, theoretical essays from parallel fields, and so forth. The *Studio as Book* Series will later be accompanied by a Studio Pamphlet Series for design studios of a shorter duration.

I wish to acknowledge the contribution of the following in bringing this project to fruition: Lindsay Bremner, Director of Architectural Research, who was the driving force behind the series, Mark Boyce, author of *Sizes May Vary, A workbook for graphic design* (Lawrence King, 2008) – and the designer of *Studio as Book*, and Filip Visnjic, designer of the series' web site, http://www.studioasbook.org.

Harry Charrington
Head of Department of Architecture
University of Westminster

ARCHITECTURE, ENERGY, MATTER

DS18: 2013-2015
EDITED BY LINDSAY BREMNER + ROBERTO BOTTAZZI

STUDIO AS BOOK
NO. 01

DEPARTMENT OF ARCHITECTURE
UNIVERSITY OF WESTMINSTER

CONTENTS

INTRODUCTION

LINDSAY BREMNER AND ROBERTO BOTTAZZI

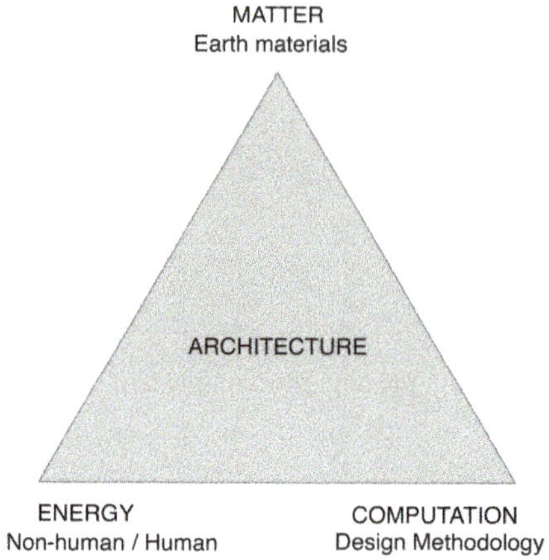

What we have then, is a kind of 'wisdom of the rocks,' a way of listening to a creative, expressive flow of matter for guidance on how to work with our own organic strata. (01)

This edition of *Studio as Book* comprises a collection of essays and an edited selection of the work produced by Design Studio 18 (DS18) tutored by Lindsay Bremner and Roberto Bottazzi at the University of Westminster, 2013-2015. The aim of the studio over this period was to approach problems of energy, energy infrastructure and resource extraction as architectural questions i.e. as political, cultural and aesthetic problems, as much as technological ones. Computational tools were used to simulate material processes and to enlist, visualise and enliven data in the service of design. The essays included in the book are by critics or interested colleagues who interacted with the work of the studio, and locate its concerns in wider intellectual debates about big data, computational design, the politics of resource extraction, architecture, energy and urbanism.

Between 2013-15, DS18 was interested in pursuing intersections between architecture, urbanism and geology, framed by debates taking place in science and the humanities about a new era of geological time, the Anthropocene. According to theorists like Bruno Latour, Timothy Morton and others, the discovery of the Anthropocene, first proposed by chemist Paul Curzon in 2002 (02) has transformed much of what we have previously assumed to be true: science has become natural history, our ideas of scale no longer apply and it is no longer possible to behave as if nature were one thing and society another. Humans have mobilised earth materials, minerals, water, air, and energy in ways that have altered the earth's oceans, atmosphere, surface morphology, and future stratigraphy to such an extent that the very idea of what geologists think of as geological systems, what oceanographers think of as oceanic systems and what architects think of urban systems or buildings has changed. Dr. Jan Zalasiewicz, geologist at the University of Leicester and chairperson of the International Anthropocene Working Group, goes so far as to suggest that buildings and cities are geology, even if, by geological standards their time scales are catastrophically fast and their structural nature chaotic. (03) What new ways of thinking and making architecture do these arguments open up?

Links between architecture and geology are not new: from Baroque grottos, to John Ruskin's and Viollet-le-Duc's Alpine studies or Buckminster Fuller's geological diagrams, architects have long been fascinated by the formal, material and structural properties of geological formations. However, the notion of the Anthropocene takes concerns with the geologic to new levels. The idea that the geological constitution of the planet is markedly anthropogenic is yet to be fully absorbed into our collective social imaginaries. Recent philosophical tendencies, including new materialism, realist ontologies, object orientated ontologies and post-humanist theories try to come to terms with the new reality: they argue that geology, anthropology, nature and culture, animate and inanimate earth systems can no longer be seen as categorically distinct, as they were in the 19th and 20th centuries. The question becomes how architecture engages these shifts and finds new ways of working imaginatively, spatially, temporally and materially with the earth, its processes and systems.

Between 2013-15, DS18 explored new conceptual frame-works, new forms of expression, new processes and new methods for designing and making architecture that acknowledged the deep interconnections between architecture, the earth, its material processes and the late-capitalist conditions (economic, political, cultural) that shape these. These ranged from the obvious and practical, such as the responsible sourcing, use, and re-use of materials, to the radical rethinking of the relationship between human-made and natural worlds. We saw an architecture that engages with the Anthropocene as not only environmentally and socially responsible, but latent with new cultural, intellectual and aesthetic possibilities.

Between 2013-15 then, DS18 was framed by three interrelated ideas:

The first, as argued above, was the idea of the Anthropocene, the idea that the human species has interfered with the earth system to such an extent that new geological conditions have emerged. One of its consequences is that it is no longer possible to distinguish between what used to be called society and what used to be called nature, humans and non-humans. Instead we now live in the anthroposphere, (04) a complex, emergent system of energies, materials and information flows, evolving according to its own non-linear logics in ways we do not control and barely understand.

From this followed the second idea behind the studio - that architecture, infrastructure and urbanism are deeply entangled with the earth system and earth materials. Buildings, infrastructure and cities are geological agents, mobilising earth, air, fire, and water and speeding up geological time. Buildings in fact, are geology, not built up slowly in discreet and recognisable layers and patterns, but muddled, mixed-up accretions of earth materials, some found, some transmogrified by prior industrial processes. How can this intimacy between architecture and the earth be made visible through design? How can design intervene strategically and instrumentally to reshape or redirect flows of geological matter where current systems are failing?

The third idea behind the studio was a methodological one - to defer thinking about architecture, infrastructure and urbanism as form, geometry or typology but rather, to think about them as data, energy and matter. Buildings, infrastructure and urban systems were seen as intensities within material systems of energy and data, mobilising and being mobilised by them in ever more complex ways. The computer enabled us to analyse, simulate and visualise these processes and intervene in them through design.

During 2013/14, DS18 began its investigations by researching hydraulic shale gas fracturing, the process commonly known as fracking. At the time, largely through Josh Fox's movie *Gasland*, (05) awareness of the negative impacts of fracking had just started to surface in the public imaginary. In July 2013, a successful protest against the drilling of a test well in Balcombe, West Sussex, UK, brought the dangers associated with hydraulic fracturing to the front of the public agenda in the UK. This gave the first iteration of DS18 topicality and urgency.

Hydraulic fracturing is a conflict-mobilising business that emerges in complex networks of geology, politics, power, economics, energy, regulation, infrastructure and land use, with a built-in tendency towards friction and failure. It is a process of micro mining that pumps a cocktail of water, sand and an undisclosed number of chemicals down a concrete well shaft, shatters geological strata at least 2 kms beneath the earth's surface and draws methane gas dislodged from the strata by the explosions up to surface. This is then fed into compressor stations and pipes for further distribution. This produces a vast urbanism of sorts, an emergent system of clearings, holes, caps, pipes and roads and other less visible actants, spreading virally and violently across landscapes in ways that are seemingly oblivious to older agricultural, arboreal or urban logics. This operation has transformed remote hinterlands into the city's doppelganger, a toxic other, its template derived not from agrarian, urban or industrial ideals, but from the fracking process itself.

In 2013/14, DS18 investigated the complex networks and operational logics that underpin hydraulic fracturing and the site-specific ecologies and geometries with which it intersects in the USA and UK. We generated ways of visualising these topographies using computational tools. We then researched fracking's inbuilt tendency towards friction and failure (deregulation, corruption, toxic contamination, pollution, loss of wildlife, loss of farmland, loss of income, medical conditions etc.) and students developed individual design agendas to respond to these conditions.

Behind the race towards and opposition against hydraulic fracturing across the globe is the transformative role it has played in reshaping remote landscapes and in altering the brittle geopolitics of global oil and hydrocarbon economies. In 2014/15, a second applied research and design studio on fracking, focused its attention on the contestation over the issuing of fracking licenses in the Karoo, South Africa. The Karoo is a vast, arid interior region of the country where a recent study by U.S. Energy Information Administration had proposed that notable shale gas reserves were to be found, turning attention to the region as a potential alternative source of energy for South Africa. We developed critical perspectives on the South African energy economy and endorsed the role of design in mapping out alternative infrastructural and social scenarios for the country's energy future.

In addition to presenting these two studios and the work produced by students, the book contains writings by a number of scholars developing new understandings of the anthropocene, new materialism, big data, post-humanist theory and how architecture might find new ways of working imaginatively, spatially, temporally and materially with the earth and its systems. David Chandler's 'Big Data: Affect Theory for Architecture in the Anthropocene' explores how computational tools, Big Data and the Internet of Things can be used to enlist, visualise and enliven data in the service of design. This is followed by Roberto Bottazzi's essay 'On Computer Simulations in the Age of Hyperobjects, which lays out the studio's computational agenda and argues for the deployment of computer simulations, particularly fluid dynamic packages, as design tools under the conditions examined by the studio. In keeping with the theme of the studio, Etienne Turpin and Tom Holderness' conversation with Kiel Moe, 'Design for Dissipative Systems,' republished with kind permission of the participants, probes Moe's conceptualisation of architecture and the city as thermodynamic i.e. non-linear systems. Lindsay Bremner's 'The Urban Hyperobject' explores the city through Timothy Morton's notion of the hyperobject, an object massively attenuated in space and time, and reflects on what this implies for design. Two contributions, Nick Axel's 'Bakken Treatise' and Lesley Green's 'Statement opposing the licensing of shale gas exploration in the Karoo' explore ways in which the law sanctions fracking and parliamentary avenues to oppose it in the United States and South Africa. We are enormously grateful to these authors for their contributions to the book and for the rich conceptual field they have opened up alongside the studio's investigations.

Notes

(01) M. DeLanda, 'Nonorganic life', *Incorporations, Zone no. 6*, New York: Zone Books, 1992, p. 143.

(02) P. J. Crutzen, 'The anthropocene', *J.Phys. IV France* vol. 12 no. 10, 2002, pp 1-5.

(03) J. Zalasiewicz, 'Buried Treasure', in G. Manaugh (ed.), *Landscape Futures*, Reno, Nevada, Nevada Museum of Art, 2013, pp. 258-261.

(04) P. Baccini and P. Brunner, *Metabolism of the Anthroposphere*, Cambridge, MA., MIT Press, 2012.

(05) J. Fox, *Gasland*, New Video Group/HBO/International WOW Company, 2010.

BIG DATA, AFFECT THEORY FOR ARCHITECTURE IN THE ANTHROPOCENE

DAVID CHANDLER

The concept of the Anthropocene postulates a new geological epoch defined by overwhelming human influence upon the earth and the biosphere, such that environmental problems such as global warming or extreme weather events can be understood as far from natural processes but as anthropologically driven: as the unintended consequences or feedback loops of the modes of design of our modern everyday existence. Architectural design in the Anthropocene is thus becoming increasingly aware of the limits of modernist framings of nature as fixed or immune from our activities in the world and fundamentally questions forms of knowledge generation that seek to externalise or universalise knowledge or to apply it on the basis of linear causal assumptions. The process of knowing ourselves and our world thereby needs to become more interactive, fluid, multiple, generous and open. This much is clear from climate scientists and from critical theorists working with a variety of approaches and intellectual traditions.

As Lindsay Bremner notes, the Anthropocene forces architectural designers to consider the wider ontological, epistemological, political, methodological and representational questions which the problematisation of the human/environment divide raises. There is seeming agreement upon the vital question facing humanity: How can we develop new ways of designing, compatible with being in a world of multiple feedback loops and complexity? But, as yet, there has been little in the way of conceptual approaches offering a way forward. Key to the answer to this question is the multi-sided and recursive relationship between design and knowledge: awareness of processes of change and emergence and its consequences for design as both the outcome of this process and, crucially, for enabling it. Key to this problem is the question of how computational tools, Big Data and the Internet of Things can be used to enlist, visualise and enliven data in the service of design.

McKenzie Wark's latest book, *Molecular Red: Theory for the Anthropocene*, captures well the need to rethink forms of knowing and acting in ways which are sensitive to the impacts of human being in the world as a complex process of interaction with secondary effects and unintended consequences "beyond any master-thinker or grand plan, beyond the magic of the market or computer modelling." [01] Key to the book's title is the need to move beyond the molar world of representational thinking and fixed entities and causal connections and towards knowledge of a molecular order of "flows, becomings, phase transitions and intensities." [02] Design practices thus have to be reorganised closer to the flux of reality rather than being abstracted from its flows of becoming; knowing has to become more akin to sensing than modernist views of causal understanding. Similar themes have been pursued by many authors in recent years, perhaps foremost amongst them has been William Connolly, who has raised the question of how we might theorise the power of affect "in a world of becoming"; affect working at the subliminal level of bodily senses before being organised into conscious perceptions, feelings and reactions. [03] Affect is here seen as a way in which the human body is sensitive to and absorbs the energy or vitality of its environment as part of the process of interaction with the world at a preconscious level.

Connolly's point, and the one which I wish to expand upon in this short piece, is the role of architectural design in enhancing the power of affect to enable a greater sensitivity to the multiple possibilities of the virtual, the emergent, the world of becoming. The key aspect is that this sensitivity of affect is not so much attuned to prediction or seeing the future but rather, in Gilles Deleuze's phase, the capacity to "grasp it at the time", the cultivation of a sensitivity of our being in the world. [04] Unlike the focus of many theorists concerned with complexity and emergent causality (seeking preventive or predictive uses of technological innovation), the problem here is seeing or being sensitive to what already exists but cannot be easily grasped by representative and conscious forms of thought (trapped in cultural and ideological modes of thinking informed by past habits and experiences). Affect theory for architectural design in the Anthropocene would thereby involve a consideration of how the ability to be sensitive could be enhanced to enable better forms of knowing and acting in the world in which both human and non-human agencies interact with uncertain effects.

Affect theory, particularly as expanded in the work of Brian Massumi, Nigel Thrift, Patricia Clough and others, has been vital in opening up the understanding of bodies as mediated through the experience of the world rather than as autonomous entities mediated only through culture and representation. These approaches have fundamentally challenged modernist binary constructions of the material and the immaterial, the integrity of the individual and the externality of its environment, the living and the non-living, and constructions of the natural and cultural. As the editors of *The Affect Theory Reader* note, affect is an inter-relational capacity: in this field of sensory mediation between the body and its environment: "lie the real powers of affect, affect as potential: a body's capacity to affect and to be affected." [05] Other authors, such as Bruno Latour, also focus on the body as an "interface" through which it is possible to "learn to be affected', to be "moved by" and to be "sensitive to" more and more of the human and non-human processes we are embedded within. [06]

As Clough has famously pointed out, however, affect theory often focuses on the capacity of the human body to be open to the world in embodied, experiential ways while neglecting the ways in which affect enables us to understand matter itself as more interactive, communicative and lively; often giving affect theory a problematic subject-centeredness which seems to limit the development of these approaches to human emotional capacities and the potential of non-representational forms of responsiveness. [07] Clough, in following Hansen, focuses on how technology depends upon and enhances the human body's capacity for indeterminate and contingent responses to affect and to be affected. Humans thus are becoming transformed through technology into being more sensitised to the world. However, Clough as much as those preceding her, focuses on the powers of affect as inherent capacities of the body or of matter itself: the focus is upon "the human body's experience of technology generally and the specific importance of bodily experience to digital technology." [08] This is unfortunately typical of biologised understandings of information and communicative system interactions, which were heavily influential in affect theory in its development through work inspired by Silvan Tomkins' psychobiology of bodily drives and Deleuzian takes on Spinozan vitalism.

However, the individual human body as interface with the world would leave little room for the wide scale transformations in design and social organisation necessary to address the global feedback loops of the Anthropocene. As Wark argues, any new forms of knowledge need to work up to multiple levels to capture the complex interactions of global humanity. If affect theory is to be developed for the Anthropocene it needs to go beyond the critical positions focused on understandings of how technology and science relate to embodied experience and consider how affect (as a relation of being in-between and beside others) can be artificially enhanced to enable the collectives of humanity to be sensitised to their environments at every level. Rather than focusing on the body as the productive and disruptive interface of affect, affect could be articulated in terms of a wider network of human/non-human assemblages of mediation between the body and its environment.

The 'interface' would then no longer be the body itself but the relations through which the human access to the world is extended and transformed through digital and technological means, which have much less to do with either psychobiology or vitalist views of the interchange and communicative capacities of matter or of life per se. The interface understood not as the (digitalised) body itself but as multiple digitalised assemblages sensing both humans and the non-human environment would enable affect to become an important aspect of knowledge production without the subject-centeredness of either early affect theory's focus on the biological underpinnings of conscious thoughts and feelings or the modernist concern with cognitive processes and with representational forms of being. Thus, affect theory could become a project of future-orientated forms of design for interactive being and becoming.

The discussion of the rise of affect in the form of sentient objects or 'sense-ables' offers a way forward as an example of artificially enhanced assemblages of mediation, where the body itself can no longer be described as the interface with the digital or the environment. As Ravi Sawney argues: "Senseables are, in essence, devices that enable the merging of self, the world, and data about one's place in that world." (09) Whereas wearable technology – such as health-trackers, watches, glasses and clothing with biosensors - provide little information about the self, and become more like a 'glorified notebook,' sensors embedded throughout the environment can technologically mediate human interaction within society and the environment without the need for cognitive decision-making. Here, the devices – or, in fact, the ever growing network of digitalised technology, the Internet of Things – become the sensing mediation between the quasi-human and the quasi-environment. The interface between the body and the external world can neither be grasped as a momentary fragment of time before conscious thought kicks in nor as a micro-level piece of tissue or nerve-ending but rather is an enormous technological network of sensors and computational power.

Of course, shaped by the concerns of neoliberal corporate capitalism, sense-ables - sensing cities, buildings, homes, furniture, cars and other everyday appliances - are concerned with how our world can become more time-saving and efficient at the same time as providing new products and markets for capitalism in crisis. For Sawney, it is market productivity and bodily efficiency which will drive these technological transformations: "When distracting quotidian decisions are removed thanks to sensors, think about how our minds will be freed up, fewer petty worries and concerns will put us on the track to becoming healthier and happier people." (10) However, the real time feedback that enables sensor-driven technological assemblages to adapt to human moods or being (location in a house for heating and light use or in traffic streams for adaptive approaches to congestion problems), indicates that affect - less mediated forms of interaction based on real-time embodiment - actually enable greater interactive sensitivities to both human, societal and environmental needs. This less mediated form of real-time responsivity is obviously - as Latour has long argued - premised on the greater strings of mediated attachments of human and non-human assemblages of sensors, screens, code etc.

Sense-ables work the other way too, as authors like Noortje Marres and Annika Skoglund have drawn out, everyday environments can be digitalised or datafied to enable human users to be sensitised to broader economic, social and environmental needs. (11) Thus, affect, understood as the capacity of human and nonhuman assemblages to be sensitive to our embeddedness in the world, need no longer be seen as separate to rather than as a transformation of modernist ways of knowing and being. The Anthropocene is doubly a product of modernist science and technology, in both the production of its conjunction of both human and non-human futures and, importantly, in the scientific discovery of its impacts and the urgency of new forms of organising knowledge and designing social practices. It has been the natural scientists as much as the critical theorists who have brought the importance of affective sensitivities rather than cognitive responses and decisions to the forefront of social and political concern (as noted by Latour). (12)

Affect theory seems to underlie the fascination with Big Data approaches as a way of generating increasingly sensitive real-time responses to problems in their emergence. Here modernist forms of knowledge, involving chains of causal understanding, and the modernist binaries and separations involved in governance as decision making from above appear outmoded: problems, from environmental degradation to humanitarian crises, are increasingly reinterpreted as emergent processes which need to be sensed and responded to through ongoing forms of localised knowing and agency. (13) Big Data, as a methodology for the Anthropocene, has become central to policy and academic discussion of urban governance and urban planning: in discourses of 'smart', 'intelligent', 'resilient' or 'sentient' cities. (14) The increasing focus on cities that understand themselves and thereby govern themselves is driven by the technological possibilities of Big Data, where cities are understood as industrial and social hubs of complex interconnections, which through datafication can produce real-time knowledge of themselves. This reflexive awareness of cities' own 'vitality' – their own 'pulse' – then enables a second order of reflexivity or of artificial intelligent 'life': "Perhaps one way in which we might consider this question is precisely through looking at how vitality develops when computational things are explicitly included in the contours of experience. Then it becomes clear that it has only gradually arisen, line by line, algorithm by algorithm, programme by programme. Cities are full of a whole new layer of emergent entities which, because they are underpinned by code using data as fuel, might be thought of as akin to sentient beings, in that they are able to produce some level of transference through correlation and measurement." (15)

The view of Big Data as empowering and capacity-building relies upon the reconstruction of society as self-governing, as self-reproducing or autopoietic. (16) However, this approach to self-government appears to be very different to modernist approaches of top-down governance, based on cause-and-effect understandings of policy interventions. In this framework, in which Big Data methodologies and understandings are central, the power of self-governance and autonomy does not stem from a development of liberal forms of power and knowledge but from their rejection. 'Smart', 'resilient' or 'sentient' cities, for example, are not successful because of a development of cause-and-effect understandings, which can then be operated upon by centralised authorities. The 'conscious' or 'cognitive' self-awareness of the 'sentient' city is understood to be very different from that of human cognition or self-awareness.

Here, Big Data materially changes the way the world is and how it is understood and governed. For Thrift, new technologies 'make this kind of relationality easier to initiate and conjugate', they are enfolded within emerging processes and essentially turn abstract constructions of relational ontologies into a perceivable social reality. (17) Bruno Latour's work points in a similar direction, where he suggests that Big Data enables access to a much 'flatter' reality, where the modernist divisions between quantitative and qualitative methods no longer needs to apply and that the 'statistical shortcuts' that constituted the 'fictive division' between the two levels of micro-interactions and macro-structures are no longer necessary. (18) This two-level or dualist approach, which has traditionally dominated social theorising, worked well, according to Latour, to describe emerged phenomena but not for grasping phenomena in their emergence, in real-time. The need for abstractions at the higher level of the 'general', 'collective' or the 'social' disappear as the real-time interactions and connections can be assembled to enable the study of the concrete and the individual to encompass ever larger collectivities or assemblages (both human and non-human). (19)

If we are to accept the need for a new theory of design for the generation of knowledge and the reorganisation of social practices in the epoch of the Anthropocene then the starting point cannot be the body as interface but the technological extension of our sensory boundaries not merely to the human collective (the experiential viewpoint of labour, advocated by Wark) but also to the extension of sensory experience to the human and non-human assemblages of mediation which begin to transform the thin and fleeting interfaces of affect theory into interfaces which can increasingly blur and disassemble modernist categories and divides and enable new forms of transformative agency. Big Data approaches with their focus on the importance of molecular interfaces and mediations rather than grand schemas of causal knowledge and molar forms of representation would seem an ideal starting point for an understanding of the limits and possibilities of these new forms of sensory knowledge production.

Notes

(01) M. Wark, *Molecular Red: Theory for the Anthropocene*, London, Verso, 2015, p. xv.

(02) M. Wark, 2015, p. xvi.

(03) W. Connolly, *A World of Becoming*, London, Duke University Press, 2011, p. 150.

(04) W. Connolly, 2011, p. 158.

(05) G. J. Seigworth and M. Gregg, 'An Inventory of Shivers', in M. Gregg and G. J. Seigworth (eds.), *The Affect Theory Reader*, London, Duke University Press, 2010, p. 2.

(06) B. Latour, 'How to Talk About the Body? The Normative Dimension of Science Studies', *Body & Society*, vol. 10, no. 2-3, 2004, 205-206.

(07) P. Clough, 'The Affective Turn: Political Economy, Biomedia and Bodies', *Theory, Culture & Society*, vol. 25, no. 1, 2008, pp. 1-22.

(08) P. Clough, 2008, p. 6.

(09) R. Sawney, 'Note to Designers: Forget Wearables, Tackle Senseables', *FastCoDesign*, no. 22, April 2015, http://www.fastcodesign.com/3045304/note-to-designers-forget-wearables-tackle-senseables (accessed 1 July 2015).

(10) R. Sawney, 2015.

(11) N. Marres, *Material Participation: Technology, the Environment and Everyday Politics*, Basingstoke, Palgrave Macmillan, 2012; A. Skoglund, 'Homo Clima: the overdeveloped resilience facilitator', *Resilience: International Policies, Practices and Discourses*, vol. 2, no. 3, 2014, pp. 151-167.

(12) B. Latour, *Facing Gaia: Six Lectures on the Political Theology of*

Nature, The Gifford Lectures 2013*, pp. 77-8, https://docs.google.com/file/d/0BxeTjgod3jSSSXZHTU9Yb3FIYms/edit (accessed 3 July 2015).

(13) See for example V. Mayer-Schönberger and K. Cukier, *Big Data: A Revolution that will Transform how we Live, Work and Think*, London, John Murray, 2013); R. Kitchin, *The Data Revolution: Big Data, Open Data, Data Infrastructures & their Consequences*, London, Sage, 2014.

(14) N. Thrift, 'The "Sentient" City and What it May Portend', *Big Data & Society*, vol. 1 no. 1, 2014, p. 8.

(15) N. Thrift, 2014, p. 10.

(16) See D. Chandler, 'A World without Causation: Big Data and the Coming of Age of Posthumanism', *Millennium: Journal of International Studies*, vol. 43, no. 3, 2015, pp. 833-851.

(17) N. Thrift, 2014, p. 7.

(18) T. Venturini and B. Latour, 'The Social Fabric: Digital Traces and Quali-quantitative Methods', in E. Chardronnet (ed.). Proceedings of *Future En Seine 2009: The Digital Future of the City, Paris*, Cap Digital, 2010, pp. 87-101.

(19) See for example B. Latour, P. Jensen, T. Venturini, S. Grauwin and D. Boullier, '"The Whole is Always Smaller than Its Parts" – a Digital Test of Gabriel Tardes' Monads', *British Journal of Sociology*, vol. 63, no. 4, 2012, pp. 590-615.

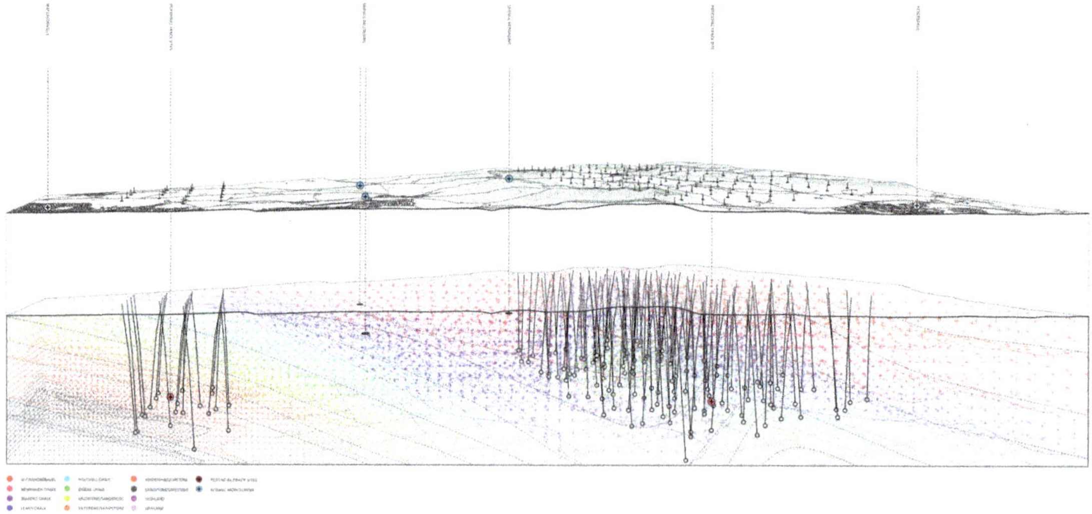

A speculative image of fracking wells drilled in Petersfield, UK

The image is a computational simulation of a seismic monitoring system. A series of three recording devices detect and monitor local seismicity induced by fracking and determine the epicenter and magnitude of the event.
Source: http://www.usgs.gov/
Image: Michael O'Hanlon

ON COMPUTER SIMULATIONS IN THE AGE OF HYPEROBJECTS

ROBERTO BOTTAZZI

Prologue

"So if the computer says it, it's true!" If you are interested in digital design or simply perceived to be by colleagues, comments such as the one opening this essay has probably been used to characterise your work or that of your students. DS18 students have not been an exception to this and the following remarks are intended as both a reply to such comments as well as preliminary notes on the use of computer simulations in an architecture studio committed to the study of the impact of the Anthropocene on architecture and urbanism. (01) Though DS18 employs computer simulations at a variety of stages in the development of students' theses, it is at the very beginning of the research that the most simulation-heavy work is undertaken, as material processes, site conditions and design scenarios are first modelled and played out in time.

Introduction

Any conversation on the nature and role of simulations seems to be forever caught into a web of entangled and even contradictory ideas and definitions. As early as 1979 Pritsker already managed to gather 21 different definitions of computer simulations. (02) More recently, Ören increased this number to first one hundred and then four hundred incontrovertibly demonstrating how far we are from a general consensus in this field. (03) Despite its complexity, resulting from an accumulation of knowledge and experiments spanning over a considerable number of decades, computer simulations in architecture are still relatively novel and, almost unavoidably, seem to quickly polarise our community between detractors and devotees. Both equally ineffective, albeit in different ways, these factions suffer from the common tendency of approaching the problem too narrowly and, consequently, conjure up a "premature metaphysics" of computation. (04) The former group stubbornly resists incorporating computer simulations into the set of techniques designers employ. Intimidated by the technical aspects, they struggle to engage with the profound conceptual and design issues at stake. As we shall see, the relevance of computer simulations in architecture goes well beyond their purely technical aspects to challenge well-established definitions for matter, scale, and representation. The latter group, on the other hand attribute to computers, and particularly to computer simulations, such degrees of novelty and internal coherence to mistake their outputs for quasi-real objects able to legitimately replace any other form of experimentation. This mistake arises from applying a misconceived idea of simulation derived from scientific research. Contrary to scientific disciplines, architects utilise these tools mostly for creative, projective purposes, thereby removing the fundamental characteristic of most scientific simulations: that of attempting to replicate real, existing phenomena, "as an alternative means to achieve some identical end." (05) Without such clear aims to attain e.g. the growth of a plant, architectural simulations cannot be validated in the same way scientific ones are. As we will see later however, the lack of an objective process of validation does not necessarily imply a lack of rigour. The three basic operations (06) performed in any simulation are still utilised even in creative simulations. However their evaluation and legitimacy relies upon a looser, more intuitive set of criteria than objective ones.

It is worth pointing out that even in the sciences, computer simulations are never really understood as exact replicas of real phenomena. To succinctly summarise the literature on this topic, we could conclude with Varenne that computer simulations can be utilised as either a kind of experiment, an intellectual tool, or as a real and new means of learning. The conversation taking place within DS18 occurs within the boundaries set by these definitions, often by moving between these three frames of reference. (07)

General definitions

Broadly speaking, simulating a phenomenon involves setting up a system (made up of variables and constants) to manage the simulation itself. The system will display varying behaviours that will give rise to useful, or not, information about the system itself. We can categorise the different elements of a simulation as: the target of the simulation – what is literally tinkered with in the process; the motivation – the larger aim behind the investigation; and the model which we define as the conceptual steps linking target and motivation. For instance, in the case of weather forecasting, the system is represented by a set of mathematical equations mimicking physical laws. These will be implemented in a digital environment – the target system – in order to forecast the evolution of weather over a certain geographical region – the motivation. The set of procedures employed to utilise the computer projection as a valid estimate of how the weather will vary are the model. Framed as such, the model unavoidably retains some elements of abstraction that can be representative of or a representation of the target. In the former case, the target is literally a part of the whole constituted by the motivation, which presents the same qualitative properties; the latter instance requires an additional layer of interpretation/abstraction to link system and motivation.

Computer simulations as representational tools

It is this last definition that best fits the nature of simulations in DS18. By including computer simulations within the milieu of representational tools, we radically change the terms of the discussion about design. Computer simulations are part of the vast set of techniques architects have been employing to represent things. It goes without saying that representational tools are not the same as the reality they represent. However, we have indulged in this confusion for centuries. Think of perspective, perhaps the oldest and most important contrivance in architects' repertoire, which allows us to represent three-dimensional space by replicating how our eyes see. Its fundamental principle, stated by Alberti around 1420 and still valid today, is that light travels following rectilinear trajectories. We now know that light, on the contrary, moves in sinusoidal waves that we approximate to lines, making the whole edifice of perspective still valid and possible to be performed by an individual on a piece of paper (incidentally, computer-generated perspectives also retain this imprecision). Perspective is both a representational and a projective method. Architects employ it not only to reproduce an existing reality but also to imagine a new one (this last point is so engrained in our way of thinking that perspectival views also have legal value and are often required documents to evaluate designs). Regardless of these approximations and errors, perspective has played and still plays a central role not only in the design process but also in articulating our relation with reality. Hence, Panofsky's elevation of perspective to a symbolic form.

Simulations, as much as perspectival views have the ability to change how we represent things and changw our "semantic alphabet".(08) So, if architects are actually used to dealing with representations, approximations, proxies, why are computer simulations still seldom included in discussions on representation?

On Data

An element causing scepticism amongst detractors of computer simulations in architecture, is the nature, or rather the role, of data. Criticism is here levelled at the abstract and self-referential role of data, whose numerical values are generated by and only make sense within the strict logic of the software managing the simulation. This line of argument is perhaps even more insidious than the previous one as it ultimately constructs an image in which computation and matter are pitched one against the other. As I will discuss later, though characterised by precise qualities, computation is only the last iteration of the complex relation between matter and ideas. The role of data in articulating our cultural and political attitude towards reality and representation cannot be underestimated. One powerful example of this is provided by the development of climatological studies. Here, as Paul Edwards is quick to affirm in his survey of the history of computational studies of climate, "without models, we have no data." (09) Data are not neutral quantities predating the use of a certain model to make sense of them. Data are actively generated by the very way in which we approach the problem. Edwards describes how apparently simple operations such as temperature measurements (type of the instrument employed, standards of measurements etc.) are tangled in complex techno-political issues having to do with the kind of model adopted. Again, the model is a conceptual and cultural device capturing the totality of the intellectual activity dedicated to teasing out patterns and controlling matter. Models, data, and matter sit at the corners of a triangular diagram in which they produce and reproduce one another. To quote from Stanford Kwinter, "to oppose computation to the real is not only incorrect but also dangerous as it obscures the complex entanglement of technological, social and political issues always at work in computers." (10)

Hyperobjects

So what are computer simulations good for in architecture, and how can we tell a good simulation from a bad one? The answers to these questions are not to be found in technical manuals or in purely philosophical conversations. Rather they emerge from repositioning architecture vis-à-vis large ecological issues that demand us to confront the impressive scales and timeframes posed by global warming. These transformations are making received notions of site, type, and material ineffective. DS18 positions itself right at the centre of these still-uncharted territories. The lack of reference points to guide navigation is to be understood both literally and conceptually. We question the very ground, again, both literally and conceptually, on which students project their designs. The semi-desert landscape of the Karoo is therefore understood as a piece of architecture in its own right, even before any speculative intervention. Its scale and temporality is massively larger than that of architecture: profound geological forces have been shaping it for millennia and equally large forces are now threating its future. Half of the work carried out concentrates on deciphering, mapping, modelling its spatial qualities. This approach also partly explains the misplaced criticism that students spend too long researching rather than designing. The massive scale of climate change, and of geological forces are invisible using traditional architectural tools. Seeing them first and foremost implies developing tools able to catch these processes, to isolate them, and foreground them. Computer simulations offer an incredible tool to enter this territory. They allow us to engage matter, its temporal and spatial characteristics, in a direct and experimental fashion.

Climate change is perhaps the clearest and most powerful example of what timothy Morton calls Hyperobjects; (11) objects whose dimension is finite (global warming is roughly matching the scale of the earth) and yet whose temporality and scale radically exceeds what our minds can grasp. Morton links the birth of the hyperbobject called climate change with that of the steam engine and projects its effects to last for the next couple of millennia. How to study them? Hyperobjects do not exist without computation, in the same way in which we would not really have debates on climate change without a substantial and prolonged computational effort to understand and simulate the climate of the planet. Computer simulations here are essential tools able to cope with the complexity of the phenomenon studied. They are epistemological devices to see, albeit through partial and distorted images, what large material configurations could be like, what properties, behaviours, etc. they might present. The aim

[top left} Simulation of solids acting as liquids under pressure
[top right] Simulation of the movement of solids interactive through applied pressure (close-up angle)
[bottom left] Simulation of the movement of solids interacting through applied pressure
[bottom rght] Simulation of the density of solids deforming under pressure
All images by Emma Swarbrick

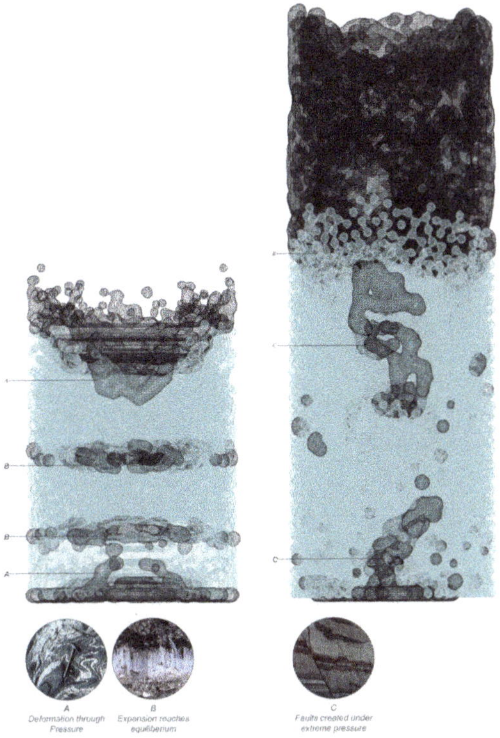
A
B
B
A
C
A
Deformation through Pressure
B
Expansion reaches equiliberium
C
Faults created under extreme pressure

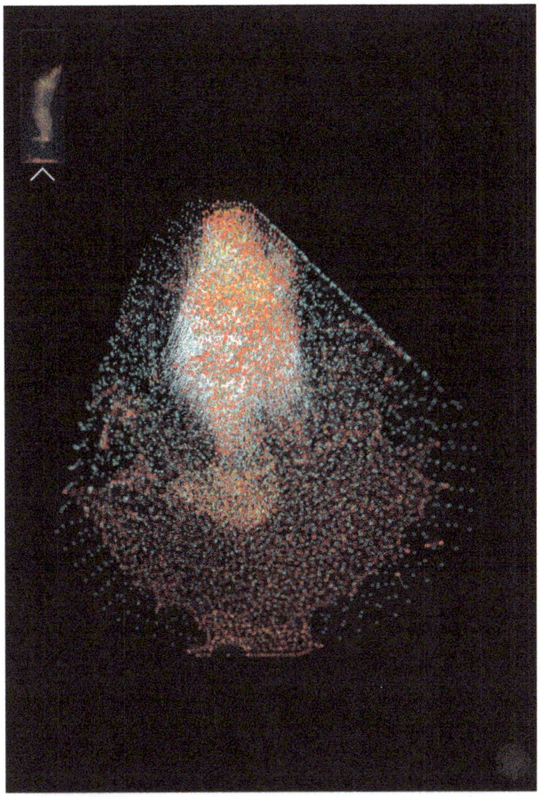

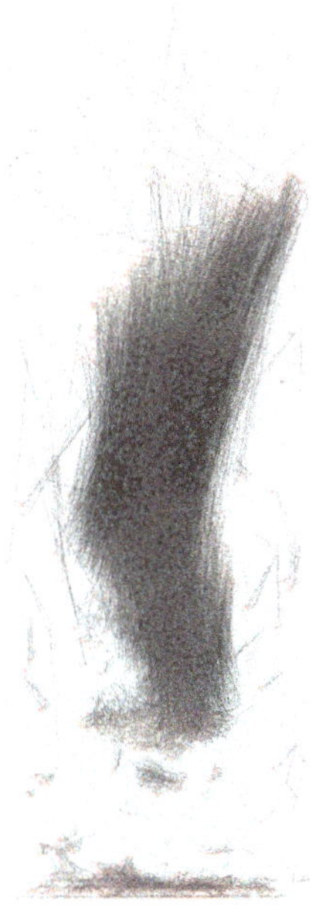

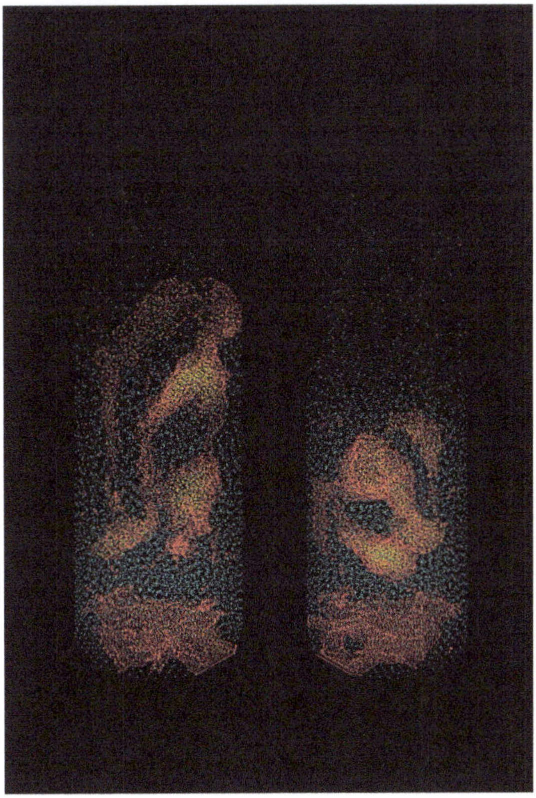

is not to imitate or even reify what is anyway a massive approximation of a real phenomenon, but rather to develop an attitude, a sensibility to deal with fluid, slow, enormous systems on which architecture is supposed to be immersed in. Hyperobjects, in their expanded meaning, act here as models; they provide the conceptual apparatus to speculatively link target and motivations.

This also highlights the great pedagogical value that computer simulations can have. Unlike experiments carried out in real labs, background noise can be removed to foreground the elements of interest. Variables can be easily and endlessly manipulated and increased to levels that cannot be reproduced in reality, thus studying a certain phenomenon under extreme conditions, "systematically and repeatedly testing variations in the 'forcing'" (12) (the variables that control the system).

Kwinter captures the possibilities enabled by computer simulation when he states that "the most powerful and challenging use of the computer (aside from the obvious benefit of automated number crunching in purely numerical domains such as accounting) is in learning how to make a simple organisation (the computer) model what is intrinsic about a more complex, infinitely entailed organisation (the natural or real system)." (13)

On Software

The choice of tools is therefore essential as they articulate the relation between systems and targets. Most of the work done in the studio in this area is not done by utilising CAD software but rather fluid dynamics packages. They present an environment much closer to the conceptual agenda sketched above. These are time-based software. Albeit at a drastically accelerated speed compared to actual phenomena, they allow us to conceptualised time and start thinking dynamically about space, matter and architecture. They are also not organised typologically. In these software packages the attribution of precise geometrical figures is deferred; rather than starting with so-called Euclidian shapes to then operate a series of geometrical [Boolean] operations on them, fluid dynamics packages ask the operator to design the initial scene by populating it with forces, frictions, materials properties, and behaviours which are eventually set into motion to interact with one another. This undoes hierarchies of matter to reduce them from pre-established geometries to particles, or voxels, endowed of physical properties. Not only is this type of spatial organisation much closer to scientific theories such as chaos and emergence, but it also forces the designer to model how simple forms of material organisation can be cultivated, aggregated and combined. These operation necessitate the intervention of one's intuition and sensibility. It has been pointed out how step-by-step simulations replace deductive processes with notions of observation and measurement.(14) Conceptually, this approach slows down the work flow: it allows us to critically analyse outputs, change initial or intermediary conditions and update the results. And then to repeat it. Finally, contrary to some CAD packages, physics-based software does not have strict or predefined scales to operate at. This again leaves open how one might redeploy these simulations. Next Technologies package RealFlow has played a central role in these investigations allowing students to handle large numbers of particles to simulate geological, environmental, and physical phenomena.

On Site

The picture that begins to appear before our eyes is therefore no longer one marked by rigid contrapositions between static categories. It is rather a fluid and complex one in which computational, mathematical, cultural and political notions and values constantly interplay one another. The ground on which this dance is performed is the physical territory of our investigation, what we commonly call the site, on which speculative proposals will be placed altering its dynamics.

To conclude, by critically employing computer simulations we can understand more holistically the relation between data and site. Site, today perhaps more than ever, is also data. Not only do we mean that sites produce data, but also that they can be seen as datascapes, as the physical embodiment of the very datasets we artificially extract from them. It is this complex relationship that allows us to perceive, as climatologists have, the average earth temperature that has risen by 0.75 Celsius over the past century, a phenomenon whose entity, scale, and duration make it completely imperceptible to human senses. The definition of site fully includes that of the technologies to model it. It is not by coincidence that Buckminster Fuller's World Game accurately charted techniques of communication and data recording as an essential element to plan at the scale of the planet. (15) It was a matter of survival.

Notes

(01) See L. Bremner, 'The Urban Hyperobject', in this volume.
(02). A. A. B. Pritsker, 'Compilation of definitions of simulation', *Simulation*, August 1979, pp. 61-63.
(03) T. Ören, 'The many facets of simulation through a collection of about 100 definitions', *SCS Modeling and Simulation Magazine*, no. 2, 2011, pp. 82-92, http://www.scs.org/magazines/2011-04/index_file/Files/Oren(2).pdf (accessed 11 August 2015). See also T. Ören, 'A Critical Review of Definitions and About 400 Types of Modeling and

Simulation', http://www.scs.org/magazines/2011-07/index_file/Files/Oren-July-11(2).pdf (accessed 11 August 2015).
(04) The idea of premature metaphysics is taken from F. Varenne, 'The Nature of Computational Things: Models and simulations in design and architecture', in F. Migayrou (ed.), *Naturalizing Architecture*, Orleans, Editions HYX, 2013, pp. 96-105.
(05) I. Peshard, 'Is Simulation an Epistemic Substitute for Experimentation?', 2010, http://philsci-archive.pitt.edu/9010/1/Is_simulation_an_epistemic__substitute.pdf (accessed 10 August 2015).
(06) Peschard, 2010, succinctly defines these three steps as: (i) preparing the system S in a certain state, by fixing initial and boundary conditions, and selectively putting under control the parameters that have an effect on the outcomes of measurement, the active parameters; (ii) letting the system evolve. The evolution of S is characterized by the evolution of a set of physical quantities characterizing the state of S, the state variables; (iii) recording theevolution of S through a sequence of states when the values of some of these parameters are varied, analysing the results.
(07) For an accurate and yet succinct summary of reasoning behind these three positions see F. Varenne, 'What does a computer simulation prove? The case of plant modelling at CIRAD (France)', in N. Giambiasi and C. Frydamn (eds.), *Simulation in Industry, Proceedings of the Thirteenth European Simulation Symposium*. Ghent, SCS Europe Byba, 2010, pp. 549-554.
(08) F. Varenne, 2013, p. 103.
(09) P. N. Edwards, *A Vast Machine: Computer Models, Climate Data, and the Politics of Global Warming*, Cambridge, MA, MIT Press, 2015, p. 140.
(10) S. Kwinter, 'The Computational Fallacy', *Threshold*, no. 26, spring 2003, pp. 90-92.
(11) T. Morton, *Hyperobjects: philosophy and ecology after the end of the world*, Minneapolis: University of Minnesota Press, 2013.
(12) P. N. Edwards, 2015, p. 140.
(13) S. Kwinter, 2001, p. 92.
(14) F. Varenne, 2013, p. 99.
(15) A short description of The World Game can be found at https://en.wikipedia.org/wiki/World_Game. The charts mentioned in the text are mostly found in: J. McHale, *World Design Science Decade 1965-1975: Phase I (1965) Document 4 The Ten Year Program*, Carbondale, Southern Illinois University World Resources Inventory, 1965.

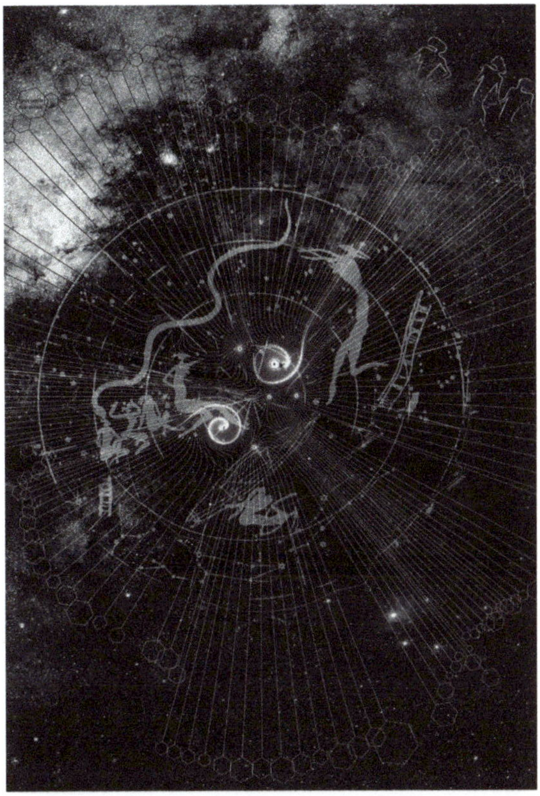

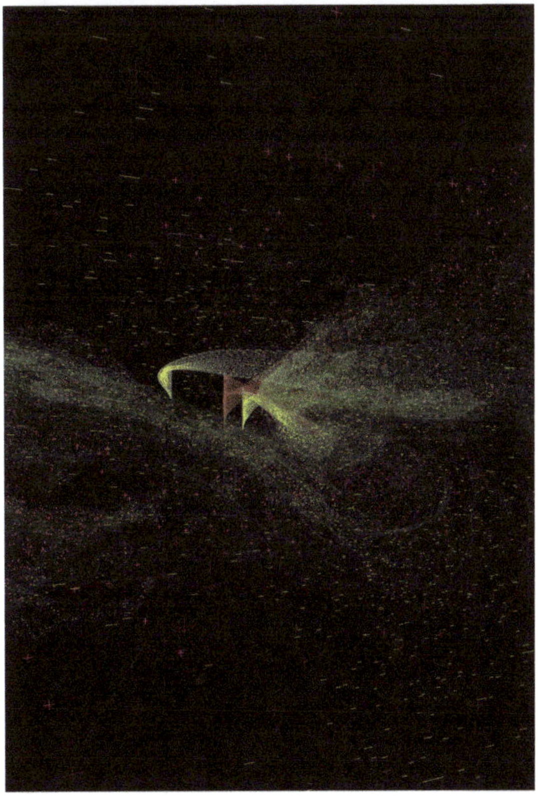

[top] Invisible Light
Simulation of transfer of energy from solar power satellites to receiving antennas via low frequency radio waves, Nieu Bethesda, Karoo
Iulia Stefan
[bottom] Simulation of Aurora Borealis as seen from the Karoo
Cheryl Choo

IMAGE CAPTIONS

[024] Simulation of Air Pollution (Plan)

[025] Simulation of Air Pollution (3D)

By setting up a stream of polluted and clean air particles, the aim was not to analyse the process or scientifically model the way in which the two interacted but to gain an understanding of properties of air, usually invisible, in two different states.

Andrew Baker-Falkner

[026] Simulation of the Melting of a Wax Cylinder (Section)

[027] Simulation of the Melting of a Wax Cylinder (Plan)

Michael O'Hanlon

[028] Simulation of a Sandstorm (Sections)

[029] Simulation of a Sandstorm (Detail)

Cheryl Choo

[030-131] Simulation of the Movement of Sand Particles in the Ocean

Iulia Stefan

[032-033] Simulation of the Entropy of Wave Energy in Collision with Tetrapod Coastal Defense Modules

Ben Pollock

Software: Exel, Grasshopper, Illustrator, Realflow, Rhinoceros.

SIMULATIONS

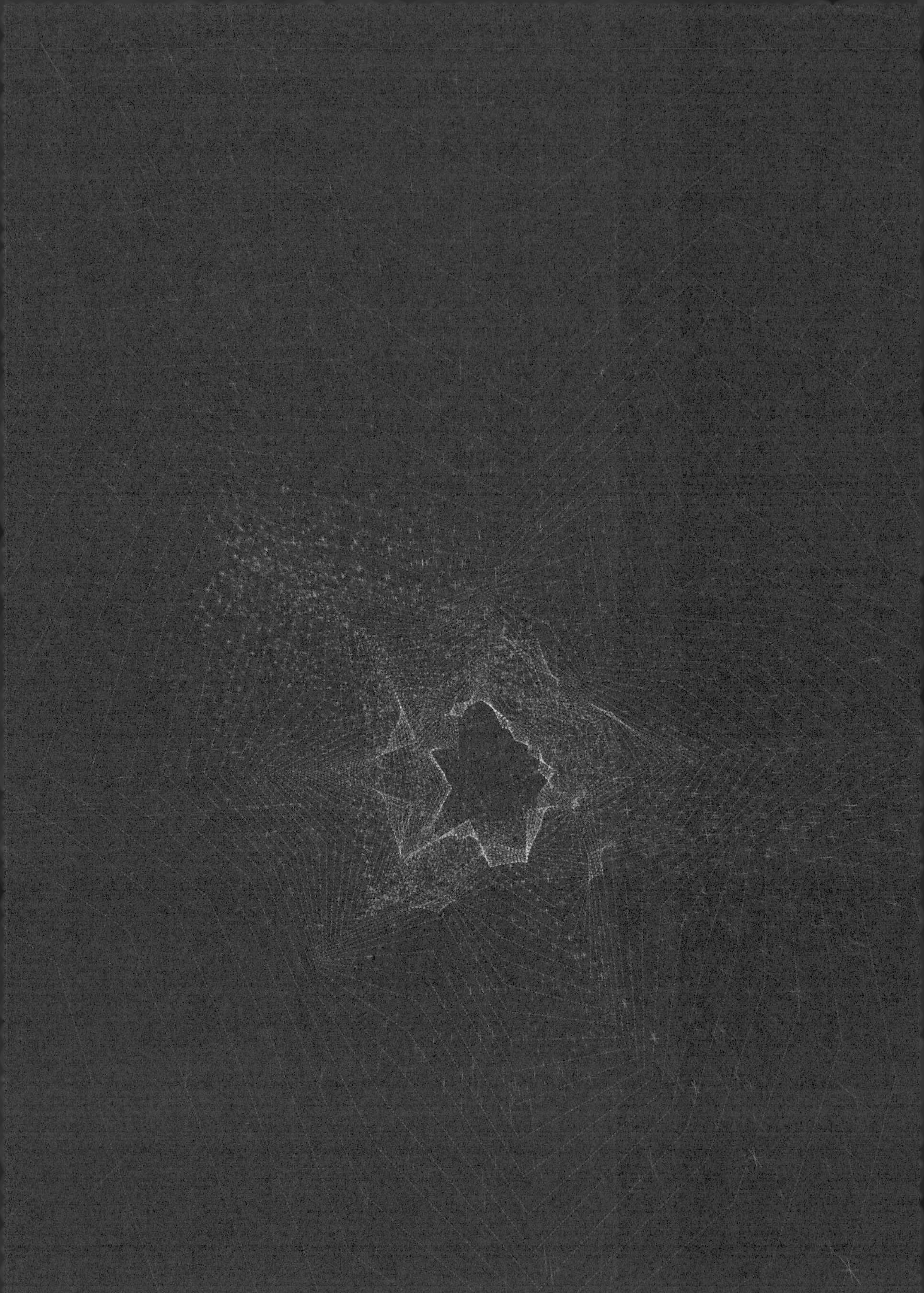

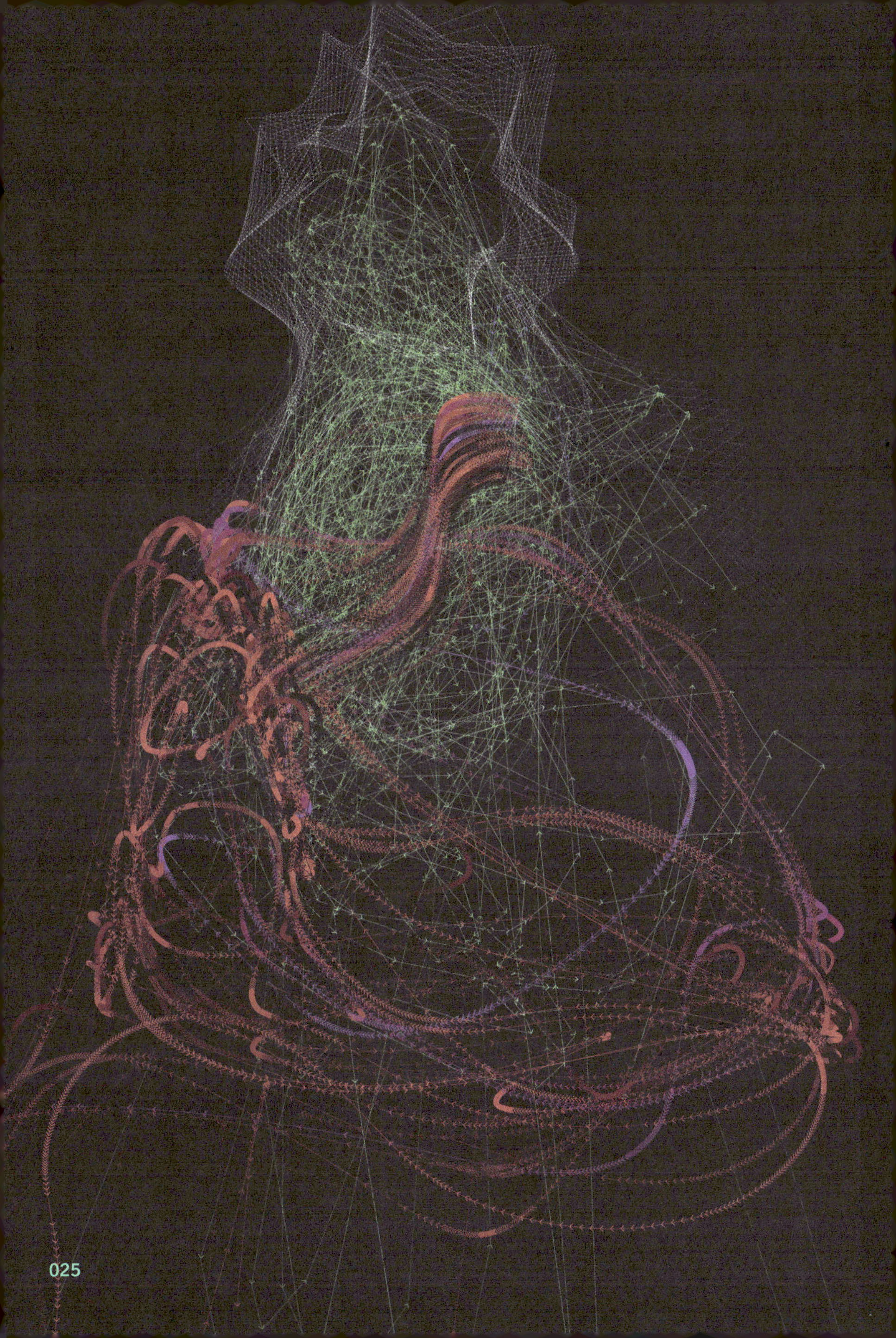

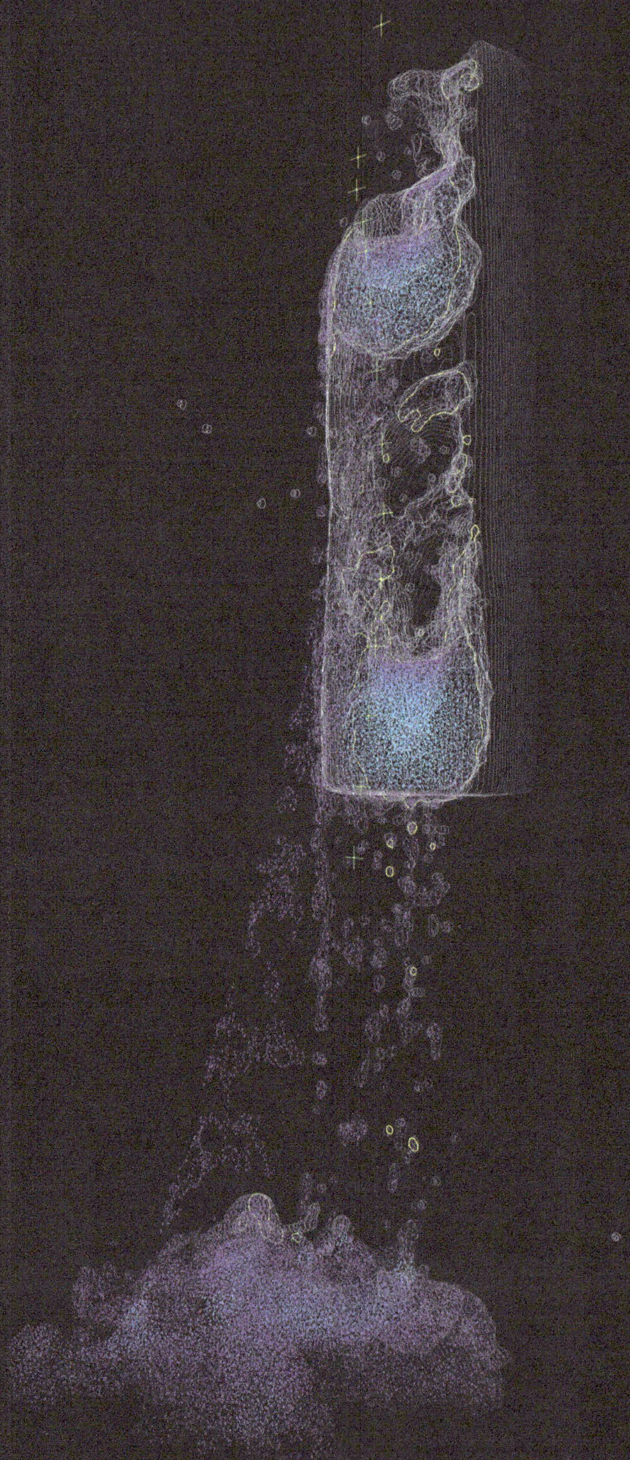

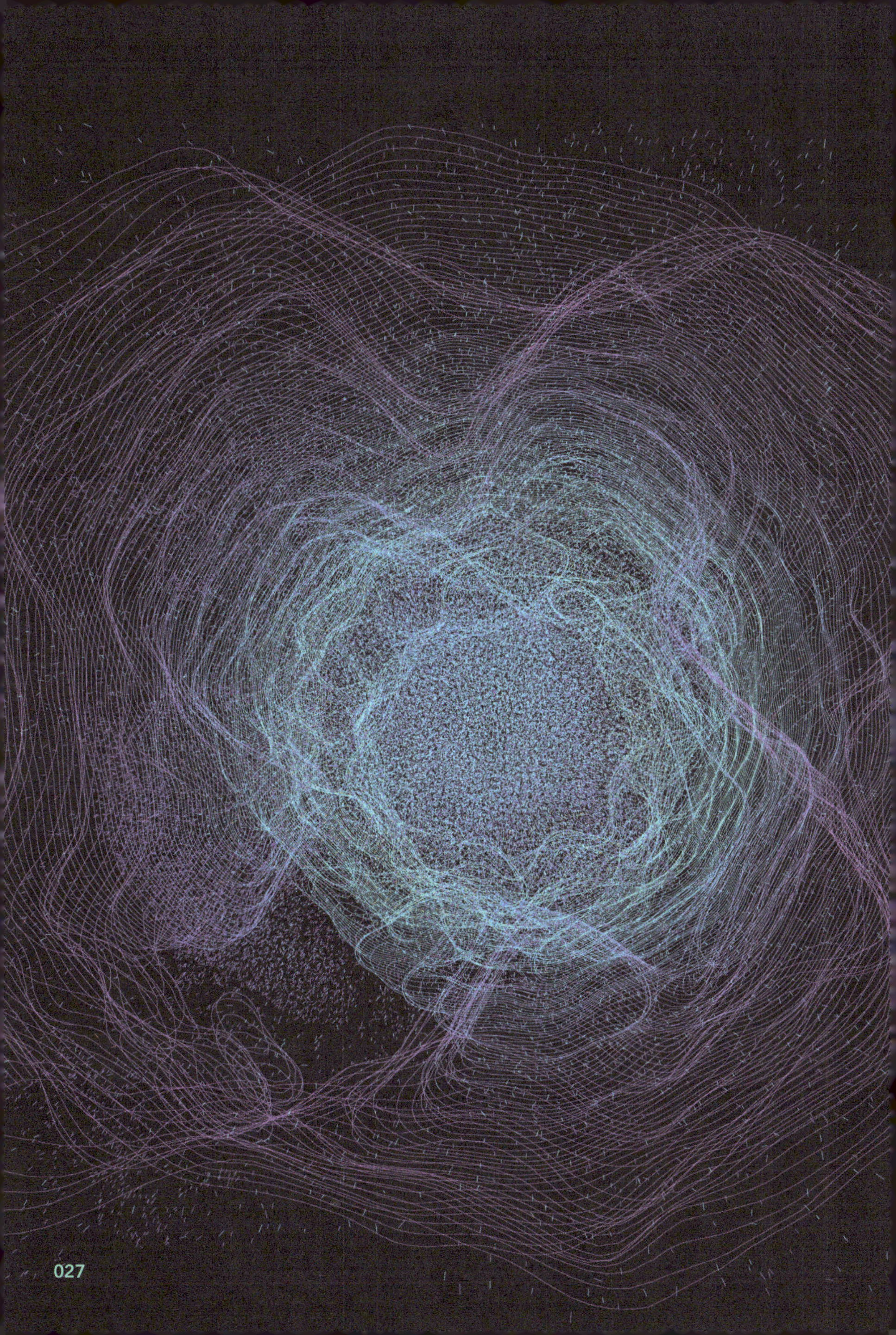

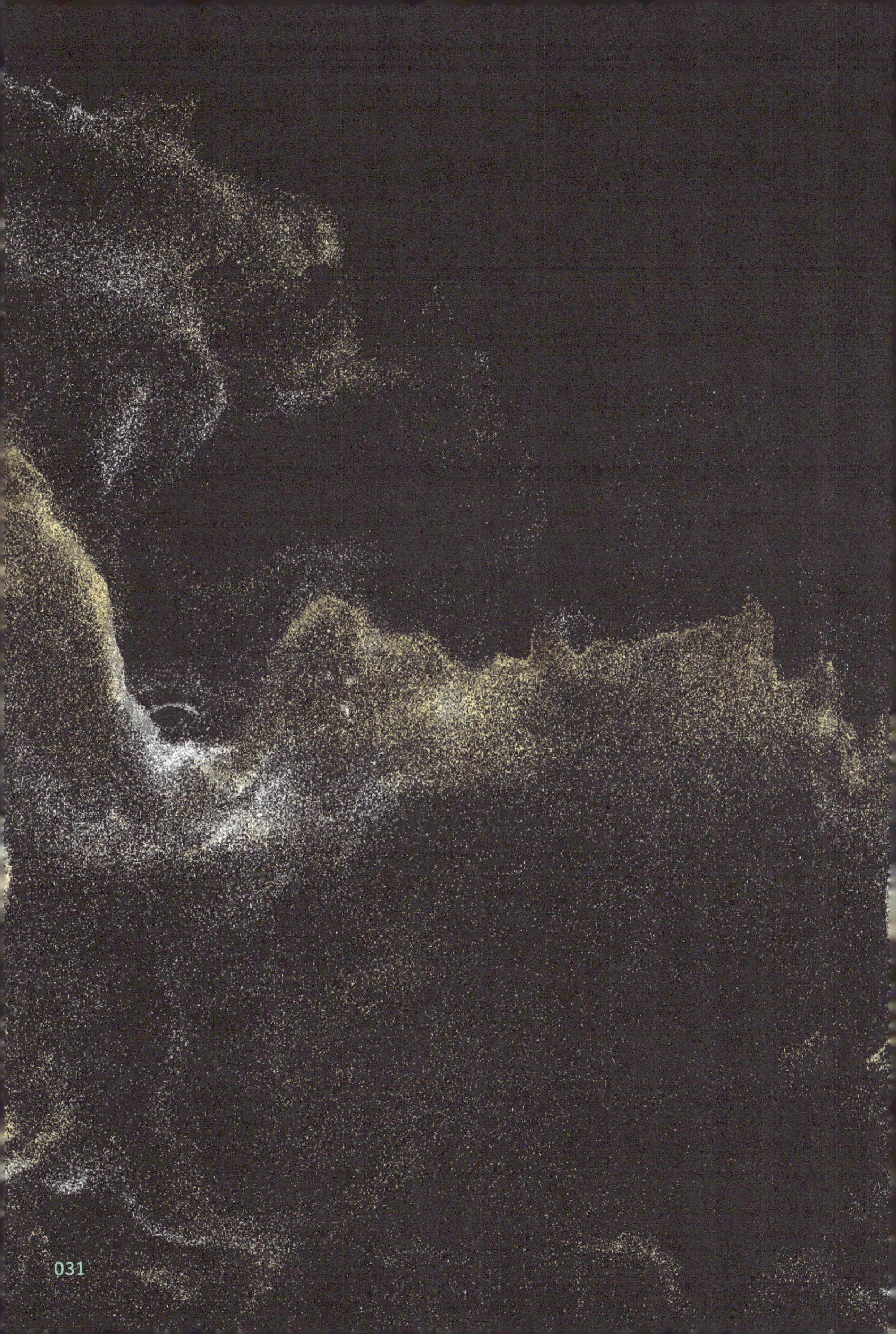

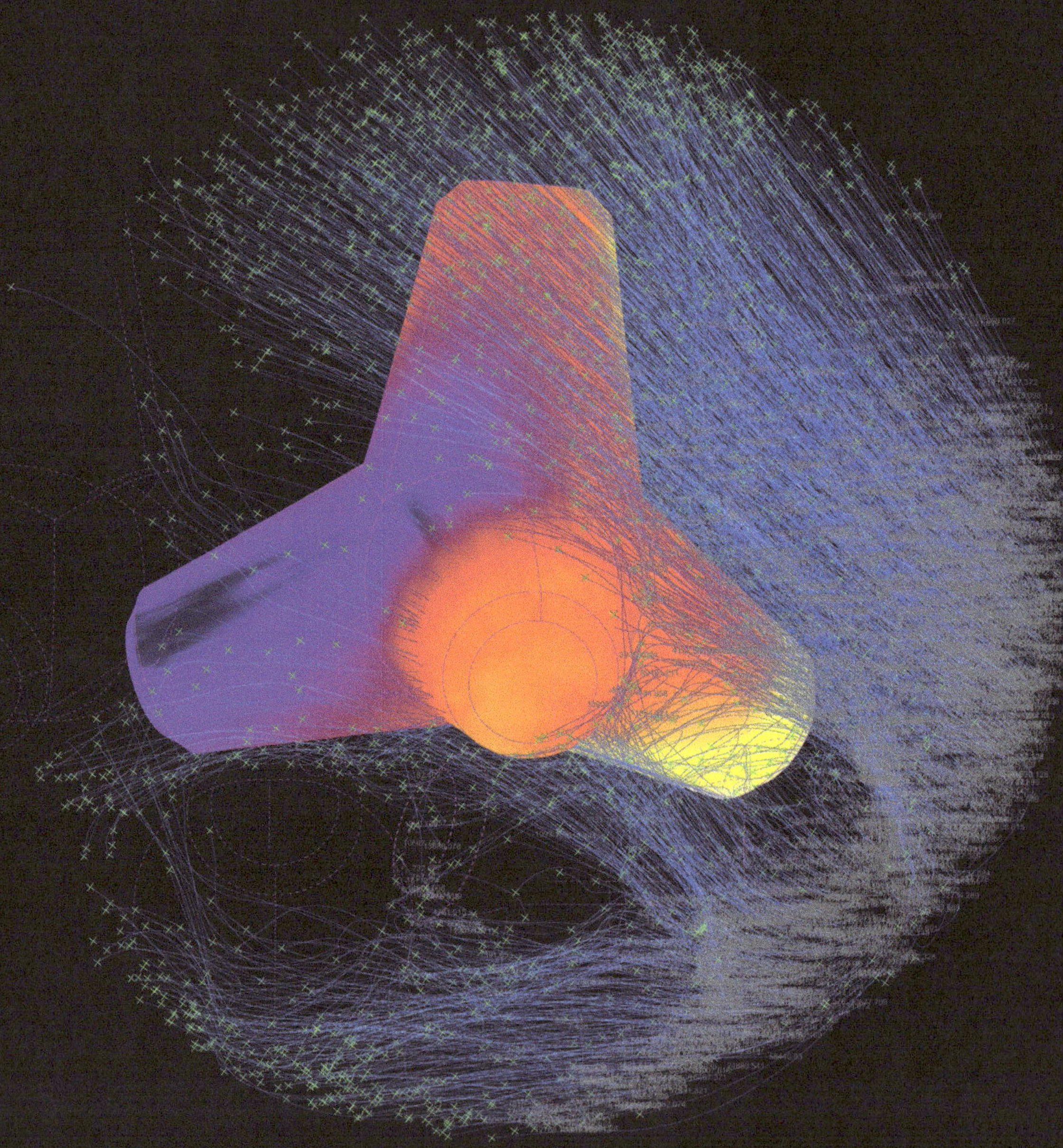

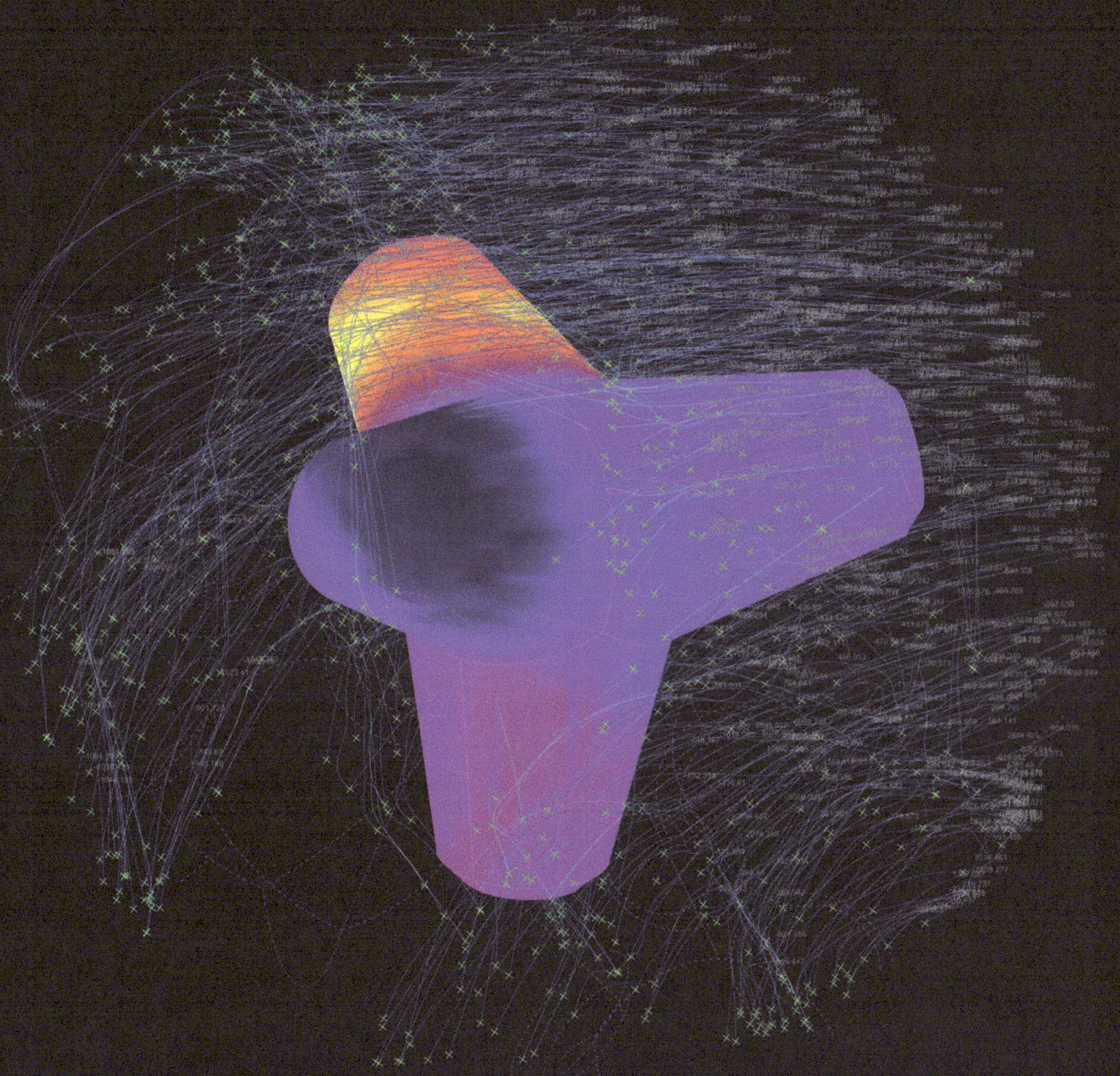

DESIGN FOR DISSIPATIVE SYSTEMS

KIEL MOE IN CONVERSATION WITH TOMAS HOLDNERNESS AND ETIENNE TURPIN (01)

At least since the publication of Ilya Prigogine and Isabelle Stengers' *Order out of Chaos*, the concept of thermodynamics in non-equilibrium systems has provoked imaginative new research in science, art, and philosophy. (02) Yet, over 30 years after the appearance of this path breaking work, the discipline of architecture remains stubbornly committed to ideas of energy that have yet to incorporate the lessons of thermodynamics. Despite the work of theorists and philosophers to address the thermodynamic imagination within design, (03) the proliferation of net-zero building pathologies and energy restrictive building systems persist. In his recent work, Kiel Moe has confronted this disciplinary blind spot with a cogent and compelling argument to re-imagine the role of energy in architecture. (04) Addressing with equal measure and attention the specificities of building science and the conceptual underpinnings of thermodynamics, Moe's research calls for a radical approach to energy that addresses the looming energetic crisis of the Anthropocene while also foregrounding the pleasure and habitability of architecture. (05) What follows is an edited transcript of our conversation with the architect and author about his theory of energy, design, and dissipative structures.

Etienne Turpin Can we begin with how you became interested in the problem of energy in architecture? You suggest, in your most recent book, *Insulating Modernism,* (06) that there is a fundamental misconception of energy, and specifically of thermodynamics, in architecture. What is the problem and what are some of the consequences?

Kiel Moe Understanding this problem took me a long time … until I was in grad school. I had a course with Sanford (Kwinter), and one with Richard Forman, an amazing landscape ecologist who is here at the Graduate School of Design. I had these courses in parallel and they had exactly the same vocabulary for form, but they manifest the program of two completely different thinkers. Still, Kwinter and Forman were both talking about dynamical systems. For me, that was the start of a bibliography about technology and systems. It took over ten years to congeal. It took me a long time to figure out - beyond a kind of trivial, alibi-for-a-studio course sort of way - how complexity discourse and science and technology discourse actually fit into architecture practice.

ET Is that because Kwinter's approach to complexity didn't quite reach the architectural scale?

KM Let's say that when I joined the faculty here at Harvard, I told Sanford (Kwinter) at one point, "Look, nobody ever understood what you were saying. It's still incomplete. You said it, but they didn't get it." That's part of what triggered our class together.

ET Your story of how architecture, as a discipline, isolated itself from science by beginning increasingly beholden to the product industry is actually a very compelling, powerful narrative. Ultimately, you are suggesting that architecture decided to insulate itself quite early on from science.

KM It is a book about insulation, but it is actually a book about isolation. It was the idea of isolation, whether it was tacit or quite overt in early architects' minds that was really powerful for modernist architects and professors. Whether they knew it or not, the idea of isolation set up the epistemological and methodological approach to energy that they used when designing buildings and cities. That is the essential problem.

ET Does a multidisciplinary approach to design relieve any of the problems in architecture? It seems to be universally promoted but then almost immediately condemned as not enough architecture.

KM I would argue that this model of multi-disciplinarity tends to trivialise both architecture and engineering. Multi-disciplinarity ontologically maintains the separation of the disciplines as its premise.

We need post-disciplinary thinkers and doers to see around our modernist corners and see the latent power of architecture and urbanisation in thermodynamic and ecological terms.

Tomas Holderness Let's cut to the question of urban thermal research and the challenge of scaling. Even if we achieve a certain comprehension in thermodynamics of buildings and cities, the difficulty is always scaling.

KM One thing I really try to push with my students is the need to understand the orders of magnitude that we are actually dealing with.

Architects end up spending a lot of time optimising something that has absolutely no significance. Understanding the relative magnitude of energy involved in making a chunk of concrete versus what that does thermally, and how to balance human comfort in all that - it is incredibly complex and non-linear. But then, if we keep zooming way out to talk about radiation at the edge of the universe, what are all the scales of energy that we need to be talking about? There is a lot to learn from each of those scales, but we should be able to have some sense of the relative order or hierarchy among them.

I think that hierarchy is different for a scientist doing global thermal imaging than it is for an architect operating at the scale of a house, but they both need to understand how the scales relate.

TH How does the architect understand the energetic movement from the building to the neighbourhood?

KM That question makes me think of Steven Chu, the Secretary of Energy appointed by (President) Obama. He came out and told everybody to paint his or her roofs white. People thought it was a version of Jimmy Carter's "put a sweater on" approach to energy. Fox News, and almost everybody else, made fun of him for this suggestion. But, in terms of radiative forcing, it would actually have a huge effect. It's this seemingly stupid thing that could actually be of some consequence.

ET I think Tomas is also asking about the tension between the design decisions that are optimized at the level of the individual house and the design decisions that would be optimal for an urban conglomeration of houses; perhaps this is a line drawn between the commitments of architecture and those of urban design.

KM Yes, but that is why the white roof and the right kind of spacing between buildings are important. And, that is where that conversation about maximum entropy systems - like the old forest that has the coolest thermal image - really tells you a lot about what a city ought to be doing. (07) You can think of some pretty direct analogues for how that could be done. As it stands now, though, we are just dumping a lot of excess heat, so we are clearly not that mature.

ET According to this energetic analysis, the most highly developed city would have the least heat loss?

KM Right. Let's say that for two cities of the same latitude, the one that has the coolest thermal image is probably the most sophisticated in thermodynamic terms. So, that's why white roofs are a good first step, but there is much more work we could get out of that thought, if we were more clever in design.

ET That is really your point, I suppose, that we can more work out of design, but from a different functionality. It is really annoyingly obvious to say, but this argument is still completely at odds with the construction of the discipline. I am curious about your strategy here; did you think about how to approach the discipline of architecture to develop this issue?

You are not just making a critique of the discipline; you are calling for a complete overhaul of its conception of energy, and therefore, of design itself.

KM I don't know the answer to that. For now, I teach and I write; I try to get it out that way. I think I'm getting ready for a slightly larger scale of conversation, but it's not an easy mind-set to shift. Net Zero, for instance, isn't about energy or ideas; it is a business proposition. It is a whole apparatus of products, consulting, and related sales materials. These people hate the idea that design could be guided by simple, common, abductive sense if we just understood a few basic thermodynamic principles. It is not in their interest to listen, much less acknowledge this alternative. I encounter that resistance all the time at lectures. When I was giving a presentation in Seattle, there was an audience member who was just furious; he adamantly refused to accept my claim that a zero energy house is just an insane marketing device. It turned out that his practice was called zero-energy houses or something. So, yes, there is going to be a lot of that sort of pushback, but you can't let it be a defining obstacle.

ET But, you admit, there are all the consultants whose livelihoods that are threatened by your proposition?

KM I think a lot of people are relieved when I say upfront: "Don't confuse these things with your motivations. What you've been taught or how you've been trained to think about these topics is the problem. Your good intentions have been poorly served by your education and the current discourse. Don't get that confused ... you're still a good person." I would say decoupling design intentions from the problematic concepts helps considerably. But to your question, I would say that all of our livelihoods are actually on the line, not just their cynical, neoliberal sustainability consultancy practices.

TH So, have you really thrown this Thermodynamics 101 explanation at the Net Zero set? Or is this confrontation just getting started?

KM I've approached this in multiple ways. We had the preeminent materials scientist from Harvard present at a meeting recently about a proposed zero-energy building at Harvard. This scientist and I agreed that zero-energy is preposterous claim, but the conversation still went nowhere even at Harvard. Basically, classically trained building scientist have a whole set of really well developed ideas for isolated systems. Maybe the universe as a whole is an isolated system, but everything else is in constant exchange, emitting or receiving energy. So, unless a house is at zero degrees Kelvin or outside of the universe—all of the houses we deal with are in the universe, however—we need a completely different approach. And, I understand it: it is actually really easy to misunderstand the first and second laws of thermodynamics. I completely understand why one might think that a house is a system unto itself and that if you don't make it efficient, it will somehow dissolve into nothingness! But the idea is absurd and has not basis in thermodynamics.

TH And, then when you try reconcile those laws with what is simultaneously going on at the other scales, with electrons and behavioural changes, etc., things get quite complex.

KM Sure, but I think even for those kinds of operations, once they are scaled up to a building or a city, that's where it gets so interesting. It is much easier to understand how a city behaves than some electrons jumping around. There are some rules and indicators and orientations that are very consistent throughout urban history. Louis Bettencourt has that great piece about scaling laws for cities. (08)

ET Do you believe in a universal physics of urban scaling?

KM I'm doing a project with Jane Hutton that maps all of the materials that have gone through the Empire State Building site over the past 200 years - it was first a forest, then a farm, then some houses and mansions, then the Waldorf Astoria Hotel, and finally the Empire State building. One thing that shocked me was that the construction materials for the Waldorf, an old load-bearing masonry building, and those for the Empire State Building, weighed the same, within about two per cent of each other. So I am starting to think that there could be a weird sort of scaling law for what is operative in a city. Through that analysis, you could start to figure out what is ecologically or ubanistically powerful from a thermodynamic point of view. The Empire State Building and the farmhouse work; their scales are well understood, but anything in between is a whole order of magnitude lower in terms of it's power and the work that it's doing. Anyway, that kind of stuff is starting to blow my mind. Through that analysis, you can start to figure out a good "velocity" for building materials, from their extraction point through to their use. It might be 20 years; it might be 100.

ET So there is an urban "sweet spot," so to speak, that can be quantified according to the same values across any geography?

KM (Hutton) is working on Central Park. She is much more concerned with the material and political geography of those projects. Issues of labour and sourcing are big parts of her work. They are both doing essentially the same task, but what we get out of them is slightly different in each case.

ET Okay, to clarify this point, let's go back a little bit. I'd you to outline what you see as the foremost consequences of ignoring the science of thermodynamics within architecture.

KM Building science systematically produced a subject without any qualities, a sort of abstract, naked person in a room. How comfortable are they? These sciences reduced the problem down to a couple of variables - air temperature and surface temperature, for example. In a few cases, air movement. They built a whole apparatus around that idea, say, midway through their evolving air conditioning research. It was all based on air, on amplifying the use of air ..., which makes no sense! There are many other physiological properties and qualities of our bodies that relate to comfort. Building science ignores these. They didn't grasp how this was a political act: to conceive of a subject with basically only air moving around it. It completely destroys all kinds of other relations. It completely changed how buildings were constructed and how materials were used. And, I should add, none of this came from an architect. It all came from an industry, and all architects could do was figure out how to incorporate these industrial products into their design practice. Very few architects stood up and said, "No, I've actually thought about it and these are the materials we should use." Or, "This is the mode of heat transfer we should focus on." There are one or two examples where an architect has fought that kind of fight, but that's all.

ET You use the example of the fireplace as one obvious relationship between the human body and radiant heat. This example connects us to Vitruvius, by way of Reyner Banham's fable of the kindling in the forest. As you note in the book, Banham's thought experiment asks, basically, if a "savage tribe" finds a pile of timber in the forest, how they decide whether to use them to build shelter from the wind and rain, or to burn them for heat. (09)

KM I use Vitruvius, Loos, and Banham to produce the narrative. I always use Banham's famous parable when I am teaching; it's absolutely essential. But still, it matters where those sticks are! Are they in northern Germany or the Sahara? That's the thing Banham left out. His point of view, in the end, was that you should just burn the sticks. Banham loved driving around in huge cars; that was cool then. It's not clear if he grasped the full thermodynamic depth of the parable. Anyway, it was a great thought experiment.

All of these parables about the origin of architecture, from Vitruvius to Alberti to Loos, come down to some version of fire and how human beings congregate around it.

That consistency is productive, and we still have a lot to learn from them.

TH The geography of the fable is critical. What you would want to build depends entirely on the context of the construction.

KM Exactly, the vernacular traditions of architecture are, in a way, totally obvious things that suggest powerful ways to build; of course, you find them everywhere in the world. I talk about this in the book as well. There are non-modern, non-isolated buildings out there. We had a few thousand years of people experimenting with what works in every region and climate in the world. Then we just abandoned all that knowledge and started training people to do research in isolation. A completely different epistemology emerged from that isolated system; some of it is valid, but much of it is not. I'm not advocating going backwards or taking a recidivist position, but I do think that the 21st century should be a non-modern century. We know how flawed the 19th and 20th centuries were and we can learn from these mistakes; but we can also learn from the previous 10,000 years of design.

ET That sensibility won't make any architects famous! It seems to me that the agenda of contemporary architecture couldn't be more at odds with your conclusions.

KM This is an important point. Let's bring it back to the scaling laws for a moment. For argument's sake, let's say that there is a scaling law for buildings—their openings, proportion, massing, and an ideal state in specific environments. All that can be clear, but when Bejan uses a scaling law for athletes in the Olympics over the last few decades, it gets really interesting because the law itself bends. Usain Bolt was not in the scaling law; he was such a freak, and was so tall, that he just ran faster. The point is that the scaling laws aren't necessarily deterministic; there is an incredible amount of novelty permissible within these scale systems. We know that the meta-systems are always changing. We tend to associate creativity and novelty with architects; the trouble is that their creativity often doesn't respond to these energy systems.

ET So how do you stack dissipative structures and make energy consumption pleasurable and productive?

KM And, in such a way that amplifies consumption elsewhere, though in a productive way. It is a little mindboggling, but it's the right question to be asking.

TH I would argue that with this idea of scale, we don't yet actually know what to measure or how to measure it; there's a juxtaposition between air temperatures used to measure human comfort for building design and the satellite-derived estimates of surface temperature, which we use to measure our cities. (10)

KM That's why I think that the research project on thermal image mapping is so incredible. (11) It actually gets at some of those questions: measuring a parking lot versus a quarry, or a recently cut field versus an intact forest. It is just a gradation of roughness. The project is a bit abstract, but it at least gives us a sense or a starting point of how or where to look. And, from there you can start to understand what's wrong and what's right as a path. For now though, no, we don't really know where to look. But I think there are adjacent fields and studies that talk about how to nest dissipative structures in a powerful way.

ET Your research is fundamentally a challenge to the normative values placed on energy, or to how the expenditure of energy is itself evaluated. You end up, at least in *Insulating Modernism*, with a remarkable provocation: we should being using a lot more energy.

KM The planet as a whole uses solar income in all kinds of ways; if you take some of that away from one process and invest in pure hedonism, then something else isn't working, probably.

In any case, the idea that we should use less energy is actually not a valid thermodynamic response.

Again, if you're dealing with an isolated system, then that's fine, but you're not. And we're not just talking about pure hedonism! We should be interested in a hedonism that has productive adjacencies. An integrated hedonism, we could say. That is where [Georges] Bataille and [Alfred J.] Lotka and a lot of other interesting people, including [Ludwig] Boltzmann, have ended up. It is reductive to say that we should be using more energy. We should intake, transform, and feedback a lot of energy. We use a lot of energy, but in totally boring and cynical ways.

ET Eduardo Kohn talks about rainforest morphodynamics in this way, where structural logics of forest systems pattern the processes of their own exploitation. (12)

KM Values are totally a part of it. Basically, there is an economic valuation to energy right now, and that's why it's construed in all the ways that it is today. But, if there was a donor valuation of energy or matter—what can it provide me and what can I do to relay that value—then it would be totally different situation. We would have systems like the forest, with an abundance of interaction between its parts. That makes me think of the role of tree aerosols: the feedback or cycles we share with trees is so intense and it's a great way for us to get at some of these scale issues. We only have the most basic sense of how these cycles operate; only the most sophisticated researchers have even a clue about how those feedback cycles are operating. But that's the model, anyway. Maximum power systems, systems that survive, according to Lotka, Boltzmann, or

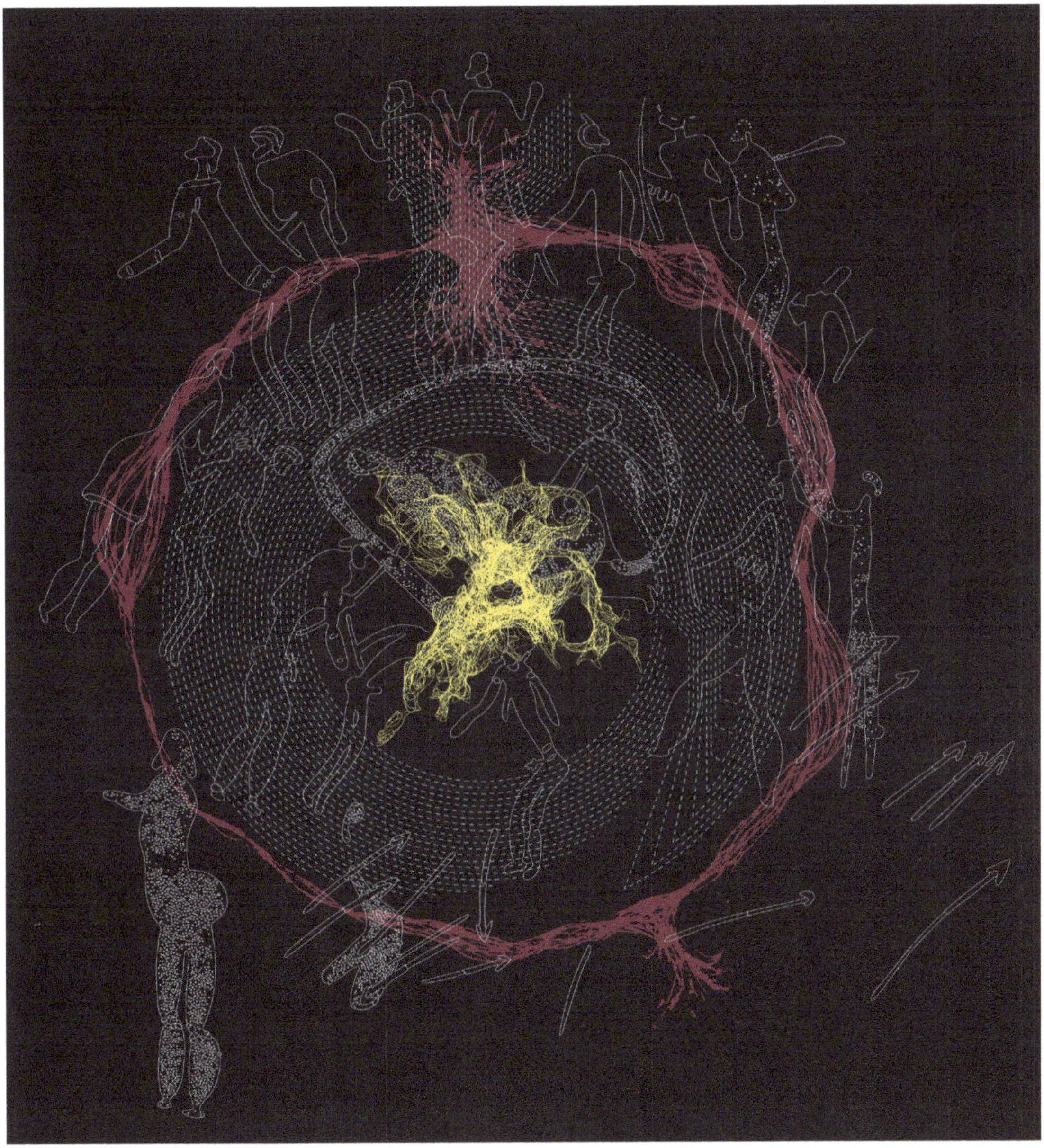

The San Trance Dance

Inspiration for the arrangement of an electromagnetic transmitter coil in Sciencity (see pages 266-275)

Cheryl Choo

[Howard T.] Odum, and develop an abundance of feedback systems. And, for us, we are not talking about recycling programs. Just because you recycle some newspaper or whatever ... I mean that's absolutely nothing in the whole picture.

ET The oldest, most integrated forests have the smallest energy loss in thermal imaging, is that right? They are dissipative entropically and yet very powerful.

KM Right, the system has gotten wealthy from using that income.

ET I'd like to talk about your discussion of the metabolic rift near the end of the book. Our colleague Seth Denizen has developed a new taxonomy for soil science that recognized cities as soil producing machines. (13) Thinking of this work I am reminded of how the processes moderns attributed to the city have been highly selective. This selection is, at least in part, what you try to address in your discussion of metabolic rift.

KM I bring up the "metabolic rift" to help architects understand that there is a huge gap between their specifications for buildings and what they're actually specifying from a thermodynamic perspective. It is very instructive to use Marx to articulate this point because it helps students and architects understand that this is a deeply intellectual and political problem that they are engaging; it is not just abstractions having to do with "boring" specifications. I think it is essential to point out this fundamental rift; it's not just a matter of concern for architects and their practice, but this impacts our lives in many ways. Few students have any idea about where all the energy in this room comes from. That's a problem because if you don't know where energy is coming from there is simply no way you can feed back into it or divert it. You don't know what the energy hierarchy is. First, you have to admit there's a problem, which is the rift. Then, there is understanding the prodigal superabundance of energy—or the metabolic "gift"—that is more or less at our disposal if we think about it in the right way. And, then there's the shift. Once you understand the rift and the gift, then you can also move, metabolically, into something that's more interesting, and probably into a better life.

ET Ecologically better, I can see that. But how does it improve the work of the architect? Because the architect is the agent, and this proposal takes thinking. It's a big ask. You are demanding a different kind of architect, not just a technician. What are the opportunities for new architects with a renewed thermodynamic imagination?

KM What I got into architecture to do, what I thought architecture was ... well, it was so disheartening going through architecture training. I was told: "that is not the proper way to build or heat or cool something." A really basic premise, for me, was to figure out how I wanted to build. When I look at 10,000 years of construction from all over the world, I'm more interested in these nonmodern works than the current paradigm, where architects are basically personal shoppers for institutional clients, downloading Revit models and plugging them in and spitting out stats and absorbing all the liability. All that has so little to do with what I think architecture is. None of the trainees say to themselves, "Oh, I'm becoming a pawn of capital that just absorbs liability for other people." Right? That's not an interesting proposition whatsoever. I think, for most architects, what they do everyday is quite remote from their original motivations or passions for joining the field in the first place. So, we need to ask what forms of knowledge, and what kinds of practices, does one need in order to pursue what I understand to be architecture and urbanization, and how do these feed into civilization. The current paradigms don't have an image of civilization.

ET The unintended consequences of the moderns' approach to energy are coming back to us now as a terrifying inheritance. What is the lineage of thought and practice in an architecture that would productively meet challenges associated with climate change? We need new images and ideas, but how to we rethink power and energy conservation to re-imagine the practice in continuity with previous experiments?

KM I think this is really the next project: to situate these concerns more squarely in the frame. If you take this line of thinking seriously, and you ask these kinds of questions about climate change, I agree very much with Erik Swyngedouw that we fetishize carbon. (14) Carbon is the troublemaker and it somehow must be the solution. But, of course that's just playing out in a neoliberal economic model.

ET Or maybe carbon is the "middle class" of climate change. Politicians play to the carbon!

KM Right. If you look at the causes and the dynamics of radiant forcing, carbon is actually a huge chunk of that. But there are many other variables that are important and are probably quite easy to address. When you make climate change as a carbon issue, it creates all kinds of problems and limitations. That's why Swyngedouw claims that the fetishization of carbon is depoliticizing; as soon as you challenge the centrality of carbon, you're automatically vilified as an anti-environmentalist. There's no politics in climate change anymore and that's a disaster. As soon as we stop asking how we should live, we're fucked.

TH It's interesting to consider the differences in response of governments between the global ban on CFCs to protect the ozone layer, and the debate on carbon emissions. But, also, as Cynthia Rosenzweig points out, many urban centres are now working at a city-scale to take a lead on CO_2 after a lack of political leadership at a national level. (15)

KM I think John May is also really good on this. He calls it "the managerial surface." (16) He's done incredible work on how this gets produced, and a lot of it deals with thermal images and how those representations are constructed. He shows that—this is how it's understood by 'the managers' - that if we have the right sort of managerial image, then we believe ourselves that we can manage the problem. But, it's not a management problem, or a representation problem; it's a question of how we should be living.

ET The Asbestos-complex?

KM Well, yeah. They thought they had something fireproof on their hands and they just went for it! It's a great example of isolated knowledge.

ET It is big business for Canadian exports. But this brings up another critical angle that you develop: the conversion of materials to products. While this is not a problem specific to architecture necessarily, it has driven any number of decisions within the field of design. Can you explain this product-economy and its influence on architecture?

KM A product is almost impossible to understand; having a complete understanding of a product is really difficult. Materials are complex unto themselves; by the time you get to products, there are as many corporate properties as there are material or thermodynamic properties. I have a hard time keeping all of that straight. When you see a chunk of insulation … how do you get your head into that? But, when you give architects six inches of wood or concrete or masonry, they can start get their head around that, at least. That's why I advocate for buildings that probably look like they could have been built 10,000 years ago! You can get at the metabolic shift with those. I think architects can actually have some legitimate formal knowledge about energy feedback loops and deep materiality. If you gave somebody six inches of wood to contemplate for their whole life, they'd have a whole lot to say about that material. And I think these nonmodern buildings are more interesting to be in, and are often more comfortable.

TH But, further than material, is it not our energy production that got us into trouble? Are we are forgetting 10,000 years of managing thermal energy? Not that it was perfect back then, before products, but there is a lot to learn from history, especially in terms of anthropogenic heat from energy production, etc. (17)

KM The basic stance of confusing fuel for work is definitely at the core of this problem as well. Fuel for work, one obscuring the other. If you talk to anyone about energy, they're thinking about it in terms of fuel, and not in terms of work. If energy was conceived of as work, then I think you'd very quickly get the right kind of systems.

ET Shiv Visvanthan has suggested, in line with what you are saying, that the second law of thermodynamics has a colonial ambition because it ignores biomass. By ignoring the energy of work, it emphasizes the most "productive" form as a fuel abstracted from its context. (18)

KM By the time that it's co-opted by Calvinists or Calvinist colonialists, sure. But, if you read one of the best explications of the second law—Boltzmann's—the key is the struggle for entropy. That's why that forest wins out in the thermal imaging contest. It struggles to create the most entropy. Hopefully the colonists, in this century, will get there too. But it doesn't look promising.

ET Your work is fundamentally about emancipating architecture from a colonial image of energy. In your teaching, what is the largest impediment or habit of mind that you see obstructing the emancipation of architecture?

KM I think it is the Calvinist legacy, which leads to very specific ideas about work construed economically rather than thermodynamically. That's a huge problem; it sets up a kind of technocratic, positivist, or managerial posture for dealing with these questions. That just sends us down the wrong path. Students who are curious, interesting, and talented designers, who are interested in open systems, look at that stuff and just get the shivers. So, there's a huge divide.

I am just trying to re-imagine what energy actually is.

This is a more interesting way of talking about this problem than bemoaning Calvinism.

Notes

(01) This interview to appear as 'Design for Dissipative Structures: On Architecture & Thermodynamics', Kiel Moe, Tomas Holderness, and Etienne Turpin, in K. Moe and S. Kwinter (eds.) *What is energy and how (else) can we think about it?*, Barcelona, Actar, 2016.

(02) I. Prigogine and I. Stengers, *Order out of Chaos: Man's New Dialogue with Nature*, New York, Bantam, 1984.

(03) See, most notably, S. Kwinter, *Far From Equilibrium*, Barcelona, Actar, 2008 and M. Delanda, *Intensive Science and Virtual Philosophy*, London, Bloomsbury Academic Reprint Edition, 2013.

(04) K. Moe, *Insulating Modernism: Isolated and Non-isolated Thermodynamics in Architecture*, Basel, Birkhäuser, 2014.

(05) For a discussion of pleasure in the architecture of the Anthropocene, see F. Roche, 'Matters of Fabulation', in E. Turpin (ed.), *Architecture in the Anthropocene*, Ann Arbor, MI., Open Humanities Press, 2013, pp. 197-208.

(06) K. Moe, 2014.

(07) Moe discusses maximum entropy systems with reference to the work of Alfred Lotka, Jeffrey Luvall and H. Richard Holbo in the chapter 'The Architecture of Dissipation', K. Moe, 2014, pp. 228-271.

(08) L. Bettencourt, 'The Origins of Scaling in Cities', *Science* no. 340, 2013, 1438-1441.

(09) See K. Moe, 2014, p169; for the original passage, see R. Banham, *The Architecture of the Well-tempered Environment*, London: The Architectural Press, 1969, p. 19.

(10) J. Voogt, J. and T. R. Oke, 'Thermal remote sensing of urban climates', *Remote Sensing of Environment*, no. 86, 2003, pp. 370-384.

(11) J. C. Luvall and H. R. Holbo, 'Thermal remote sensing methods in landscape ecology', in M. G. Turner and R.H. Gardner (eds.), *Quantitative Methods in Landscape Ecology*, New York, Springer-Verlag, 1991, pp. 127-152.

(12) E. Kohn, *How Forests Think: Toward an Anthropology beyond the Human*, Berkeley and Los Angeles, University of California Press, 2013.

(13) S. Denizen, 'Stratophysical Approximations', in S. Springer and E. Turpin (eds.), *Land & Animinal & Nonanimal*, Berlin: K. Verlag & Haus der Kulturen der Welt, 2015, pp. 50-74.

(14) E. Swyngedouw, 'Apocalypse Forever? Post-political Populism and the Spectre of Climate Change', *Theory, Culture, & Society* vol. 27, no. 2–3), 2010, pp. 213-232.

(15) C. Rosenzweig, 'All climate is local', *Scientific American* no. 305, 2011, pp. 70-73.

(16) J. May, 'The Logic of the Managerial Surface', *Praxis* no. 13, 2012.

(17) E. J. Chaisson, 'Long-Term Global Heating From Energy Usage', *EOS, Transactions*, American Geophysical Union no. 89, 2008, pp. 253-254.

(18) S. Visvanathan, 'Between Cosmology and System: The Heuristics of Dissenting Imagination', in B. de Sousa Santos (ed.), *Another Knowledge is Possible: Beyond Northern Epistemologies*, London and New York, Verso, 2008, pp. 182-218.

THE URBAN HYPEROBJECT

LINDSAY BREMNER

I will begin this short essay with three urban vignettes:

In Johannesburg in 2000, politicians came up with the controversial idea of defining an urban edge to the city as a cordon to secure its perimeter and contain urban sprawl. This took the form of a line on a map, reactivating a rather ancient idea of the city as a bounded territory, with policy infrastructure (not walls) separating inside from outside, them from us, order from disorder and defining an interior for administration, taxation and servicing purposes. In 2011, Johannesburg rescinded the urban edge as a policy instrument because it exacerbated the problems it set out to solve. It allegedly pushed up land values within the urban perimeter, relegating poor communities to live outside its boundary and the line was simply routinely adjusted year on year to incorporate earlier approved urban developments beyond it. A fixed line urban edge was shown to be unworkable. (01)

Graham Shane illustrated his essay, 'The Emergence of Landscape Urbanism,' in *The Landscape Urbanism Reader* (02) with Cedric Price's famous three part diagram of the city as an egg. (03) Price drew the ancient city as a hard-boiled egg arranged in concentric layers within its shell or walls; he portrayed the industrial city as a fried egg, its perimeter deformed and extended outwards by railway corridors, and the post-industrial city is drawn as a scrambled egg - polycentered, granular, lumpy, uneven, with a morphology of enclaves and isolated building typologies.

This version of the city remains grounded and two-dimensional. In 2012 however, the city was portrayed in three dimensions, in a brilliant graphic analysis of Hong Kong. (04) Whereas in most cities of the world, the city is a composition of figure-ground relationships and the ground plane is the datum of urban life, in Hong Kong, this is not the case. The city is a result of a combination of "top-down planning and bottom up solutions ... played out in three dimensional space." (05) A continuous network of elevated or underground pedestrian passageways, stairs, escalators, elevators and footbridges pass through malls, office lobbies, train stations, bus stations, ferry terminals, public parks etc. and replace the ground as the basis of urban life. The city is, quite literally, ungrounded, emerging as a topological continuum, where atmosphere, or "microclimates of temperature, humidity, noise and smell [06] generate urbanity, direct circulation and make place. To make sense of the complexity of this city spatially and politically requires thinking it as and through volume. [07]

In contrast to this, one of the things that has become apparent through the work of DS 18 at the University of Westminster over the past two years is that today's cities can no longer be thought about as bounded territories or even volumes at all. They are entities massively disaggregated and distributed across space and time that generate profoundly different temporalities and scales than the ones we are used to and do not fit neatly into the nested spatial model of contemporary politics. Cities today leak out of all such enclosures. They are what Timothy Morton describes as "viscous:" [08] sticky, oozy; they secrete (toxins, sewage) they belch (pollution), they devour (energy, food, water), they suck in (people, commodities) and spit out (waste) etc. They simply cannot be thought of as distinctive bounded territories any longer. Put another way, cities are punctuated, discontinuous geographies and exchanges that envelop the planet [9] and no longer coincide with notions of bounded territory or sovereignty at all [10] despite the appearance otherwise.

In what could be characterized as a condition of generalized urbanization, increasingly diffuse agglomeration patterns blend with a dense mesh of infrastructural networks and are strongly interwoven with expanding zones of production, supply, and disposal that cover the whole planet. [11]

This is the premise behind 'The City of 7 Billion,' a research project by Joyce Hsiang and Bimal Mendis of Plan B Architecture and Urbanism and the Yale School of Architecture, to "reinvent the world as a single urban entity" and "give agency to the architect to confront the global crisis of urban growth." [12] The project has developed a spatial visualization of population density for a number of cities for the period 1990 – 2015. This data was built as a physical installation at the 2011 Chengdu Biennale, in the form of a 10x10x10ft cube that visitors could walk into. Inverted three dimensional population density maps for North America, Asia and Africa protruded into the cube from the ceiling and two walls. This spikey space, surrounding visitors in three-dimensional data, conveyed a sensual experience of the urban globe inside of which we now live. [13] Our work has gone further and suggests that cities today exceed even planetary boundaries. Their mass warps subterranean strata, animates the ocean, grips and distorts the atmosphere and beyond, [14] casting its shadow back and forward across millennia of time.

A baby vomits curdled milk. She learns to distinguish between the vomit and the not-vomit and comes to know the not-vomit as self. Every subject is formed at the expense of some viscous, slightly poisoned substance, possibly teeming with bacteria, rank with stomach acid. The parent scoops up the mucky milk in a tissue and flushes the wadded package down the toilet. Now we know where it goes. For some time we may have thought the U-bend in the toilet was a convenient curvature of ontological space that took whatever we flush down it into a totally different dimension called Away, leaving things clean over here. Now we know better: instead of the mythical land Away, we know that waste goes to the Pacific Ocean or the wastewater treatment facility. Knowledge of the hyperobject Earth and of the hyperobject Biosphere, presents us with viscous surfaces from which nothing can be forcibly peeled. There is no Away on this surface, no here and no there. In effect, the entire Earth is a wadded tissue of vomited milk. [15]

How can we think and represent such a condition and how does it alter what we think of as (urban) design? I suggest that we turn to three related ideas to explore this: to Timothy Morton's idea of the hyperobject, to the idea of urban metabolism and to the idea of propositional design.

"Hyperobject" is a term invented by Morton to designate entities of such spatial and temporal dimension that they defy traditional ontology and comprehension by traditional means. They are "nonlocal," a technical term in quantum theory to describe the entanglement of particles at some distance from one another. Einstein called this "spooky action at a distance." (16) Morton proposes that hyperobjects cofound the idea of the local, as they only exist as "knotty relationships" (17) between gigantic and intimate scales. If cities are considered in this way, a blow is dealt to the idea that they, or anything else for that matter, can be considered discrete spatio-temporal objects; instead they are objects massively distributed in space and time with blurred boundaries at scales considerably larger than we used to think. So, for instance, when I turn on a light switch in London, I might tap into the fossilized remains of plants and animals sedimented under the North Sea millions of year ago; the polystyrene cup I drink my morning coffee from is itself a by-product of liquefied dinosaur bones, otherwise known as oil, and will outlive me by 400 years in a distant landfill. Hyperobjects are time-space stretched to such a vast extent that they become almost impossible to hold in mind. (18) We are only able to experience bits and pieces of them at any one time, as they phase in and out of our consciousness, at a scale 1 + n dimensions lower than their dimensionality. Hyperobjects are trans-dimensional objects that we can never know in total, but only register or plot as time and space ripple around their edges. We experience them as "little edd(ies) of metastability" (19) or haecceities in the mesh of inter-objective spatial and temporal relations in which they embedded.

Particular cities provide knowledge and experience of the urban hyperobject that is otherwise withdrawn from us. They are enmeshed in its vast and long, past and future, history of interconnected, emergent relations. They are both cause and after-effect of these relations. Thought in this way, cities, buildings and infrastructure are opened up to investigation as what Keller Easterling calls "active form," (20) records of past action and arrangements that actively do things in the present and into the future. In Smudge Studio's *Geologic City: A Field guide to the GeoArchitecture of New York* (21) for instance, the geological materials that make up New York's iconic buildings or sites are identified, traced back to their origins and placed on a geological time scale. This reveals the city as a "geological hot spot" (22) rivaling major geological events. At the same time, the authors realize that the materials and forces they encounter are not inanimate things. They are lively actors, "making things happen in the city and catalyzing events by assembling with the world of humans." (23) Geologic time and human time converge. The city's infrastructure serves as the equivalent of dinosaur footprints – it sets up a sensuous, inter-objective system that connects earth materials and geological time with the daily life of New Yorkers.

Architecture and urban design need to grasp that cities comprise and operationalize a multitude of distant but socially and ecologically enmeshed territories, processes, materials and timescales. This includes vast zones of food production, subterranean resource extraction sites, quarries, energy production and distribution networks, satellites, landfills, oceans, the logistical spaces of trade and circulation etc. One way of thinking this is as a "multiplicity of metabolic cycles operating at a series of both spatial and temporal scales, from the building to the planetary, from the daily to the geologic." (24) The concept of urban metabolism becomes a way of weaving together the diverse locations, diverse actors (human and non-human) and diverse social and ecological processes generated by cities, plotting their flows and measuring their geographical imprint to enable design to intervene in them. Design conceptualized in this way becomes a geographic or metabolic agent, unlocking geologic, thermodynamic or oceanographic imaginaries and their infrastructural potential. Designing infrastructure," says Easterling, "is designing action." (25)

The ecologies of cities flow across borderlines, continents and oceans in ways that confound discourses of territoriality, sovereignty and governance. (26)

It is impossible to look at the city as a kind of discrete entity any more given the way financial networks, ecological networks, or social networks work. These systems have much larger footprints that the actual physical or political boundaries of cities. (27)

Thinking (and representing) the ways cities mobilise resources, alter weather patterns and serve as attractors of people and things makes it clear that they operate as interconnected entities in an all-encompassing, though discontinuous, urban system. Cities can no longer be thought of as fixed or static, but are complex, flowing, topological eco- (eco-nomic / eco-logical) systems that 'slice' across geopolitical boundaries, invade bodies, disrupt weather and send real estate prices rocketing. The agents and practices that still define cities as entities (administration, city government, city planning etc.) rely

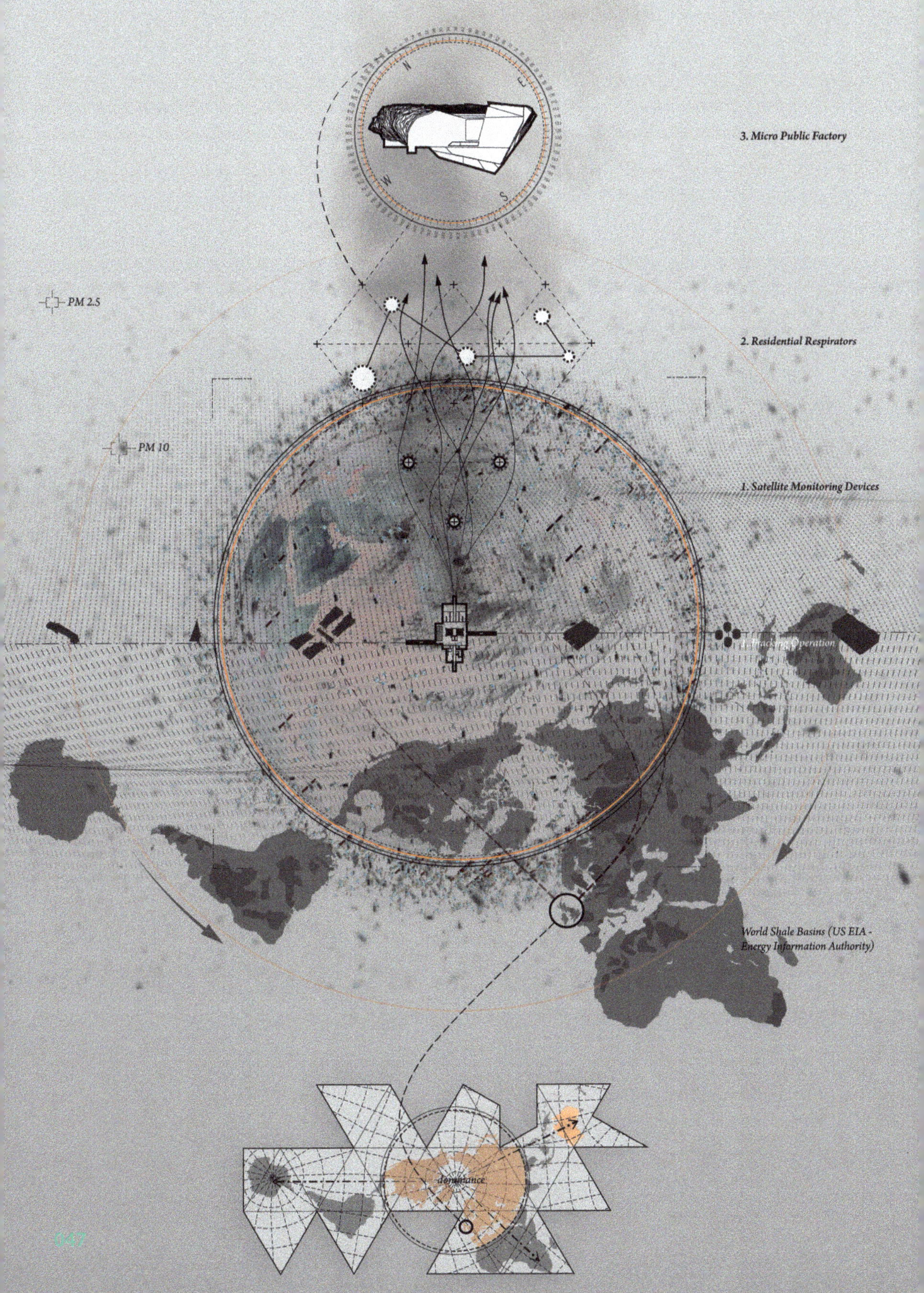
3. Micro Public Factory
PM 2.5
2. Residential Respirators
PM 10
1. Satellite Monitoring Devices
1. Tracking Operation
World Shale Basins (US EIA - Energy Information Authority)
dominance

on outside inputs, in the form of foreign investment, multinational corporations, distant energy supplies, migrant labour and so on to sustain them. At the same time, the toxins or pollutants they release into the earth, atmosphere, rivers and oceans, disperse around the globe according to any number of planetary logics. Take, for instance, the Fukushima-Daichii nuclear power plant in Japan that began operation in 1971 and was damaged beyond repair by the 2011 tsunami. Its design, construction and operation mobilized an extensive international network of state and non-state actors. It was designed by American multi-national, General Electric, with reactors supplied by General Electric, Toshiba and Hitachi. Construction was by Japanese construction company, Kajima and it was run for 40 years by the Tokyo Electric Power Company [28] After the tsunami of 11 March 2011, air- and ocean-borne radionuclides released from its damaged reactors were shown in computer simulations to have dispersed in vast sweeping circles eastwards across the Pacific till they breached the west coast of the USA and spread over the north American continent. [29] The world, as a bounded entity, all but disappeared.

Cities and their infrastructures today are elements of a porous, leaky hyperobject, being continually de- and re-territorialised by the polymorphous relations and networks that constitute it. Design can no longer afford to ignore this. This requires firstly, a new global sensibility. By this I do not mean ramping design up to a global scale, as in the 'City of 7 Billion' project, which remains committed to a modernist design paradigm. It claims that with enough data and computing power, it will be able to plan the planet so to speak, a somewhat futile technocratic exercise that cannot but fail to direct the complexity of the global urban hyperobject. No, by global sensibility, I mean the awareness that what we do at a local level is impacted by and impacts on other entities at multiple scales in complex ways. Design is always local, but embedded in relational networks that span the globe and beyond and extend backwards and forwards in time. These networks are incredibly diffuse, while being particularly local. I am arguing for design as a contextual, situated practice that takes into account its consequences at multiple scales and temporalities, many of which can be tracked and imaginatively engaged.

The second thing the urban hyperobject requires of design is an ecological awareness that extinguishes the very idea that we exist in an environment.

> *Ecological awareness is a detailed and increasing sense, in science and outside of it, of the innumerable interrelationships among life forms and between life and non-life ... [This] ends the idea that we are living in an environment. ... What it means is that the more we know about the interconnections, the more it becomes impossible to posit some entity existing beyond or behind the interrelated beings. When we look for the environment, what we find are discrete life forms, non-life and their relationships. But no matter how hard we look, we won't find a container in which they all fit; in particular we won't find an umbrella that unifies them, such as the world, environment, ecosystem or even, astonishingly, earth.* [30]

All there is an open ended, rhyzomatic mesh of interrelating, co-existing, entities, a complex system of dynamic exchanges and distributed agencies. Deleuze and Guattari describe it thus:

> *Here there are no longer any forms or development of forms; nor are there subjects or the formation of subjects. There is no structure, any more than there is genesis. There are only relations of movement and rest, speed and slowness between unformed elements, or at least between elements that are relatively unformed, molecules, and particles of all kinds. There are only haecceities, affects, subjectless individuations that constitute collective assemblages. Nothing develops, but things arrive late or early, and form this or that assemblage depending on their compositions of speed. Nothing subjectifies, but haecceities form according to compositions of nonsubjectified powers or affects. We call this plane, which knows only longitudes and latitudes, speeds and haecceities, the plane of consistency or composition (as opposed to the plan(e) of organization or development). It is necessarily a plane of immanence and univocality.* [31]

This opens up a very different problematic for design than the modernist one, where design was a conceptually driven operation a designer performed on some externality - a problem, an environment, a site etc. If there are no externalities outside of entities and their interrelationships, this means that design is no longer sovereign, but embedded within complex, relational systems. It interacts reflexively with and from inside these dynamic systems. This means admitting the agency of things as diverse as data, computational software, non-designers, earth materials, energy fields, politics etc. into design and re-conceptualising it as a complex, emergent system itself. Latour would call this, I think, a propositional approach to design. "Propositions," a term he borrows from Alfred North Whitehead, "are not positions, things, substances or essences ... but occasions given to different entities to enter into contact." (32) Propositions are associations of entities that, through contact over the course of an event, perform in certain ways, their definitions are modified and their attributes and competencies in relation to one another are played out. This seems to me to be a working definition for design at the time of hyperobjects, opening the door for all sort of entities, "strange strangers," (33) scales and times to enter the arena of design and for design itself to be reformulated.

Notes

(01) Gauteng Growth and Development Agency, 'The Gauteng Spatial Development Framework', Johannesburg, 2011.

(02) G. Shane, 'The Emergence of Landscape Urbanism', in C. Waldheim (ed.), *The Landscape Urbanism Reader*, New York: Princeton Architectural Press, 2006, pp. 55-67.

(03) C. Price, 'Three Eggs Diagram' Montreal, Centre Canadien d`Architecture, 1982.

(04) A. Frampton, J. D. Solomon and C. Wong, *Cities Without Ground: A Hong Kong Guidebook*. ORO Editions, 2012.

(05) A. Frampton, J. D. Solomon and C. Wong, p. 6.

(06) A. Frampton, J. D. Solomon and C. Wong, p. 100.

(07) S. Elden, 'Secure the volume: Vertical geopolitics and the depth of power', *Political Geography*, 2013. DOI: http://dx.doi.org/10.1016/j.polgeo.2012.12.009

(08) T. Morton, *Hyperobjects: Philosophy and Ecology after the End of the World*, Minneapolis: University of Minnesota Press, 2013, p. 27.

(90) N. Brenner and C. Schmidt, 'Planetary Urbanization', in M. Gandy (ed.), *Urban Constellations*, London, UCL Urban Theory Lab, 2012, pp. 10-13.

(10) G. Bridge, 'The Hole World. Scales and Spaces of Extraction', *Landscapes of Energy, New Geographies* no. 02, 2009, pp. 43-49.

(11) D. Ibanez and N. Katsikis (eds.), *Grounding Metabolism, New Geographies* no. 06, 2014, p. 3.

(12) J. Hsiang and B. Mendis, B. 'The City of 7 Billion: An Index', *New Constellations New Ecologies, Proceedings of the 101st ACSA Conference*, 2013, pp. 596-606, http://higherlogicdownload.s3.amazonaws.com/AIA/aa21f56a-baa6-4d44-a308-a08ffbc272db/UploadedImages/ACSA.AM.101.81.pdf (accessed 14 July 2015).

(13) E. Badger, 'What if the Entire World Lived in 1 City', *CityLab*, 07 March 2013, http://www.citylab.com/design/2013/03/what-if-entire-world-lived-1-city/4897/ (accessed 14 July 2015).

(14) The SOHO solar observation satellite for instance orbits the earth at the Langrain L1 point about 1,5 million kms from the earth at a stable point where the earth's and the sun's gravity balance.

(15) T. Morton, 2013, p. 31.

(16) T. Morton, 2013, p. 45.

(17) T. Morton, 2013, p. 47.

(18) T. Morton, 2013, p. 58.

(19) T. Morton, 2013, p. 85.

(20) K. Easterling, 'The Action is the Form', in M. Shepard (ed.), *Sentient City*, Cambridge, MA., MIT Press, 2011, pp. 154-158.

(21) Smudge Studio, *Geologic City: A Field guide to the GeoArchitecture of New York*, 2011, http://smudgestudio.org/smudge/GeoCity.html (accessed 14 July 2015).

(22) Smudge Studio, 2011.

(23) Smudge Studio, 2011.

(24) D. Ibanez and N. Katsikis, 2014, p. 6.

(25) K. Easterling, 2011, p. 155.

(26) T. Kuehls, *Beyond Sovereign Territory: The Space of Ecopolitics*, Minneapolis, University of Minnesota Press, 1996.

(27) E. Badger, 2013.

(28) Fukushima Daiichi Nuclear Disaster, https://en.wikipedia.org/wiki/Fukushima_Daiichi_nuclear_disaster (accessed 15 July 2015).

(29) CEREA, 'Atmospheric dispersion of radionuclides from the Fukkushima-Daichii nuclear power plant', http://cerea.enpc.fr/en/fukushima.html, last updated 9 Jan 2014 (accessed 15 July 2015).

(30) T. Morton, 2013, p. 129.

(31) G. Deleuze and F. Guattari, *A Thousand Plateaus, Capitalism and Schizophrenia*, trans. B. Massumi, London, Continuum, 2003, p. 266.

(32) B. Latour, *Pandora's Hope. Essays on the Reality of Science Studies*, Cambridge, MA., Harvard University Press, 1999, p. 141.

(33) T. Morton, 2013, p. 130.

MAP CREDITS

P052-053 Submarine Cable Network, 2015

Ben Pollock

Sources: https://www.telegeography.com; http://www.cablemap.info http://sedac.ciesin.columbia.edu/data/set/gpw-v3-population-count-future-estimates; http://www.dospeedtest.com/speedtest-result/country-statistics/ Maldives

P054-055 Geostationary Satellite Network, Maldives Coverage

Ben Pollock

Sources: http://opensignal.com; http://satbeams.com; http://sedac.ciesin.columbia.edu/data/set/gpw-v3-population-count-future-estimates; http://satellitedebris.net/Database/LaunchHistryView.php

P056-057 Global distribution of mined materials found in e-waste, abstractly modelling the topographical impact of this in section

Alice Thompson

Source: http://indexmundi.com

P058-059 Impacts of Climate Change and Local Stressors on Coral Reefs

Jessica Hillam

Sources: http://www.wri.org/publication/reefs-risk-revisited; http://marinebio.org/oceans/coral-reefs/

P060-061 Global Sand Flows

Iulia Stefan

Sources: http://www.win.tue.nl/~vanwijk/myriahedral/; http://www.trademap.org/Country_SelProduct_Graph.aspx; http://unctadstat.unctad.org/wds/ReportFolders/reportFolders.aspx; http://atlas.media.mit.edu/en/visualize/geo_map/hs92/import/show/all/2505/2013/; http://www.naturalearthdata.com/

062-063 Aerosol movement across the Arabian Sea from 2nd - 24th Feb 2014

These drawings provide a platform to examine the geo-political and socio-economic impacts of transboundary aerosol pollution.

Calvin Sin

Source: https://worldview.earthdata.nasa.gov/

Software: Exel, Grasshopper, Ilustrator, Realflow, Rhinoceros

MAPS OF THE ANTHROPOCENE

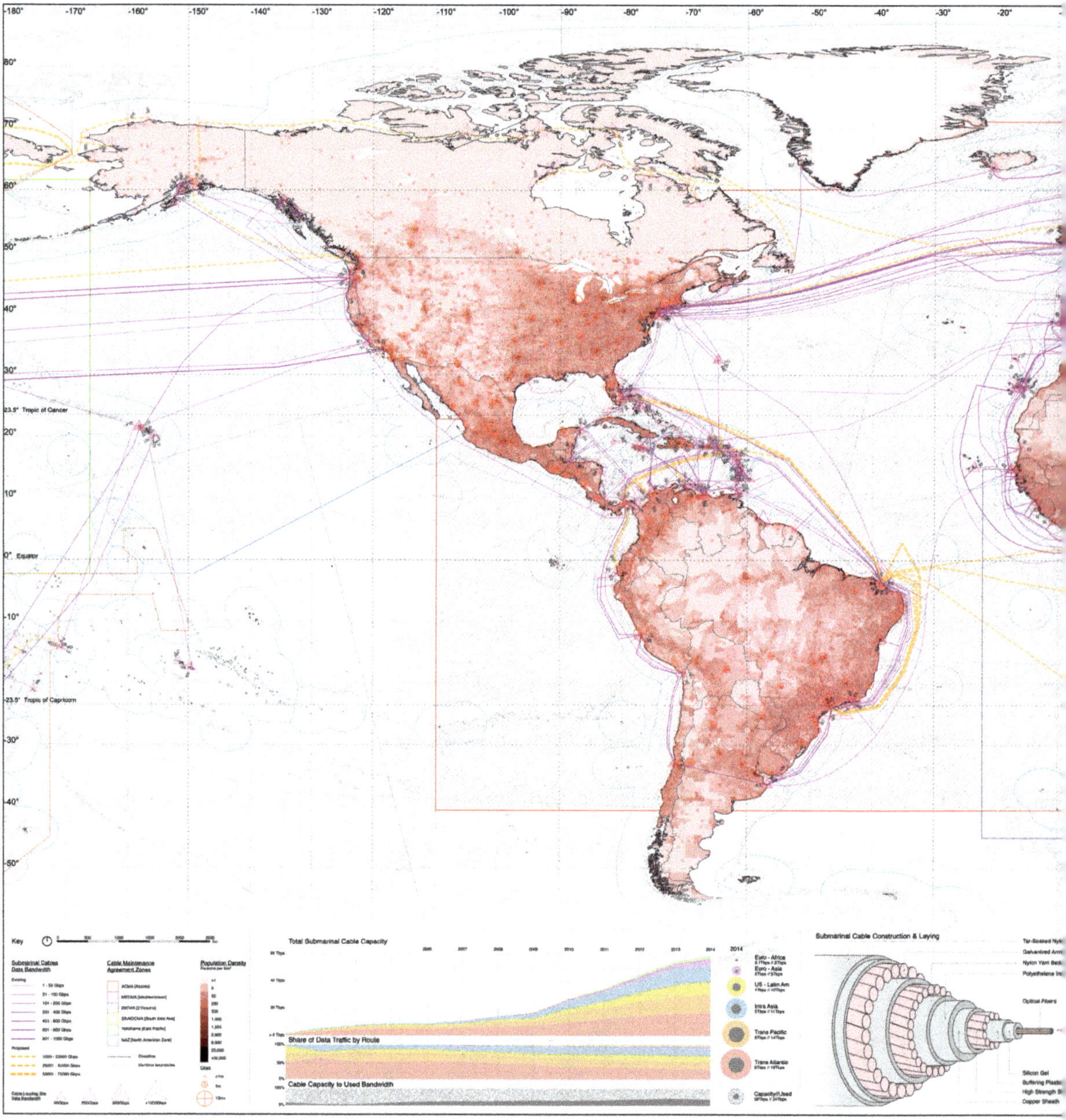
-180°
-170°
-160°
-150°
-140°
-130°
-120°
-110°
-100°
-90°
-80°
-70°
-60°
-50°
-40°
-30°
-20°
80°
70°
60°
50°
40°
30°
23.5° Tropic of Cancer
20°
10°
0° Equator
-10°
-23.5° Tropic of Capricorn
-30°
-40°
-50°
Key
Submarinal Cables Data Bandwidth
Existing
Proposed
Cable Maintenance Agreement Zones
Population Density
Total Submarinal Cable Capacity
Share of Data Traffic by Route
Cable Capacity to Used Bandwidth
2014
Euro - Africa
Euro - Asia
US - Latin Am
Intra Asia
Trans Pacific
Trans Atlantic
Capacity//Used
Submarinal Cable Construction & Laying
Optical Fibers
Silicon Gel
Copper Sheath

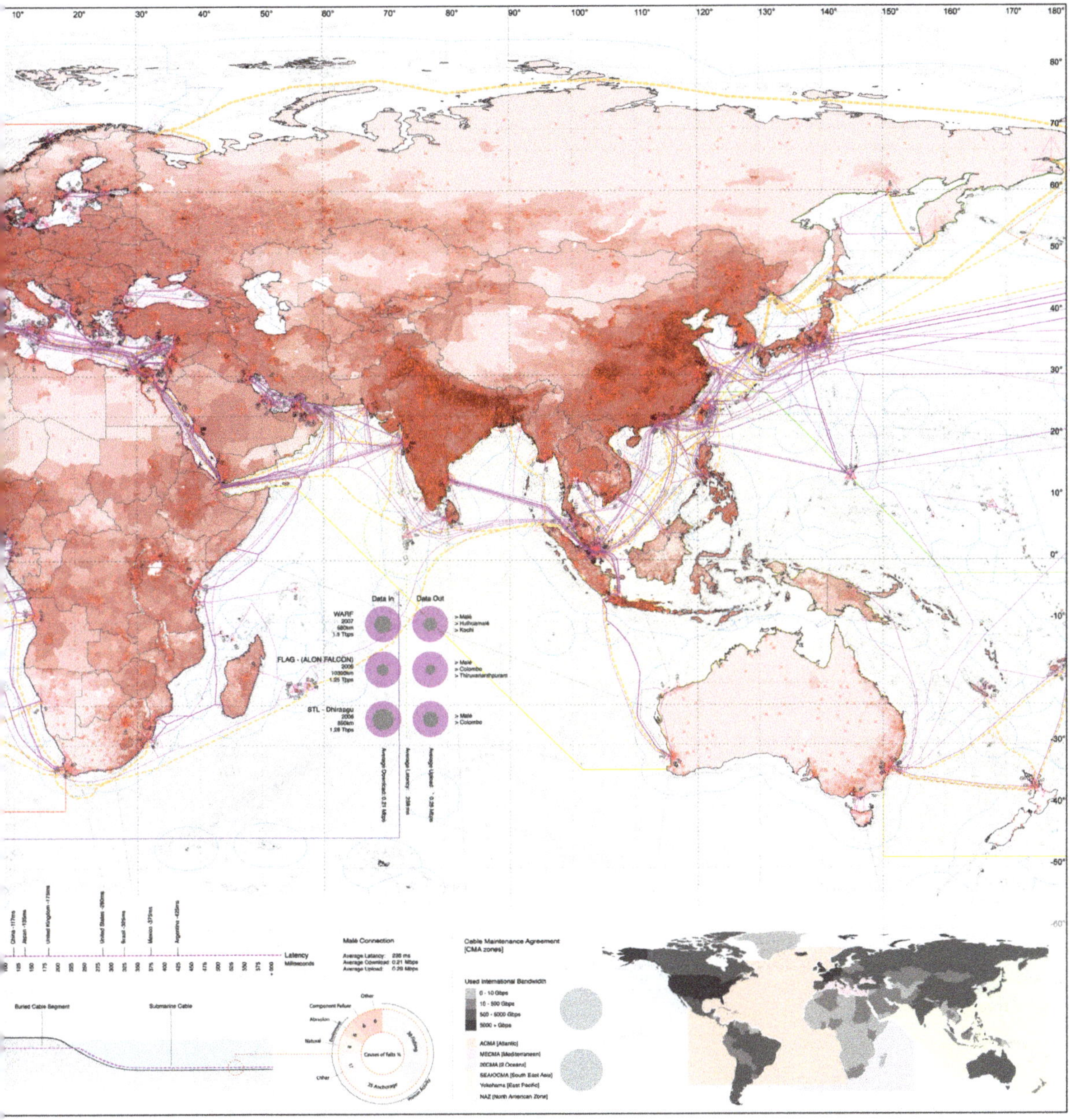

Data In
Data Out
WARF
2007
680km
1.5 Tbps
> Malé
> Hulhumalé
> Kochi
FLAG - (ALON FALCON)
2006
10300km
1.25 Tbps
> Malé
> Colombo
> Thiruvananthpuram
STL - Dhiraagu
2006
850km
1.28 Tbps
> Malé
> Colombo
Average Download: 0.21 Mbps
Average Latency: 236 ms
Average Upload: 0.29 Mbps
Latency
Milliseconds
Malé Connection
Average Latency: 236 ms
Average Download: 0.21 Mbps
Average Upload: 0.29 Mbps
Buried Cable Segment
Submarine Cable
Other
Component Failure
Abrasion
Natural
Other
Causes of faults %
25 Anchorage
Cable Maintenance Agreement
[CMA zones]
Used International Bandwidth
0 - 10 Gbps
10 - 500 Gbps
500 - 5000 Gbps
5000 + Gbps
ACMA [Atlantic]
MECMA [Mediterranean]
20CMA [2 Oceans]
SEA/OCMA [South East Asia]
Yokohama [East Pacific]
NAZ (North American Zone)

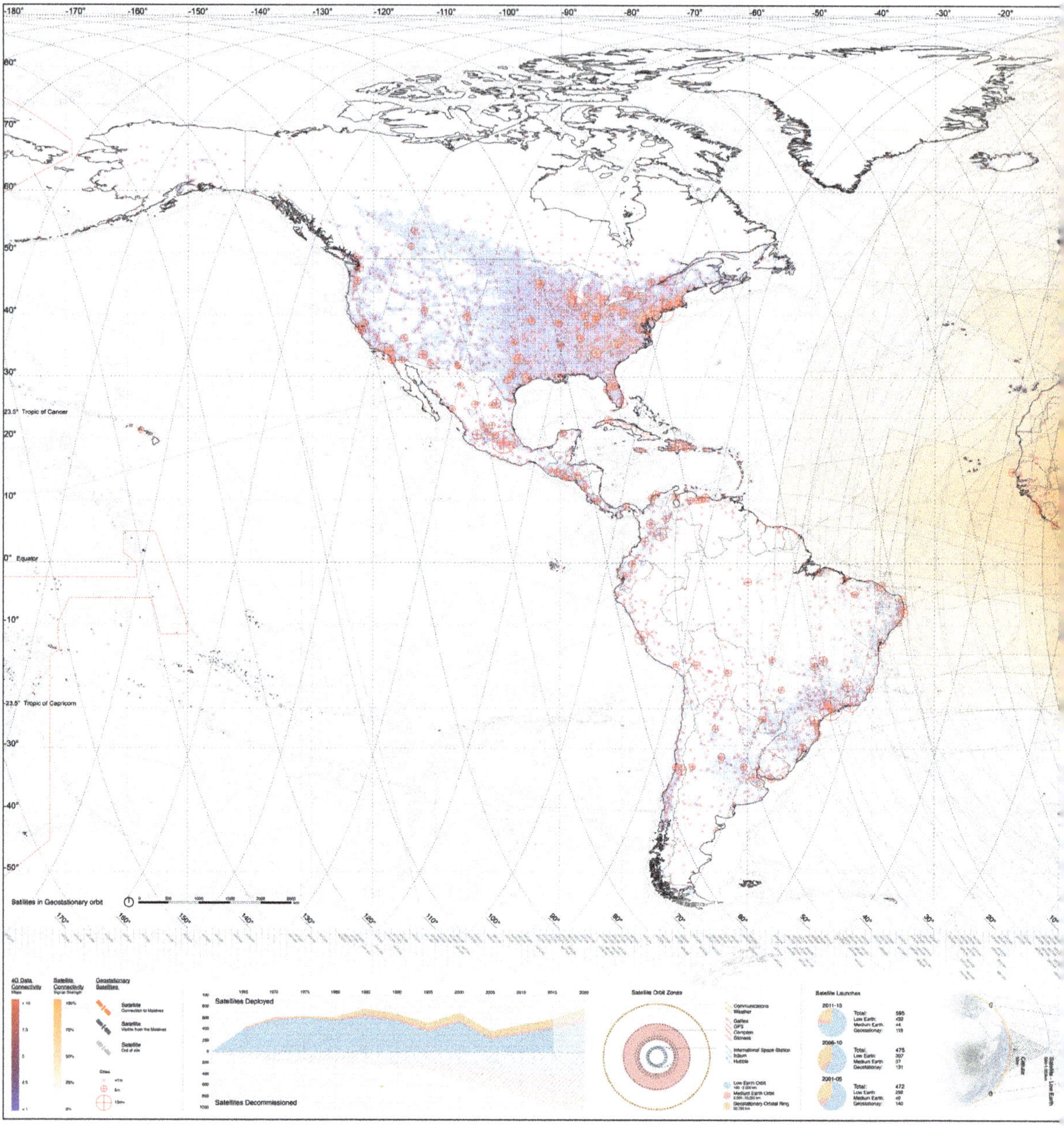

-180°
-170°
-160°
-150°
-140°
-130°
-120°
-110°
-100°
-90°
-80°
-70°
-60°
-50°
-40°
-30°
-20°
80°
70°
60°
50°
40°
30°
23.5° Tropic of Cancer
20°
10°
0° Equator
-10°
-23.5° Tropic of Capricorn
-30°
-40°
-50°
Satllites in Geostationary orbit
170°
160°
150°
140°
130°
120°
110°
100°
90°
80°
70°
60°
50°
40°
30°
20°
10°
4G Data Connectivity
Satellite Connectivity
Geostationary Satellites
Satellite
Connection to Maldives
Satellite
Visible from the Maldives
Satellite
Out of site
Cities
<1m
5m
15m+
Satellites Deployed
Satellites Decommissioned
Satellite Orbit Zones
Communications
Weather
Galileo
GPS
Compass
Glonass
International Space Station
Iridium
Hubble
Low Earth Orbit
Medium Earth Orbit
Geostationary Orbital Ring
Satellite Launches
2011-15
Total: 595
Low Earth: 432
Medium Earth: 44
Geostationay: 119
2006-10
Total: 475
Low Earth: 307
Medium Earth: 37
Geostationay: 131
2001-05
Total: 472
Low Earth: 292
Medium Earth: 40
Geostationay: 140
Satellite - Low Earth

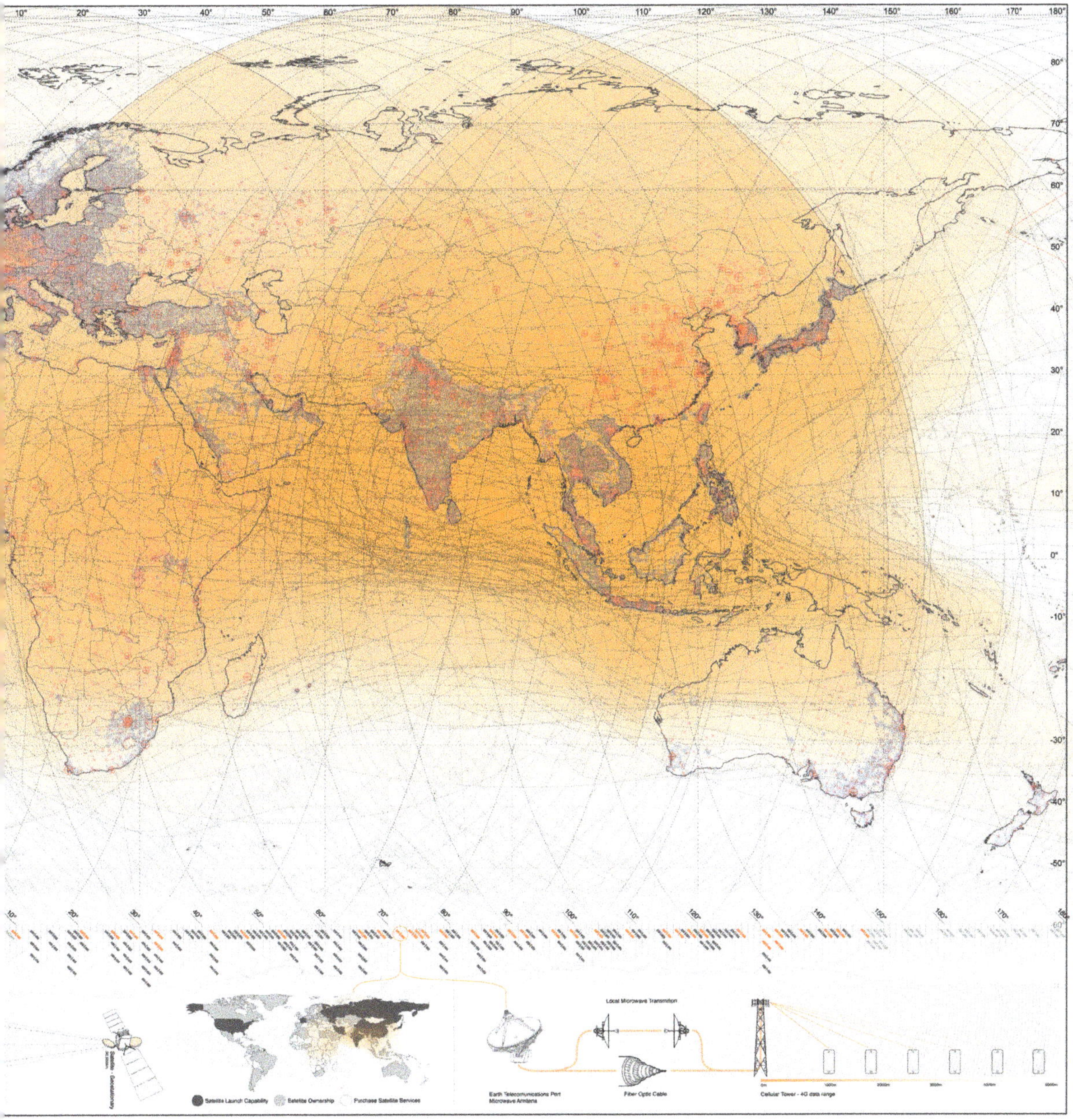
Satellite Launch Capability
Satellite Ownership
Purchase Satellite Services
Earth Telecommunications Port
Microwave Antenna
Local Microwave Transmition
Fiber Optic Cable
Cellular Tower - 4G data range

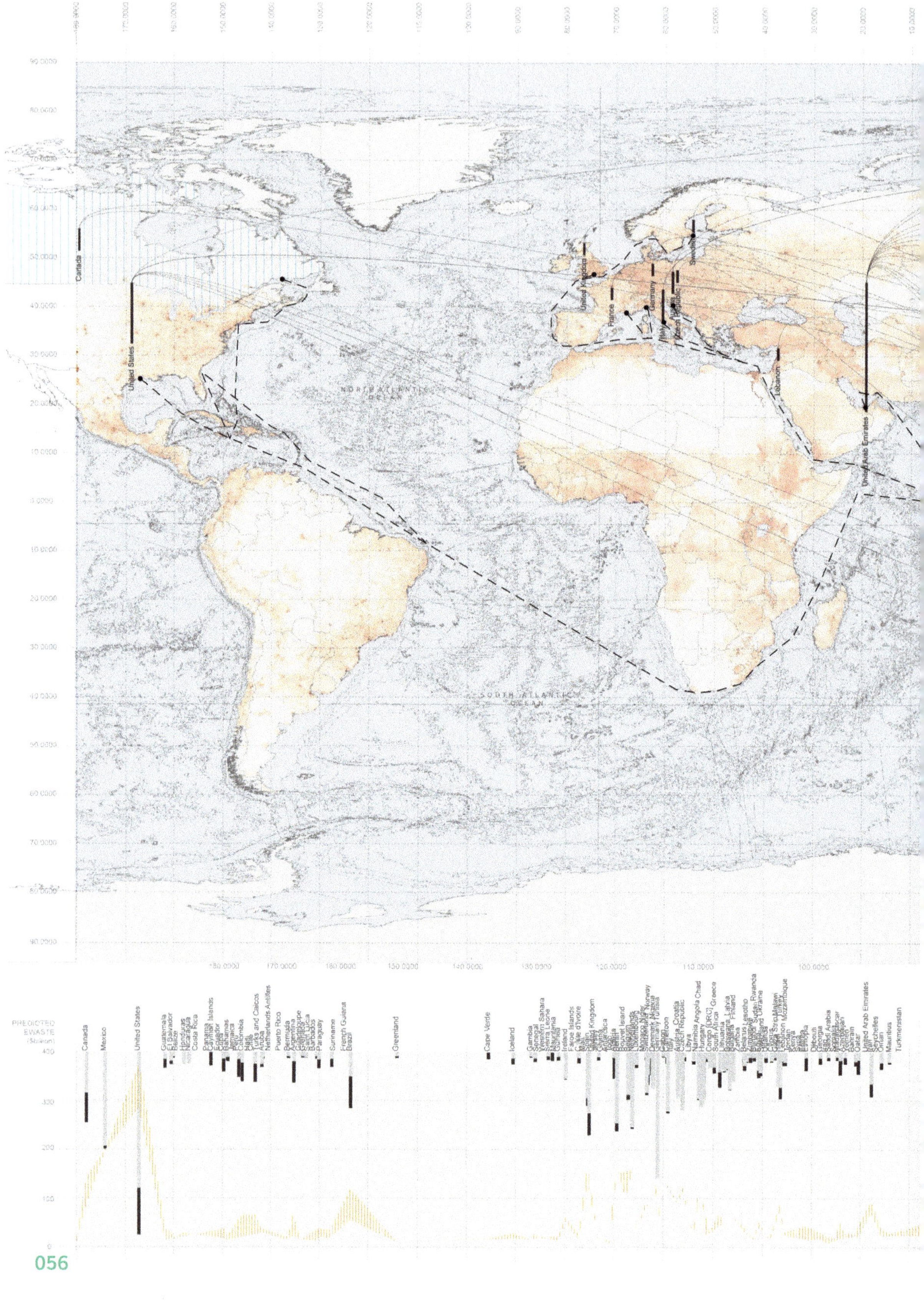

NORTH ATLANTIC OCEAN
SOUTH ATLANTIC OCEAN
Canada
United States
United Kingdom
France
Germany
Sweden
Lebanon
United Arab Emirates
PREDICTED EWASTE
Mexico
Greenland
Cape Verde
Iceland

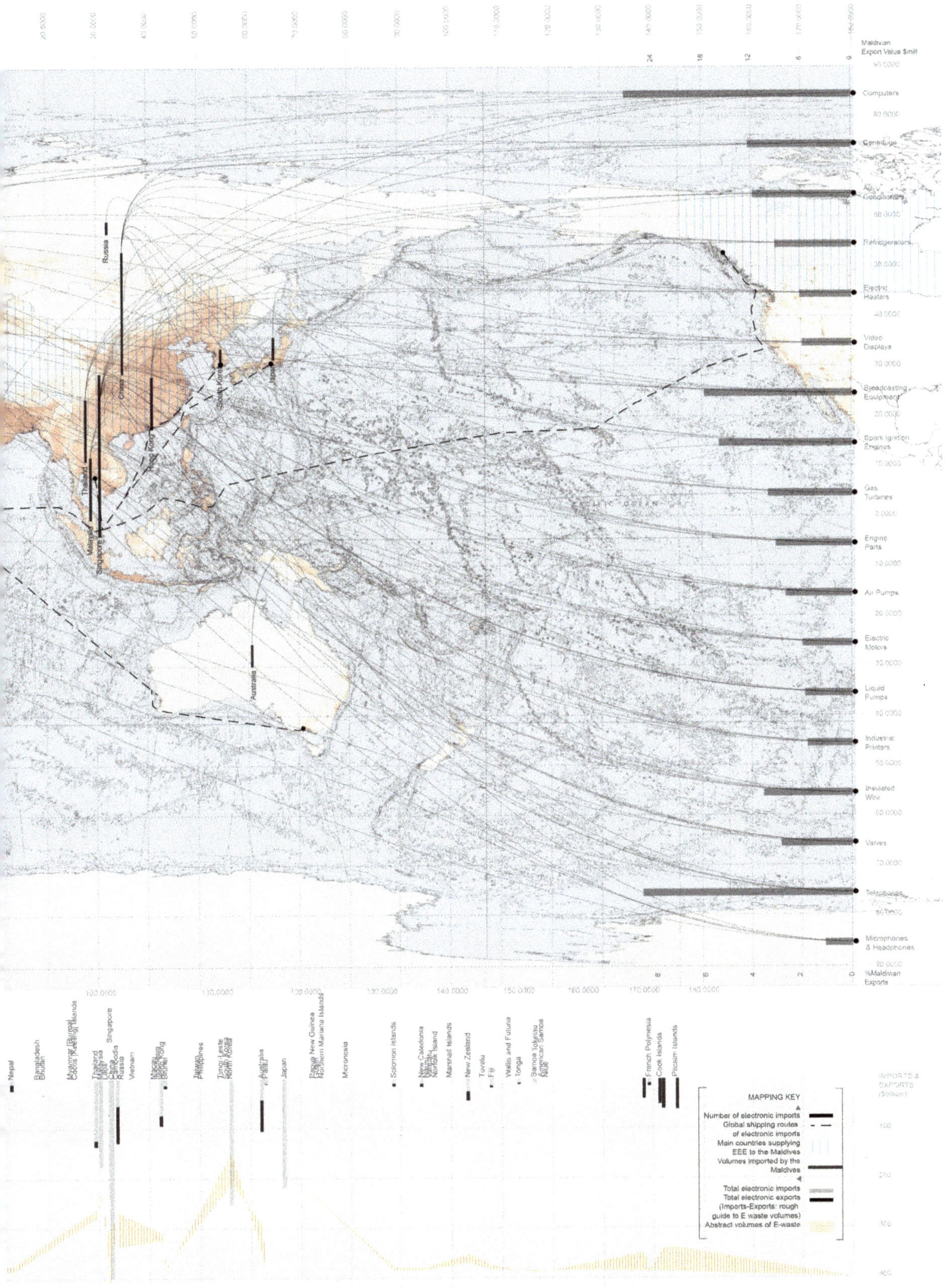

Maldivian Export Value $mil
Computers
Centrifuge
Air Conditioners
Refrigerators
Electric Heaters
Video Displays
Broadcasting Equipment
Spark Ignition Engines
Gas Turbines
Engine Parts
Air Pumps
Electric Motors
Liquid Pumps
Industrial Printers
Insulated Wire
Valves
Telephones
Microphones & Headphones
%Maldivian Exports
Russia
China
Hong Kong
Thailand
Malaysia
Singapore
South Korea
Japan
Australia
Nepal
Bangladesh
Bhutan
Myanmar (Burma)
Cocos (Keeling) Islands
Vietnam
Philippines
Timor-Leste
North Korea
Papua New Guinea
Northern Mariana Islands
Micronesia
Solomon Islands
New Caledonia
Norfolk Island
Marshall Islands
New Zealand
Tuvalu
Fiji
Wallis and Futuna
Tonga
Samoa
Tokelau
American Samoa
Niue
French Polynesia
Cook Islands
Pitcairn Islands
IMPORTS & EXPORTS ($Billion)
MAPPING KEY
Number of electronic imports
Global shipping routes of electronic imports
Main countries supplying EEE to the Maldives
Volumes imported by the Maldives
Total electronic imports
Total electronic exports (Imports-Exports: rough guide to E waste volumes)
Abstract volumes of E-waste

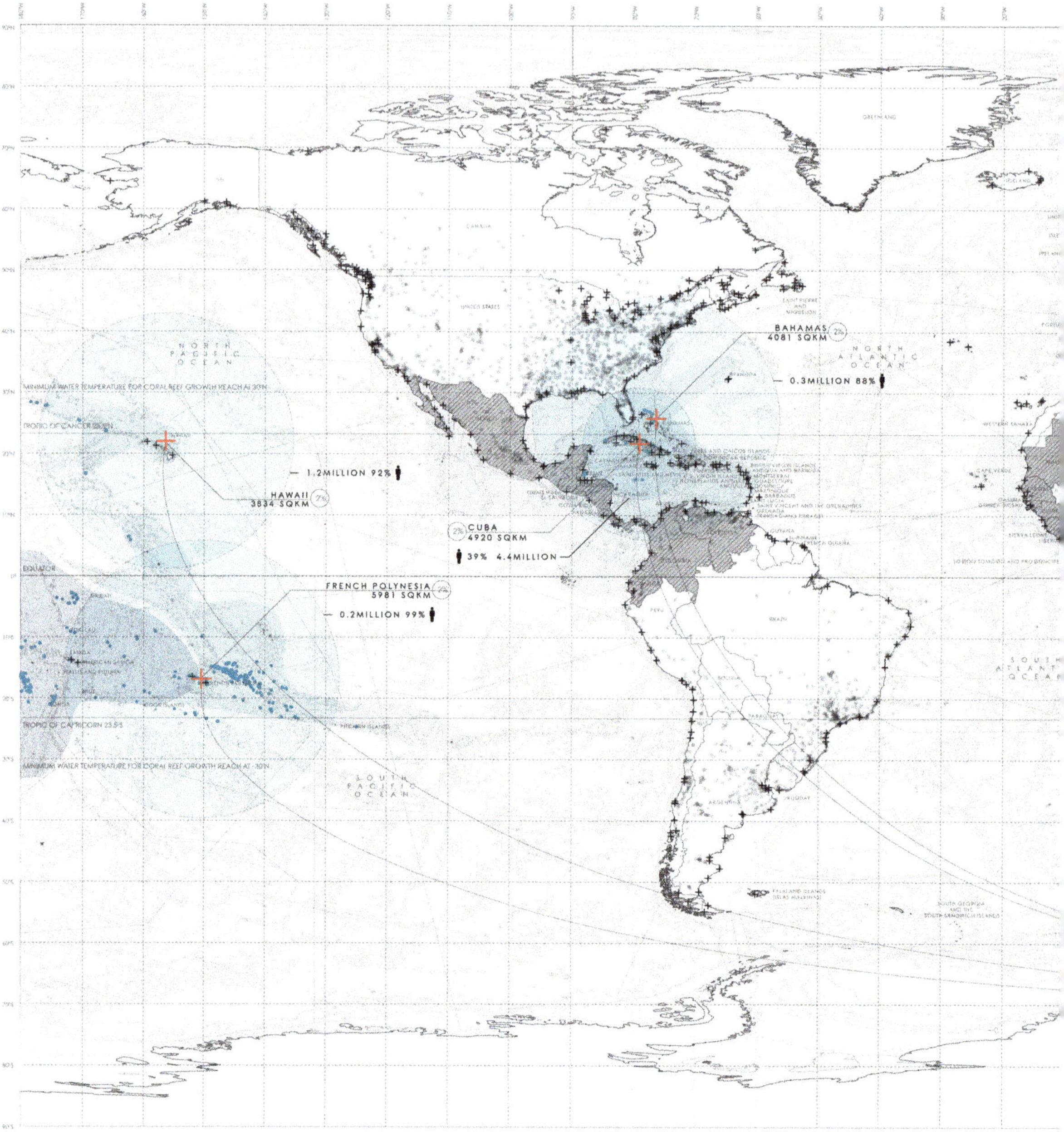

MAP KEY

Reef species

Over 500
301 - 500
201 - 300
101 - 200
50 - 100
Less than 50
Coral reef Locations
Port

MAP DATA

$350 BILLION
ECONOMIC VALUE OF REEFS

249,713 KM²
GLOBAL AREA OF REEFS

$1000 P M²
TO REPAIR DAMAGED REEFS

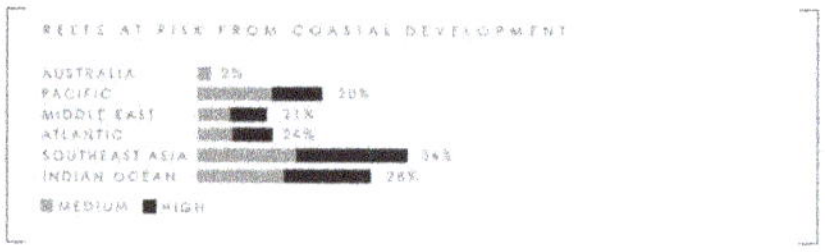

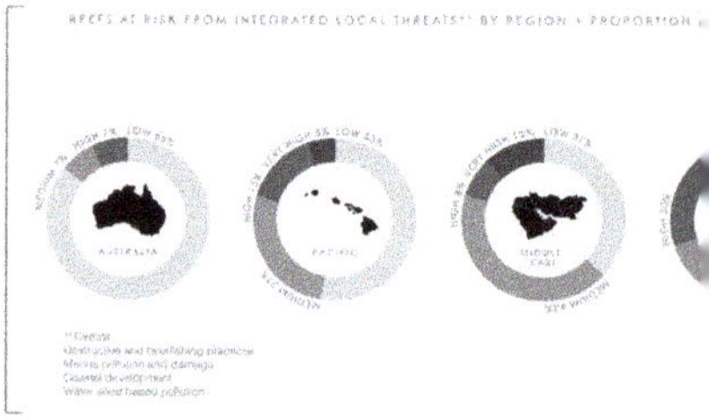

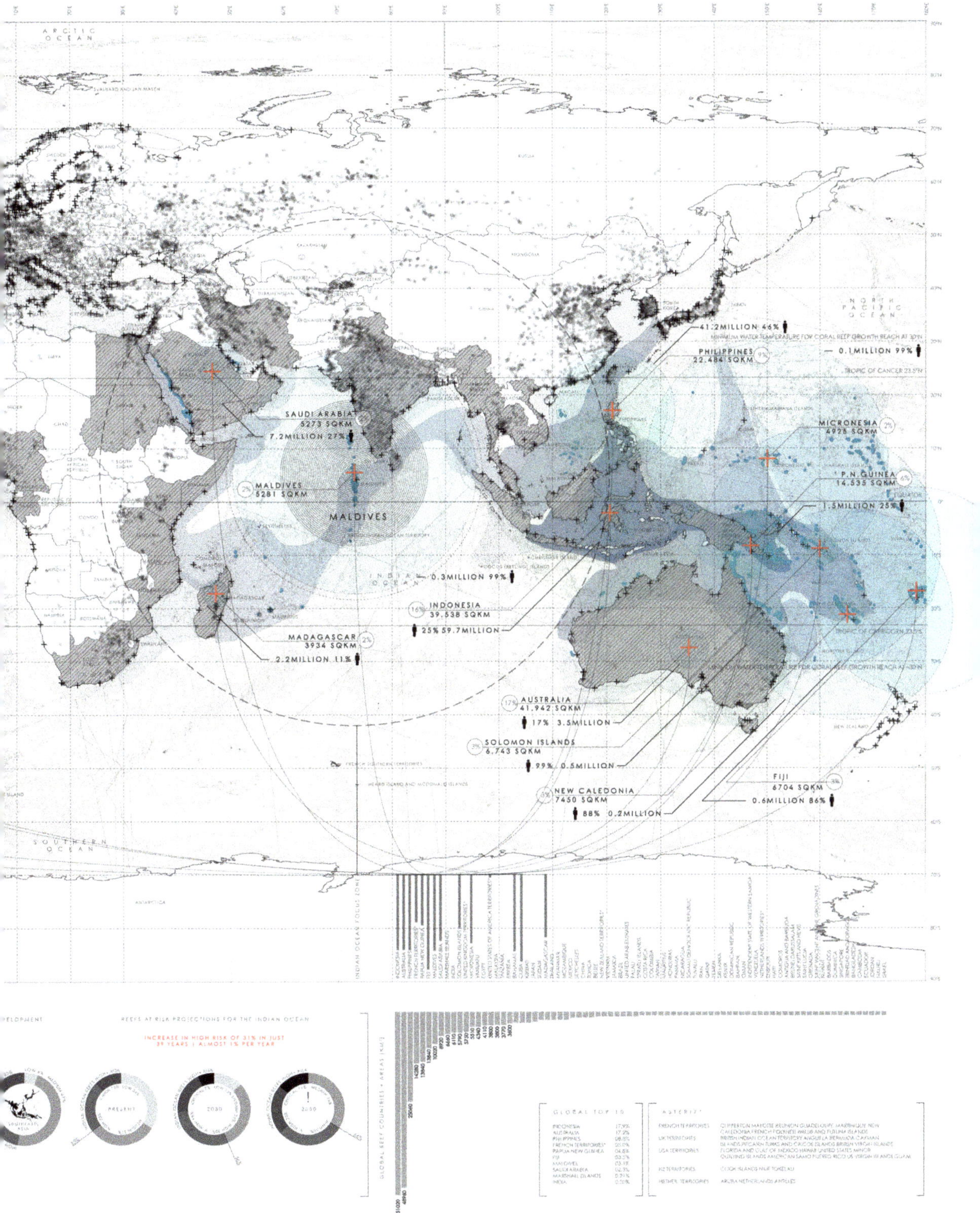
ARCTIC OCEAN
NORTH PACIFIC OCEAN
INDIAN OCEAN
SOUTHERN OCEAN
41.2MILLION 46%
PHILIPPINES
22.484 SQKM
9%
0.1MILLION 99%
MINIMUM WATER TEMPERATURE FOR CORAL REEF GROWTH REACH AT 30°N
TROPIC OF CANCER 23.5°N
SAUDI ARABIA
5273 SQKM
2%
7.2MILLION 27%
MICRONESIA
4925 SQKM
2%
P.N. GUINEA
14.535 SQKM
6%
EQUATOR
1.5MILLION 25%
2%
MALDIVES
5281 SQKM
MALDIVES
0.3MILLION 99%
16%
INDONESIA
39.538 SQKM
25% 59.7MILLION
MADAGASCAR
3934 SQKM
2%
2.2MILLION 11%
TROPIC OF CAPRICORN 23.5°S
17%
AUSTRALIA
41.942 SQKM
17% 3.5MILLION
2%
SOLOMON ISLANDS
6.743 SQKM
99% 0.5MILLION
FIJI
6704 SQKM
3%
0.6MILLION 86%
3%
NEW CALEDONIA
7450 SQKM
88% 0.2MILLION
INDIAN OCEAN FOCUS ZONE
REEFS AT RISK PROJECTIONS FOR THE INDIAN OCEAN
INCREASE IN HIGH RISK OF 31% IN JUST 39 YEARS | ALMOST 1% PER YEAR
PRESENT
2030
2050
GLOBAL REEF COUNTRIES + AREAS [KM²]
51020
48960
25060
14280
13840
12640
10020
8920
6660
6110
5790
5750
5310
4340
4110
3850
3800
3770
3600
GLOBAL TOP 10
INDONESIA 17.9%
AUSTRALIA 17.2%
PHILIPPINES 08.8%
FRENCH TERRITORIES* 05.0%
PAPUA NEW GUINEA 04.8%
FIJI 03.5%
MALDIVES 03.1%
SAUDI ARABIA 02.3%
MARSHALL ISLANDS 02.1%
INDIA 02.0%
ASTERIX*
FRENCH TERRITORIES
UK TERRITORIES
USA TERRITORIES
NZ TERRITORIES
NETHER. TERRITORIES

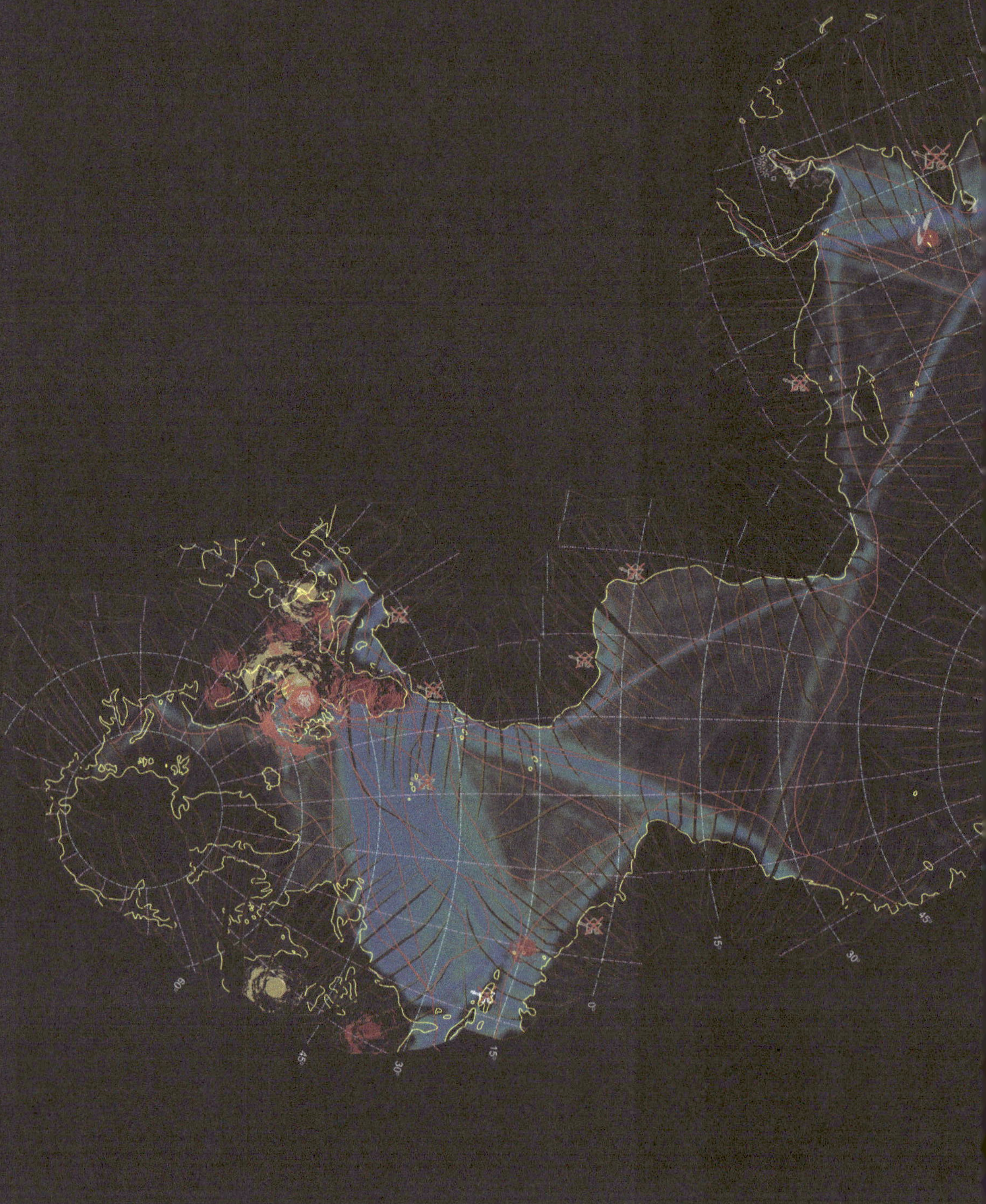
15°
30°
45°
0°
15°
30°
45°
60°

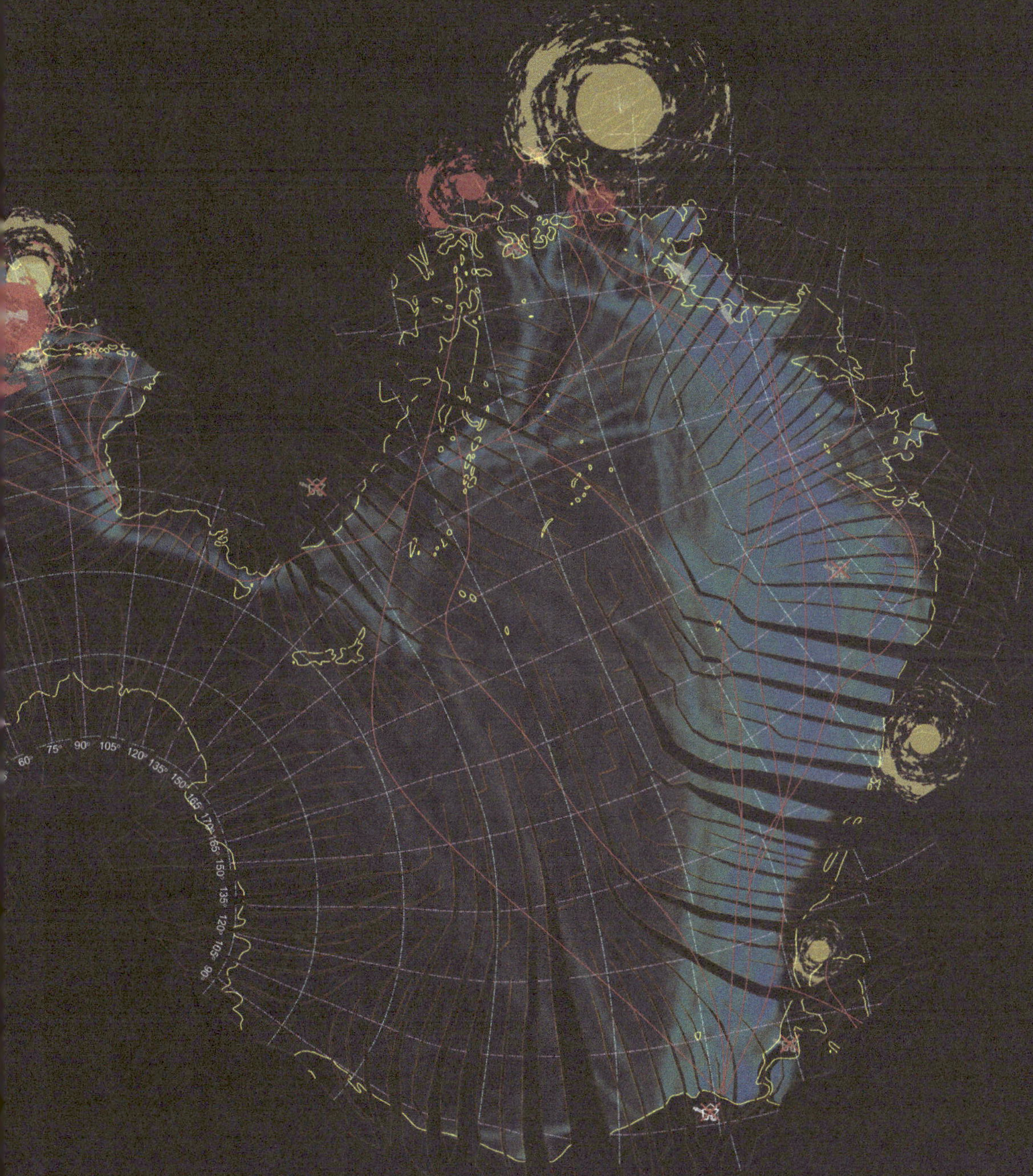
60°
75°
90°
105°
120°
135°
150°

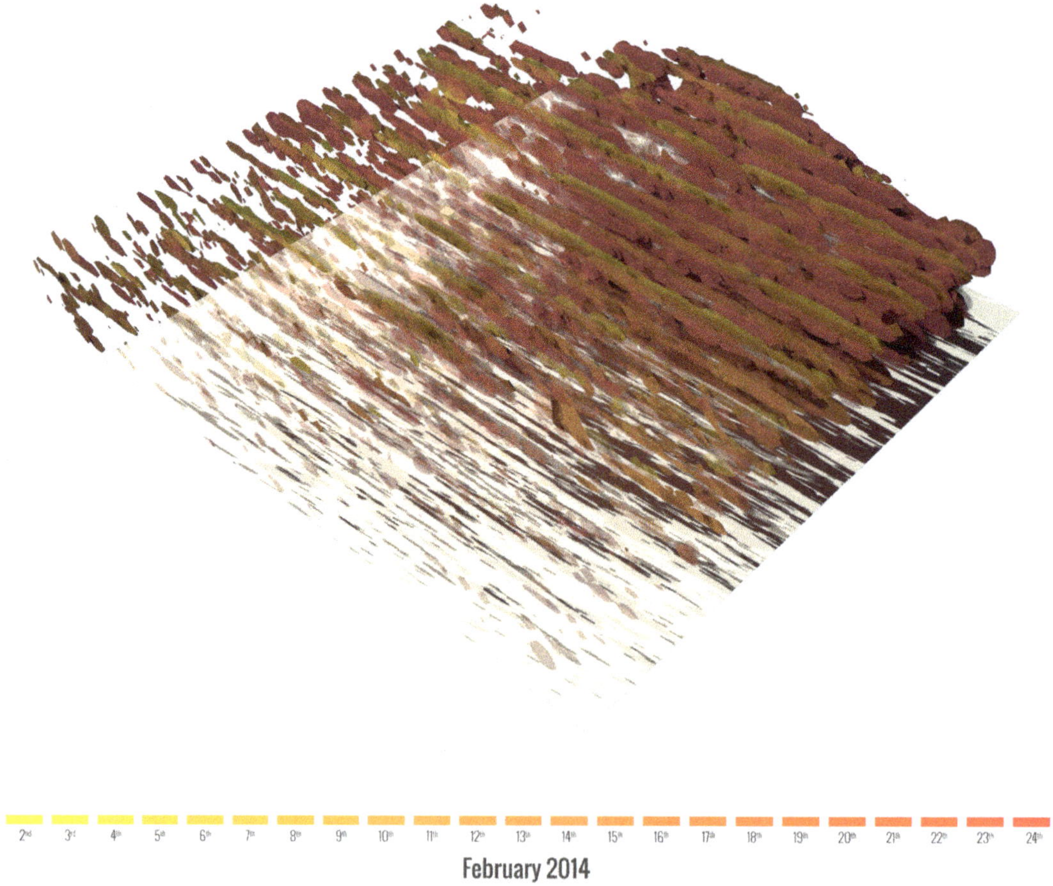
2nd
3rd
4th
5th
6th
7th
8th
9th
10th
11th
12th
13th
14th
15th
16th
17th
18th
19th
20th
21st
22nd
23rd
24th
February 2014

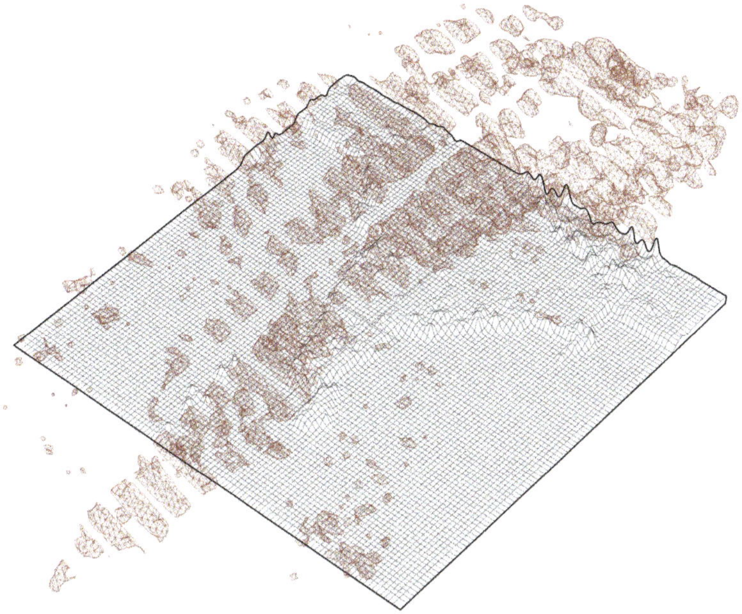

THE BAKKEN TREATISE

NICK AXEL

1.0

On August 8th, 2005, George W. Bush signed into law the Energy Policy Act, within which hydraulic fracturing', or fracking as it is commonly referred to, was deregulated. Coupled with the economic recession and mortgage crisis that began in 2007, what transpired was a nationwide hydrocarbon (oil and gas) drilling boom. While fracking has deployed infrastructural systems such as rail and pipelines throughout the country at an impressive scale, the base unit driving all else is the well, a hole dug deep into the ground to connect the surface of the earth and everything on top of it with all that lie below. In under ten years, upwards of one hundred thousand new wells have emerged across the national landscape. The proliferation of wells, a process entirely made possible by the deregulation of fracking, has reconfigured the relationship between citizen agency and land rights, and as such, the constitution of territory as a political technology in the United States.

1.1

Starting in the late 19th century in the USA, landowners began selling the rights to extract minerals from beneath their land while holding onto the property rights of the land itself. This was a gesture of profit maximization, like sharecropping, just along vertical axis. This inaugurated what became the legal norm in 1916 and is today called a "split estate" (01) which is where one person owns the surface rights and another owns those of the subsurface. Yet seeing what lies below the surface cannot be used in a conventional sense of domestic habitation or agricultural cultivation. What follows is referred to as Mineral Estate Dominance, a legal property doctrine that is in fact over five hundred years old. This states that whoever owns subsurface rights is granted the right to use the surface as is "reasonably necessary" to extract the resources that lie below. (02)

1.2

Fracking is a supplementary step in the conventional process of drilling for hydrocarbons only necessary in so-called unconventional locations, meaning where one can't just tap into a pool of liquid or gas and suck out the goods. (03) Fracking injects large quantities of a chemical fluid mixture down into a well at extremely high pressure in order to break open the rock, fracturing its geological structure and releasing the resources that were previously trapped within. By making it possible to extract hydrocarbons from where it couldn't be done before, fracking reclassified territory. Many parts previously recognized as not having oil and gas were rendered positive and subjected to a new set of legal conditions and forces of development. Thanks to the doctrine of Mineral Estate Dominance, by being able to break open the geology below, fracking effectively opened up the geography above to industry and finance. The legal deregulation of hydraulic fracturing coalesced a series of historical conditions into a land grab for subsurface property rights. (04)

1.3

In a matter of just a few short years, North Dakota became the locus of the fracking boom. People from all over the country flocked to the northwest corner of North Dakota, a relatively desolate place along the Canadian border at the geographic center of North America. They came for what lay underneath: the Bakken formation, a vast and continuous geological layer of oil-rich shale. Six-figure salaries were offered on the oilfield, and an effective minimum wage twice the national standard. Nearly 15,000 new wells were drilled after 2008. State-wide oil production levels grew by over 1000%, from under 100,000 barrels per day to over 1,000,000. (05) The State of North Dakota produced more oil than the individual output of four OPEC countries: Algeria, Libya, Ecuador and Qatar.

North Dakota experienced a notably less-intense boom just a few decades before, but one whose bust nevertheless ruined state coffers and left the built environment in a state of destitution. The infrastructural demands of this boom, from housing to highways, were therefore slow to be addressed. As a result, the city at the recent boom's center, Williston, had the highest average rental prices in the entire country – even higher than Manhattan or San Francisco. Fly-in fly-out short-term labor patterns, typical of oil and gas operations in remote areas, further altered the extreme nature of the local real estate market. Measures were put in place to encourage a more sedentary lifestyle of relocation, which are just now being brought to fruition and promoted as such. (06)

Yet as a friend recently put it to me, "the boom is over."

1.4

The price of oil is influenced by many factors, but is underpinned by a delicate global balance of supply and demand. Since the deregulation of hydraulic fracturing, global oil supply levels have grown by 9.0 million barrels per day (mb/d), or 110%. (07) During that time, national crude oil production levels in the United States grew by nearly 4.0 mb/d, or 180%, almost half of the global oil supply increase. (08) With global oil consumption unable to keep up with production levels, a glut has ensued where supply outstriped demand and production was precariously stockpiled. (09) The result has been a collapse in oil prices that has wrought havoc on national economies both outside of and in OPEC, from Russia to Nigeria and Venezuela.

The origin of the oil price collapse is often identified as 2014, when, in spite of projections of an imminent imbalance between supply and demand, OPEC maintained their collective production quota first set in 2011 of 30.0 mb/d, which they have continued to do since. (10) The event was framed as a decision by OPEC, and Saudi Arabia in particular, to lower their production quota and boost global oil prices that were already in slow decline; in short, to bend to trends in non-OPEC countries to increase production. Indeed, what was at stake was OPEC and Saudia Arabia's power over oil as a global commodity. Instead, what was decided upon was, for all intents and purposes, a war of attrition. (11) Indeed, oil is not a cheap resource to extract; to drill a hole deep into the ground and fitted with complex machinery in often quite remote locations, a sizable (though temporary) workforce is needed. Thus every location has a specific price point for oil, where the cost of extracting it is positively outweighed by its sale price. In short, the maintenance of 30.0 engenders economic Darwinism, where the market is used to return itself to equilibrium. This has already worked, at least on paper, in two different ways. First, it has instigated a contraction in non-OPEC supply, (12) and second, it has led to apparently-long-sought-after "international dialogue and cooperation efforts" (13) across OPEC's organizational borders for the first time in fifteen years. (14)

2.0

In 2013, Russel Gold, senior energy reporter for *The Wall Street Journal,* found that over fifteen million people in the United States lived within a one-mile radius of a oil or fracking well. (15) Since fracking's deregulation, there has been an intense focus of anti-fracking activism based on the fact that the specific chemical content of the fluid mixture used in fracking is proprietary information and thus cannot be appropriately safeguarded against or (re) mediated. While fracking does injecti new and unknown things into the ground that very well might be seeping into our ecological and metabolic systems, it brings along with it a host of other risks posed by drilling for oil and gas: surface spills, subsurface cracks, off-gassing, tailing pond aeration, train derailments, car crashes and road degradation are just some of the potential risks brought about by fracking.

Aside from that, fracking augments the patterns and intensity of oil and gas development, and thus the qualitative nature of the risks it distributes. Fracking wells can only extract oil from the rock it cracks open. This means that wells can be drilled at a significantly higher density than those that have to collectively draw from a single pool of resources. So for economic stability, more needs to be drilled, but that can, at least conjecturally, be done in the same area as would be needed to do so with conventional extraction methods. Yet if exposure to risk is centered in, on and around the well itself, the threat of contamination and becoming what Saskia Sassen has recently terms "dead zones" is exponentially higher: (16) whereas conventionally there might have only been one site of risk exposure, one well, now there might be twelve in the same area.

2.1

The question I set out to myself back in 2013 at the start of my research into fracking was simple: who or what is it that sanctions a well to be drilled? The word sanction is a peculiar one, belonging to the linguistic category of contronyms, meaning single words whose multiple definitions are opposites of each other. Thus to sanction means both to allow and forbid. So by asking what sanctions a well to be drilled, I sought to implicitly find the means of refusal. Clearly, as I discussed above, there are larger forces at play in the determination of whether drilling operations take place or not. Yet there are a series of complex mechanisms that must be enacted, terms and conditions that must be met and followed, in order for each and every well to be drilled. It is thus the well that presents itself as a site for intervention. The well is the constitutive site of systemic relations, and indeed, the system itself. In a form of architectural critique, I sought to locate design. I sought to locate the spaces of agency.

My starting point for research was complicity. I accepted the fact that fracking is legal, politically backed and financially viable. I sought to contort this reality against itself. The architecture of oil is built in law. I sought for the law to be a deconstructive practice. In order to do this, I learned to read legal scriptures; the architectural drawings of oil. This architecture is made of history, from history. Common law is one of precedence. The legal framework of land rights is one of the critical infrastructures of oil's architecture. Oil entered into this history recently, and was built upon the legacy that came before it. Mineral Estate Dominance, the legal doctrine cited before, was first enunciated in 1536, when the British Empire needed to pay and fuel its colonial army. Mineral Estate Dominance is the starting point for this project of architectural critique. It is the first time within the legal history of resource extraction that a landowners' sovereign power over their land, once given, was given conditions for its abrogation. The entire post-colonial history of land in the United States is built upon Mineral Estate Dominance. After independence, the federal government distributed land in such a way as to circumvent the need for the public to invoke this doctrine; citizens were simply not given land with subsurface resources on it. Yet Mineral Estate Dominance haunts settlement, no matter where.

The question I ultimately set myself out to try and answer was: What can be done? Since the discovery of oil, United States land law has progressively evolved to restrict the means by which a landowner can refuse a well. The law has evolved to favor drilling interests. As it turns out though, in the fracking boom, most people whose property held hydrocarbons and were given a choice to have a well drilled on it chose to comply. What concerned me in my research were those who said no. As an ideological position, I do not think hydrocarbons should be extracted from the Earth. I would say no. Yet as the case of Mineral Estate Domination demonstrates, the ideology of the United States' legal system supports positive rights (freedom to) at the expense of negative rights (freedom from). I would not be able to say no. While I sought something more operative in my research, I found that the law Mineral Estate Dominance dominated. The only conclusion I could come to is that it must be overturned. Territorial sovereignty must be transferred back to the landowner. Mineral Estate Dominance undermines our ground; as individuals, as citizens, as people. This is not a conservative position. This is a radical position. It is nothing more than a statement. Yet what more is the law than a series of statements, resonating in the air?

Notes

(01) The USA Homestead Act (1916) distributed only surface rights regardless of whether they were over resource-rich or resource-poor sub-surface geology.

(02) In 1568 Queen Elizabeth I of England was declared to own all mines of gold and silver, no matter who owned the land where those mines were to be found, due to the fact that these materials were needed to maintain a standing army and therefore in the public interest. This logic of sovereign exceptionality in land rights and resource ownership persisted into the formation of the United States of America, and is the basis for subsurface property rights in nearly every country around the world today.

(03) This describes the common pool situation, which is a locus of historical conflict in oil and gas perfectly embodied in the quote "I drink your milkshake" in the heated exchange between Daniel Day Lewis and Paul Dano in Paul Thomas Anderson, There Will be Blood (2007).

(04) One of the earliest legal measures against extraction is the Termination Of Oil Or Gas Interests In Land Act 42 1963, Amended, Act 519, 2006.

(05) 'Crude Oil Production', *U.S. Energy Information Administration*, 03 March 2016, https://www.eia.gov/dnav/pet/pet_crd_crpdn_adc_mbblpd_a.htm (accessed 20 March 2016).

(06) M. Adamson, 'Anthropocene Realism', *The New Inquiry*, 30 November 2015, http://thenewinquiry.com/essays/anthropocene-realism/ (accessed 20 March 2016).

(07) Global oil supply current stands at nearly 90.0 mb/d. OPEC, 'Monthly Oil Market Report March 2016', *Organization of the Petroleum Exporting Countries*, 16 March 2016, www.opec.org/opec_web/flipbook/MOMRMarch2016/MOMRMarch2016.html#46 (accessed 20 March 2016); OPEC, 'Monthly Oil Market Report August 2005', *Organization of the Petroleum Exporting Countries*, http://www.opec.org/opec_web/static_files_project/media/downloads/publications/MOMR_082005.pdf (accessed 20 March 2016).

(08) U.S. Energy Information Administration, ibid. United States oil production currently stands at nearly 9.0 mb/d.

(09) G. Smith, 'IEA Raises Estimate of Surplus Oil Supply on Higher OPEC Output', *Bloomberg Business*, 9 February 2016, http://www.bloomberg.com/news/articles/2016-02-09/iea-raises-estimate-of-surplus-oil-supply-on-higher-opec-output (accessed 20 March 2016).

(10) OPEC, 'OPEC 166th Meeting concludes', *Organization of the Petroleum Exporting Countries*, 27 November 2014, www.opec.org/opec_web/en/press_room/2938.htm (accessed 20 March 2016).

(11) Smith, ibid. The 30.0 mb/d OPEC quota does not correlate with collective production levels, which currently stands at over 32.6 mb/d. Ibid.,

(12) OPEC, 'OPEC 168th Meeting concludes', *Organization of the Petroleum Exporting Countries*, 4 December 2015, www.opec.org/opec_web/en/press_room/3193.htm (accessed 20 March 2016).

(13) OPEC, ' Opening address to the 166th Meeting of the OPEC Conference', *Organization of the Petroleum Exporting Countries*, 27 November 2014, www.opec.org/opec_web/en/press_room/3193.htm (accessed 20 March 2016).

(14) M. Sergie, G. Smith and J. Blas, 'Saudi Arabia, Russia to Freeze Oil Output Near Record Levels', *Bloomberg Energy*, 18 February 2016, www.bloomberg.com/news/articles/2016-02-16/saudi-arabia-and-russia-agree-oil-output-freeze-in-qatar-talks (accessed 20 March 2016).

(15) R. Gold and T. McGinty, 'Energy Boom Puts Wells in America's Backyards - WSJ,' accessed September 1, 2014, http://online.wsj.com/news/articles/SB10001424052702303672404579149432365326304.

(16) S. Sassen, *Expulsions: Brutality and Complexity in the Global Economy*. Cambridge: Harvard University Press, 2014.

THE LEAKY LANDSCAPE

CLAIRE HOLTON

Four frackmaps of Midland County, Texas, USA (01)

Aerial Photograph showing Fracking Pads and Roads

Notes

(01) Sources: http://www.usgs.gov/; http://www.airnow.gov/; Http://www.gisp.rrc.stats.tx.us/; http://www2.epa.gov/; www.topoquest.com

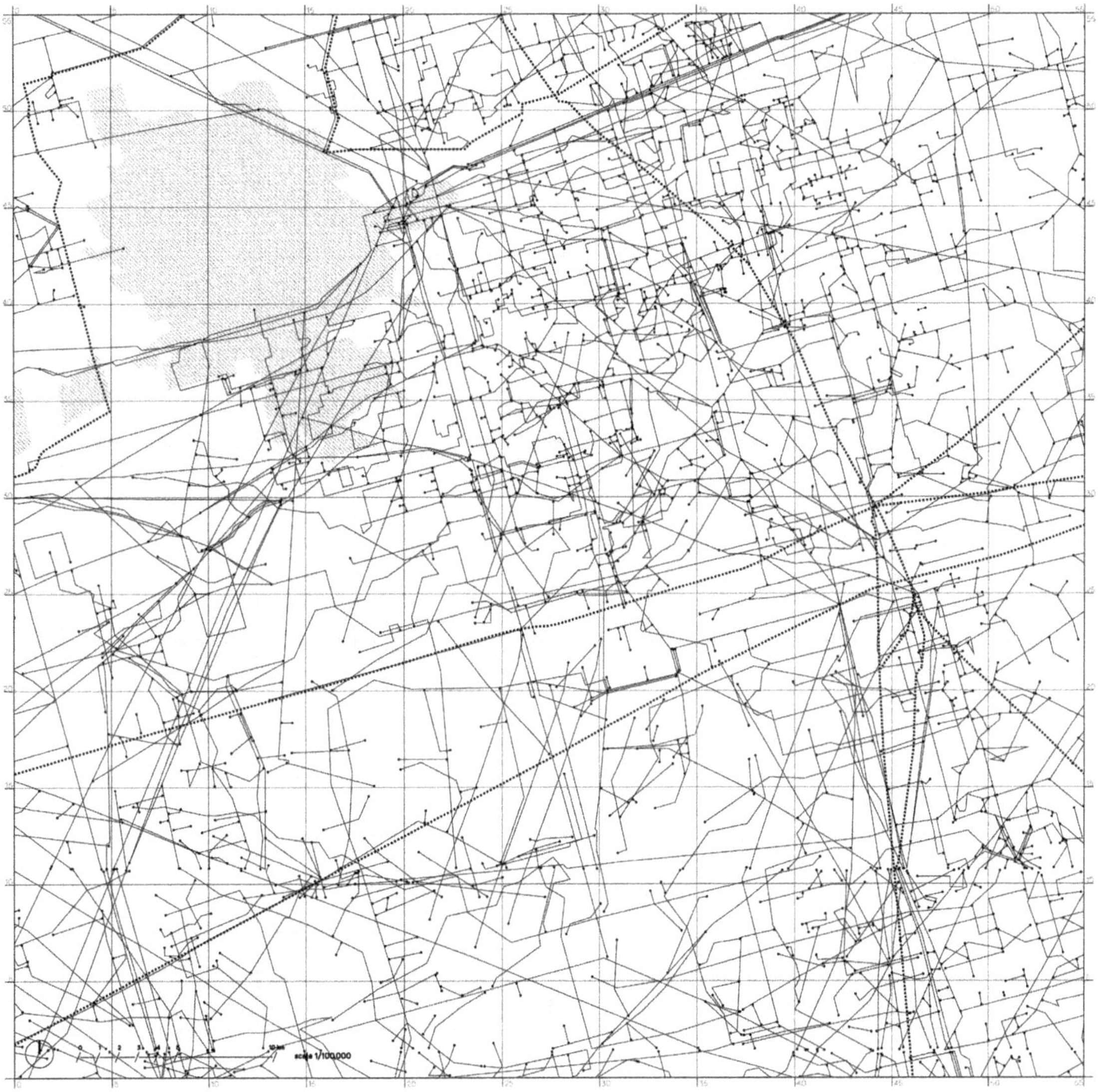

Survey of Subterranean Wells and Pipes

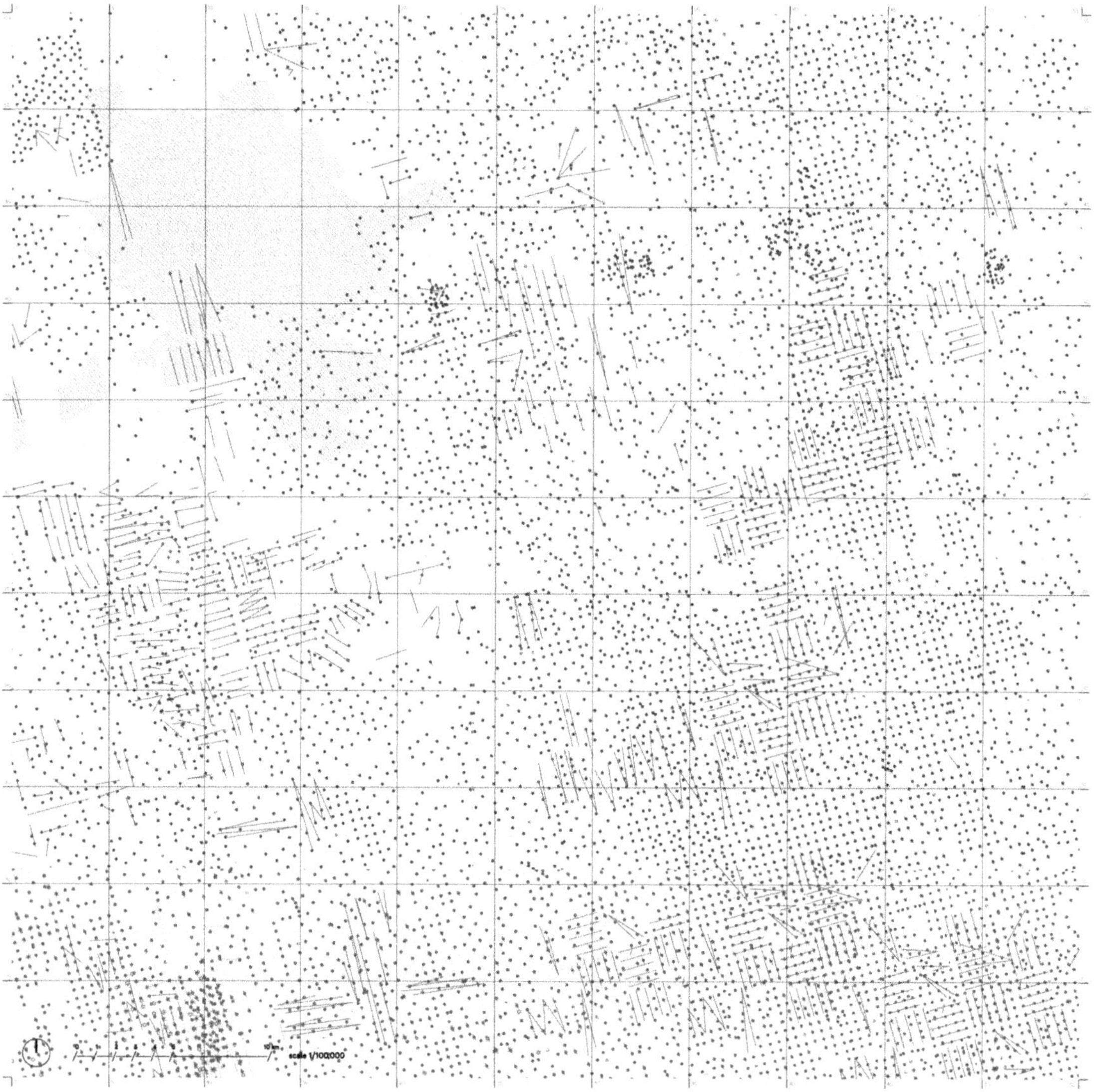

Survey of Subterranean Wells: Vertical Shafts and Horizontal Extents

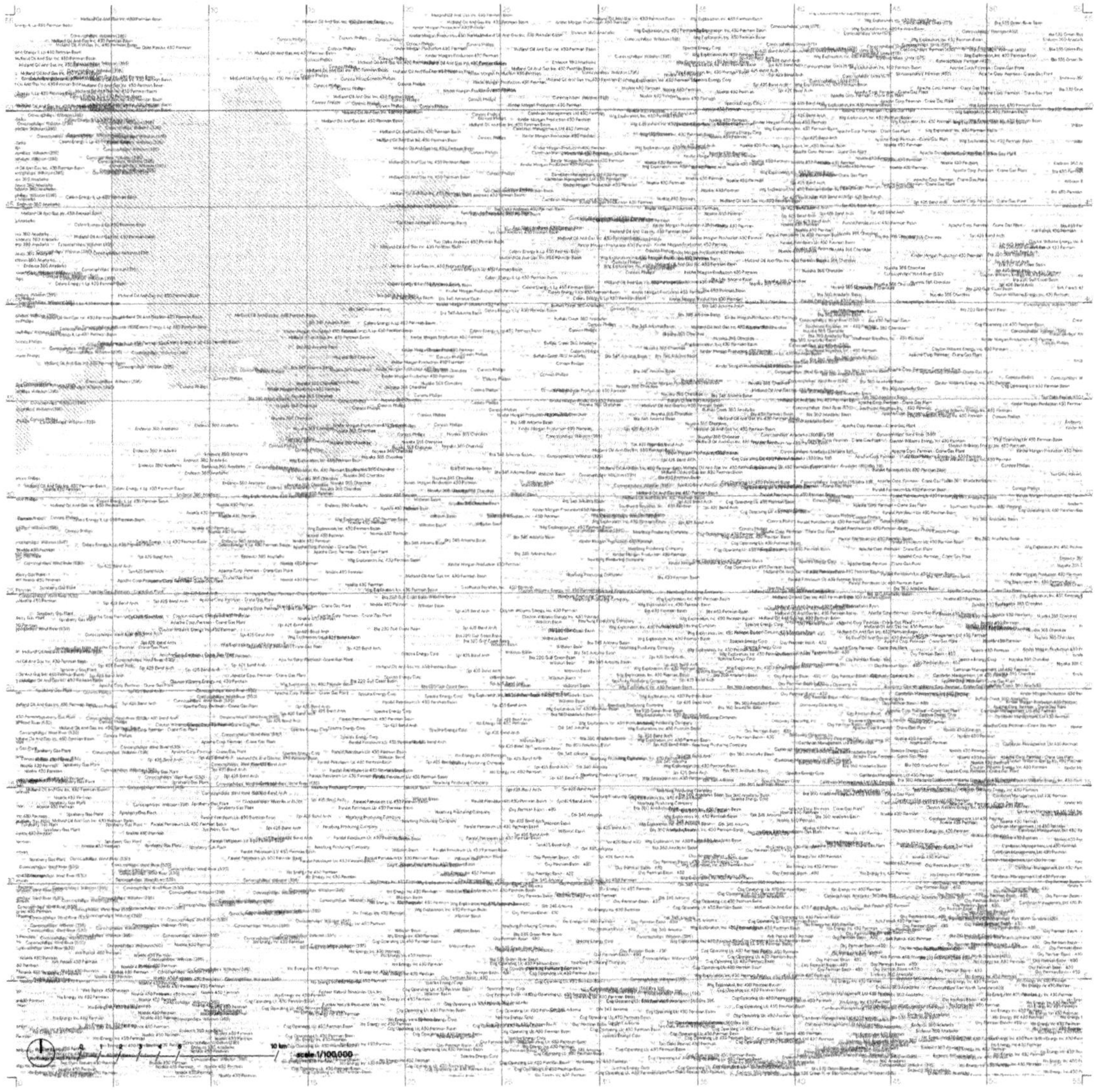

Survey of Subterranean Pipe Ownership

STATEMENT OPPOSING THE LICENSING OF SHALE GAS EXPLORATION IN THE KAROO

This statement was lodged via the Public Participatory Process on fracking in the Karoo region in the Eastern Cape on 15 February 2015, for which Bundu Gas and Oil Exploration Pty Ltd had submitted an environmental management plan. Its purpose was to reflect on the ways in which the scientific evidentiary on hydraulic fracturing had, in the United States, been systematically closed down, via non-disclosure agreements, out-of-court settlements, and a combination of secrecy, inadequate resourcing and incompetence on the part of environmental regulators. As I researched the record of those non-disclosures, it became more and more apparent that there were two additional ways in which the fracking debate was being shut down: ignoring the timeframe of environmental impact in the long duree, and maintaining the idea of a 'pro' and an 'anti' fracking faction so that there could be a neutral middle ground that states could occupy to determine that fracking could proceed. (01) The one exception, in the USA where the possibility of an artificially-created neutral middle ground has been foreclosed, was in the state of Maryland where senators voted in July 2015 in favour of a bill that declared hydraulic fracturing to be "an ultrahazardous and abnormally dangerous activity." (02) The implications of that were to set a different baseline for court decisions, so that plaintiffs did not have to prove that 'neutral' industry was harmful, they simply had to show what damages had been caused and link them to the activities near the site in question. Researching these issues underscores the value of work in knowledge studies, for democratic processes: to follow carefully, systematically, and empirically how an evidentiary is generated.

I am an Associate Professor of Anthropology at the University of Cape Town, where I lead the Environmental Humanities South graduate programme and research initiative.

I have reviewed the published scholarly literature in social science journals on the impact of fracking on soils, water, and air that support the life of both humans and non-humans, and have further studied the documents and records released by the Courts and by public regulatory bodies in the USA, notably those in Pennsylvania and California. From these documents it is clear that there is a serious mismatch between the scientific claims, in the USA, about the safety of fracking, on the one hand, and on the other those in the social science literature, in in-depth journalistic investigations, and in the recent release of classified documents by environmental regulatory authorities. These documents, released in Pennsylvania and California in the past six months since August 2014, provide evidence that has been hitherto hidden from the scientific community, from decision-makers, and from publics.

That evidence is of a very serious nature that affects all South Africans, and affects the scholarly community which is tasked with providing the knowledge base on which our democracy can proceed.

Unless that new evidence is presented and evaluated in South Africa, to publics, to the affected government Departments, and to Parliament, the process of public consultation will be proceeding on fraudulent grounds.

Not doing so, risks violating the rights, guaranteed in the South African Constitution, of a clean environment.

It risks burdening the fiscus with the costs of cleaning environments, and of restoring damaged earth processes permanently, in the future.

It is hubristic: No government is capable of protecting all future generations of humans and animals from permanent damage to fossil water sources. The time-scale of the impact of fracking on fossil water, far exceeds the possibilities of legal regulatory mechanisms.

It risks new forms of serfdom and servitude in which generations to come are dependent on the purchase of water from elsewhere. This is a neocolonial relationship, yet South Africa's post-apartheid government is bound to ensuring liberty from all forms of oppression, both now and into the future.

It undermines South Africa's democratic process, in which the secrecy surrounding the evidence of harms caused by fracking in the USA, is allowed to tamper with our capacity to make independent and informed decisions.

It risks violating the rights of the Earth, and of the community of all on Earth, as methane emissions contribute to global warming at rates that far exceed those of carbon dioxide.

It will harm the possibility of fair and just land redistribution on the Karoo, as land claimants risk receiving poisoned lands.

To go ahead with fracking on the basis of current claims about its safety, will undermine the relationship of scholarship to South African democracy.

On all of the above concerns, I submit that I am an interested and affected party, and a stakeholder in the scholarly community which has a duty to raise concerns about the framing ideas and scientific claims that are being relied upon to determine that fracking is in the interests of South Africa.

Fracking in South Africa must be delayed until there has been a full review of the scientific evidentiary that has informed government decision to date in South Africa, with specific attention to the evaluation of recent revelations by environmental regulators and courts.

Such a review panel must include scholars in the Social Sciences and Humanities, who must be empowered to present a full review of the scientific evidence, and the environmental philosophy and ideas, that have been presented to Parliament and government officials to date in South Africa as supporting the claims to the safety and economic effectiveness of fracking, and must include a review of newly released classified records in the USA and elsewhere which have shown that fracking has been damaging to (inter alia) water, earth, atmosphere, health and life, and the fiscus.

Cape Town, 15 February 2015.

Notes

(01) See Lesley Green, 'Give us Evidence!' What does it mean to require "evidence-based research" in decision-making on hydraulic fracturing?' in J. Glazewski. (ed.) *Hydraulic Fracturing in the Karoo: Critical Legal and Environmental Perspectives*, Cape Town, Juta, 2016, pp. 275-288.

(0). See Senate Bill 458 'Civil Actions – Hydraulic Fracturing Liability Act', introduced for debate on February 6 2015 and passed with amendments in July 2015, http://mgaleg.maryland.gov/2015RS/bills/sb/sb0458f.pdf (accessed 6 September 2015). See also the Maryland Attorney General's comment on that bill in March 2015, http://www.wbaltv.com/blob/view/-/31874610/data/1/-/a0f404/-/AG-letter-on-SB-458-fracking.pdf (accessed 6 September 2015), explaining that declaring fracking an 'ultrahazardous and abnormally dangerous activity' would remove from the plaintiff the burden of proving that fracking is dangerous and hazardous, but he or she would still need to show that an injury was caused by hydraulic fracturing.

HISTORY OF THE KAROO FOR THE ANTHROPOCENE

The Karoo is an almost Darwinian landscape of information about the phase transitions in the earth and animal, plant and human populations that have occurred over its long history. (01) 320 million years ago it was part of the super-continent Godwana and lay over what is now the South Pole. Its lowest geological strata was formed by glaciers to its north depositing a layer of mud into in a vast inland basin, forming a 1km thick sedimentary layer. As Godwana drifted northwards 200 million years ago, this turned into a swampy inland sea inhabited by all manner of long extinct carnivorous reptiles. The peat in these swamps turned into coal, now mined in other provinces of South Africa. Muddy deposits flowing from the mountains to the south then deposited a 5.6 km sedimentary layer that stratified into sandstones and shales. Finally, 180 million years ago, Godwana broke up and the earth exploded with volcanic activity on a titanic scale, bringing an end to a teeming reptile population, then engulfed and transformed into fossils. Lava covered the surface in a 1,6 km layer of basalt, forced under high pressure between horizontal layers of sedimentary rock solidifying into dolorite sills, and welling up through long vertical fissures solidifying into dykes. (02)

Even rocks flow: their atoms migrate along grain borders (self-diffusion), dislocation boundaries within grains move, cracks and fissures propagate. In this sense, the flow of rocks is very viscous; they constantly change, but at extremely slow speeds. Furthermore, under extreme heat and pressure, rocks may undergo a bifurcation (limestone, a sedimentary rock, metamorphises into marble). Alternatively, rock may be melted into lava and reincorporated into the convection flows driving plate tectonics. Stratification then, is in no way a terminal state: free matter and energy stratify, and the stratified destratifies. (03)

This is what gives the Karoo its distinctive geomorphology: horizontal, loosely bound sandstone and shale layers that weather easily, interrupted by hard, resistant dolorite dykes and sills that appear as flat topped hills and weather into rounded domes. Underground, dolorite dykes and sills trap water, creating water reservoirs that, when pushed upwards through fissures, emerge as springs, the attractors around which the lives of Koisan hunter-gatherers evolved, from around 20,000 years ago. (04) Koisan hunter gatherer's relationship to the land was a 'gaseous' one – they believed that their society was held together by stories that floated across the land, with waterholes as their attractors. People lived in small brush shelters, the extent of a group's territory determined by the distance they could roam and the land's ecological resources. Each group of between 12 and 25 people had a water source and a defined territory, adjusted seasonally according to availability of resources. Society was egalitarian, resources were shared through loose ties and alliances and wealth was not accumulated. People bonded with the landscape through myths and legends about its features and rock painting and engravings were used as part of ritual practices to mediate social conflict and record experiences in the spirit world. (05)

Arthur Iberall describes this form of society as a system of "weak force" and "gas-like" interactions and the phase change into a pastoralist society one of liquification or condensation around more fixed centres of population. (06) In response to external pressures (diminishing resources and external social pressures), human society became less gaseous and more viscous. Slowly, the domestication of plants and animals and new technologies made it less mobile and more stratified. This phase change occurred in the Karoo around 2,000 years ago when Khoekhoe pastoralists from the same genetic pool as the Koisan, introduced sheep, cattle, domestic dogs and pottery into the region. These pastoralists had a 'fluid,' cyclical relationship to land; their herding strategies comprised large transhumance cycles, centred on perennial springs

Typical Karoo landscape near Aberdeen
Photograph: Thomas Marincowitz

P076 Typical Karoo windmill and water reservoir, Klaarstroom, Karoo
P077 Blockhouse and railway bridge built on the frontline of the Anglo Boer War (1899 - 1902) near Laingsburg, Karoo
Photographs: Thomas Marincowitz

and rivers and exploiting seasonal ecologies. They were constantly on the move and their houses and belongings were lightweight, assembled and disassembled and carried with them. They lived in mat covered houses and travelled with their stock, carrying their possessions on the back of oxen. Wealth began to be accumulated in the form of livestock used for food and milk or exchanged and passed down from one generation to the next. Social hierarchies developed. There was constant oscillation between the gaseous (Koisan) and the liquid (Khoekhoe) social groups: Khoekhoe rarely slaughtered their stock other than for ritual purposes and, like the Koisan, lived on game and wild plants. Koisan hunter gatherers sometimes broke into Khoekhoe society by becoming servants. (07)

In the mid 17th Century however, the phase transition that was to 'crystallise' these cyclical gaseous and liquid social formations into a hard, stratified, hierarchical society began, when the Dutch East India Company (VOC) established a base in Table Bay as a refuelling station for their ships sailing to the far East. The base was surrounded by Khoekhoe pastoralists who controlled flows of trade to the interior and by the Koisan hunter-gatherers, most of who lived in the mountains. Within five years, the VOC had transformed Table Bay from a trading and refuelling post to a colony and Dutch company men were granted land to farm on the Cape Peninsula. (08)

What followed over the next two centuries was a nonlinear process of social crystallisation as the Dutch extended their colony further and further inland. They frequently faced brutal resistance from and, in turn, brought brutal pressure to bear on the two indigenous groups. Through this process, Khoekhoe society hardened close to the bottom of the social hierarchy as farm labour and Koisan society was virtually exterminated. Hundreds of men, women and children were killed, captured or enslaved; those who survived were incorporated into the strata of colonial society at the lowest possible level. (09) Today many of their descendants still live as impoverished shepherds, labourers, domestic workers, or as unemployed township dwellers.

Colonial relationships with land in the Karoo proceeded through a series of phase changes, resulted in a highly stratified, crystallised landscape. Occupation was initially based on fluid, short-term grazing licenses that gave rights to as much grazing as the licensee needed in a vaguely defined area of common land. This was replaced with a loan farm system combining mobile and fixed rights, itself replaced by fixed freehold tenure on a surveyed area, in return for an annual rent. At the end of the 19th Century, farmers were able purchase their properties and tenure was fixed. Most land in the Karoo is today under private ownership and managed on a commercial, rotational basis.

These phase changes in land tenure have been matched by evolving grazing practices. As water sources and rangeland was privatized, grazing orbits shrank and a rigid kraaling system was implemented. Livestock was herded from kraal to rangeland to water source to kraal on a daily basis. Over time, this linear system resulted in erosion and disruption to the ecology. Imported windmills and fencing in the early 20th Century, enabling rangeland to be subdivided into camps each with its own water source, produced a more flexible system, rotating flows of animals through highly striated land. Today 80 % of the Karoo belongs to private owners where sheep and goats are farmed on a rotational basis, or as game conservancies. 50% of farms are smaller than 3,000 ha, 25% larger than 6,000 ha. (10)

The British arrived at the Cape at the end of the 18th Century, laying claim to it as part of the British Empire through force. This 'liquified' the bonds of Dutch colonial society and farmers to the land. Trekboere, as they became known, resistant to British rule began following lines of flight into the interior of Southern Africa, where violent confrontations with social groups already there ensued. During the 19th Century, the Trekboere established independent forms of republican rule in the interior (the Orange Free State and the Transvaal) and the Karoo became the boundary of the British Cape Colony. But then the discovery, first of diamonds and then of gold brought unprecedented flows of energy (prospectors, railways, new technologies, new concentrations of people and money, etc.) into the Boer republics, pushing the near equilibrium conditions that had been reached into a phase change again. Two Anglo Boer Wars (1880-1881 and 1899-1902) followed, pitting the rigid might of the British Empire against the fluid Boer commando system with its gaseous guerrilla tactics (they melted into thin air). Eventually the British won the war by stopping supplies to the Boer commandos and solidifying their populations into concentration camps. In 1910, the Union of South Africa was established as a dominion of the British Empire. Key to British victory had been the railway line through the Karoo to the interior. The Karoo was the front line in the war, still visible in the British blockhouses that guarded strategic railway bridges. The war contributed to rural poverty, increased urbanisation and fanned Afrikaner and African nationalism. The British Empire had been shaken by two small republics, which were resolved in their determination to be rid of British influence and stratify society in their own interests. African people, despite being used by both sides during the war, were not given the franchise in 1910. Following this, the South African Native National Congress was formed in 1912, becoming the African National Congress in 1923.

The 20th Century is largely a history of the phase change of South African society into one rigidly stratified

under Afrikaner control and the lines of flight taken by the black majority to bifurcate it. This was eventually achieved in 1994, with the first democratically elected South African government, which liquefied the country's racially based social stratification, while crystallising society in new ways and making new bifurcations possible. (11)

This is the landscape that DS18 engaged in 2014/15. It is a complex landscape, the product of millennia of non-linear, self-organising processes of geological, human and more-than-human life. The pretext for the studio was the imminent introduction of fracking into the Karoo, as a potential phase-changing moment, when solidified, stratified geological and social formations would be disrupted, bifurcated and transformed along any number of fault-lines.

The Karoo, given its geological history and climatic conditions, has long been viewed as a reservoir of energy - sun, wind and hydrocarbons. In the 1960's, the South African state-owned oil exploration company, Soekor, drilled a series of wells for oil near the town of Aberdeen. This project was abandoned as a failure, but provided evidence of resources and fault lines the far under the earth's surface. In 1967, after drilling to a depth of 4 km, the drilling fluid in a well vanished. Six weeks later, a farmer 30 kms away noticed that his borehole was discoloured. Investigation indicated that the contamination was toxic and could only have come from the Soekor well.

The first formal interest in fracking for shale gas in the Karoo was in 2008, when Bundu Oil and Gas, a subsidiary of Australia's Challenger Energy, applied for exploration rights on a conservancy close to Graaff Reinet. Three years later, on the basis of a U.S. Energy Information Administration (EIA) assessment of 390 trillion cubic feet of technically recoverable reserves, Dutch Shell applied for an exploration licence covering more than 95,000 square km, almost a quarter of the Karoo. Such quantities, if realised, could be a phase changer an economy that has always been a big oil and gas importer. An outcry from activists, citizens and farmers and legal challenges however, led to a moratorium on the granting of licences, lifted in 2013 when President Jacob Zuma described shale gas as a game changer for the economy, saying that Pretoria would allow fracking within the framework of environmental laws. The Department of Mineral Resources released draft technical regulations on oil and gas exploration and production in late 2013 and was expected to publish regulations before the elections in May 2014, but has still not done so. Shell currently has three pending exploration license applications, Bundu Gas & Oil and another company, Falcon Oil and Gas Ltd. one each. (12)

Two kinds of human institutions are at play here, interacting dynamically in the struggle to channel, amplify and control this land and its reservoirs and flows of energy - on the one hand, the hierarchical structures of state

P078 Post-apartheid Site-and-Service Infrastructure, Umasizakhe, Karoo
P079 Anti-fracking Poster, Nieu Bethesda, Karoo
Photographs: Lindsay Bremner

institutions, on the other the equally hierarchical structures of energy monopolies, and on the other, the more or less fluid structures of farmers, citizens and activists acting without any central organ directing their processes of assembly. In the current stand off between them, the opportunity exists for to re-evaluate the energy resources and flows of the Karoo, organic and nonorganic, human and more-than-human, and, through design, to develop strategies, maps, models and prototypes to channel the dynamical behaviour of its flows in creative new ways.

Notes

(01) E. Palmer, *The Plains of Camdeboo*, London, Fontana, 1966.
(02) M. E. Meadows and M. K. Watkeys, 'Paleoenvironments', in W. R. J. Dean and S. J. Milton (eds.), *The Karoo: Ecological Patterns and Processes*, Cambridge: Cambridge University Press, 1999, pp. 27-41.
(03) M. DeLanda, 'Nonorganic life', Incorporations, Zone no. 6, New York, Zone Books, 1992, p. 143.
(04) H. J. Deacon and J. Deacon, *Human Beginnings in South Africa*, California: AltaMira Press, 1999.
(05) J . Morris, D. Rusch, and N. Parkington, *Karoo Rock Engravings*, South Africa: Krakadouw Trust, 2008.
(06) In M. DeLanda, 1992, p. 154.
(07) R. Ross, 'Khoesan and Immigrants: the Emergence of Colonial Society in the Cape, 1500-1800', in C. Hamilton, B. K. Mbenga, and R. Ross, (eds.), *The Cambridge History of South Africa*, Volume 1, Cambridge, Cambridge University Press, 2012, pp. 168-210.
(08) R. Ross, 2012.
(09) A. B. Smith, 'Hunters and herders in the Karoo landscape', in W. R. J. Dean and S. J. Milton (eds.), *The Karoo: Ecological Patterns and Processes, Cambridge*: Cambridge University Press, 1999, pp. 243-256.
(10) M. T. Hoffman, B. Cousins, T. Meyer, A. Petersen, A. and H. Hendricks, 'Historical and contemporary land use and the desertification of the Karoo', in W. R. J. Dean and S. J. Milton (eds.), *The Karoo: Ecological Patterns and Processes*, Cambridge: Cambridge University Press, 1999, pp. 257-273.
(11) See I. R. Smith, 'The Revolution in South Africa Historiography', *History Today* vol. 38, no. 2, February 1988, http://www.historytoday.com/iain-r-smith/revolution-south-african-historiography (accessed 09 February 2016) for a discussion of changes in the study of South African history.
(12) See Department of Mineral Resources, 'Full Report on Investigation of Hydraulic Fracturing in the Karoo basin of South Africa 18 September 2012', http://www.dmr.gov.za/publications/summary/182-report-on-hydraulic-fracturing/853-full-report-on-investigation-of-hydraulic-fracturing-in-the-karoo-basin-of-south-africa-.html, (accessed 09 February 2016) and Department of Mineral Resources, 'Mineral and Petroleum Resources Development Act, Act No. 28 of 2002, http://www.treasurethekaroo.co.za/images/Draft_Fracking_Regulations_20131015-GGN-36938-01032.pdf (accessed 09 February 2016).

PHOTO ESSAYS

THE KAROO

IMAGE CAPTIONS

P083 Water channel in Nieu Bethesda

P084-085 Panoramic view from site of rock engravings, Nelspoort

P086-087 The Valley of Desolation, Camdeboo Local Municipality

P088-089 Camdeboo National Park

P090-091Umasizakhe, Camdeboo Local Municipality

P092-093 Nieu Bethesda, Camdeboo Local Municipality

P094-095 House in Nieu Bethesda

Photographs: Lindsay Bremner

THE DORSET COAST

IMAGE CAPTIONS

P097 Kimmeridge Oil Shale Strata
P098-099 Anvil Point
P100-101 Durdle Door
P102-103 Old Harry's Rocks
P104-105 Lulworth Cove
P106-107 Kimmeridge Bay
P108-109 Kimmeridge Cottages
Photogaphs: Lindsay Bremner

DESIGN STUDIOS

FRACKED URBANISM

SEM 1 2013/14

INTRODUCTION

LINDSAY BREMNER AND ROBERTO BOTTAZZI

If one comes across Google Earth images of landscapes characterised by dense arrays of small white patches linked by a network of straight or wiggly lines, one has come across the clearings, holes, caps, water tanks, pipes and roads that form the most visible traces of the process of resource extraction known as hydraulic shale gas fracturing. This is an emergent system that began with the lives of tiny creatures 400 million years ago, approximately 2 kms under the surface of today's earth, and is likely to extend millions of years and great distances into space/time in future. What is fracking? What is its genetic code? How does its metabolism work? What are the inputs and outputs, stocks and flows that keep it working? Who or what are the actors and agents involved in its networks? At what scale do they operate in space and time? How are they interconnected and how can their flows be spatialised? Most importantly, what risks does fracking pose and what are its limits? What conflicts does it generate? What dangers does it pose? Where does fracking fail? Who or what does it fail?

These were the questions with which DS18 began its research into shale gas fracturing in 2013. We adopted no specific vantage point from which to investigate the issues and human and non-human actors were given the same value. Their properties, actions, and constraints in relation to the fracking process were identified and visualised (See Appendix). Through this initial research and the data it generated, spatial and architectural opportunities within the violent territorial transformations produced by fracking and its after-effects were identified. These included things like risks of increased levels of geological instability, gas explosions, air and water pollution, disease, biodiversity loss etc. How could these dangerous and toxic conditions give rise to new spatial and architectural imaginaries to e.g. prevent the risk from occurring, monitor its development, reduce its impact or exploit its potential for further transformation. For example, could the risk be limited by implementing a seismic activity monitor, by building a water treatment plant, an air filtration device, by alterations to topography necessary to direct water flow etc. Strategies such as these were developed spatially and visually at a territorial scale, alongside which students were then required to design a prototype with both dynamic and repeatable properties.

Two responses to this brief follow, John Cook's 'Remediated Landscape' and Philip Hurrrel's 'Residential Respirator.' The first investigated the contamination of soil by two widespread modes of territorial transformation, industrial agriculture and hydraulic fracturing, in Weld County, Colorado. Cook proposed using the bio-remediating plant species Helianthus annuus (commonly known as the sunflower) to restore the ground back to natural grassland. He developed a system for scattering seeds and harvesting the sunflower plants once they had absorbed the toxins in the soil, on a territorial scale. Evidence has emerged that homes near large-scale hydraulic fracturing sites across the USA are experiencing an increase in smog and health related issues linked to air pollution. Philip Hurrell investigated how to protect such homes against air pollution when residents do not have the option to move. These two projects attempted to think through the far-reaching consequences of fracking on human and non-human life and designed systems and prototypes aimed at protection and reparation. In addition to these two projects, Claire Holton drew on data from a variety of internet sources to developed four maps of Midland County in Texas to visualise the underground as a matted network of gas wells, pipes and property. These are featured on pp. 68-71.

Raye Levine, 2009.
http://www.DamascusCitizens.org

REMEDIATED LANDSCAPE

JOHN COOK

This project began by examining the impact that two widespread farming methods have had on the USA's landscape and resources: water-intensive agricultural farming and the heavily toxic subterranean mineral extraction known as hydraulic fracturing (fracking). Focusing on an area known as 'The Wattenberg Gas Field' in Weld County, Colorado (one of the first shale deposits discovered in America), the project looks to address both issues of wide spread drought and water shortages and the heavily contaminated soil conditions resulting from hydraulic fracturing. A future time when the fracking wells have ceased operation and the grassland's soil and air conditions have become too toxic for prolonged inhabitation or crop growth is imagined. A process of reversing this with as little human interaction as possible, in order to limit exposure to the toxins, is proposed. A long-term solution is proposed that employs a variety of phytoremediating plant species (such as the sunflower) to cyclically extract ground soil contaminants over hundreds of harvests, and eventually remediate these poisoned landscapes. The remediating plant crop must be harvested prior to the development of seed, or else the local fauna and birdlife will egest and spread the now heavily toxic yield. Instead, the natural seed spreading methods by wind or animal are automated via the use of large scale seed disperal units. The process of harvesting, composting and disposal is controlled remotely.

The strategy is explored and represented at a number of scales, from the mechanised seed dispersal units to the resulting wind-driven masterplan, slowly evolving and expanding over time.

P117 Platteville, Colorado - perspective of a landscape under remediation

P118 USA Shale / Agricultre / Water Resources

These maps illustrate the relationship between deep geological shale deposits, areas of intensively artificially irrigated farmland, and the fragile subterranean water aquifers that fuel both crop and mineral 'farming'.

Sources: http://water.usgs.gov/; http://www.eia.gov/; http://www.usda.gov/

P119 Plant Species, Bioremediation and Water Usage

A study of known fracking ground soil contaminants, and the bioremediating plant species that are able to absorb them. The sunflower species is investigated further in terms of its water consumption versus that of a fracking well pad, and Colorado's most heavily farmed crop, corn.

Sources: http://oeffa.org/; Colorado Agricultural Statistics Annual Report 2012 (http://www.colorado.gov/); http://www.mhhe.com/biosci/pae/botany/botany_map/articles/article_10.html; Obo, O. (2013). Hydraulic Fracturing (Fracking): Procedures, Issues, and Benefits. Petroleum Zones; Wise, D. L. (2000). Bioremediation of Contaminated Soils. CRC Press.

P120 Platteville Aerial Survey

Two farming towns, Platteville (north) and Fort Lutpon (south) sit upon the banks of the South Platte river. The river is a vital source for the townspeople and their industry, but has been prone to a number of serious flooding events in recent years.

Sources: http://www.usgs.gov/; http://cogcc.state.co.us/; http://water.state.co.us/; http://geosurvey.state.co.us/

P121 Platteville Agricultural Infrastructure Survey

The one square mile grid has shaped the landscape of Platteville. Irrigated crop circles past and present are peppered with thousands of ground source water wells - drawing water from the depleted aquifers below.

Sources: http://www.usgs.gov/; http://cogcc.state.co.us/; http://water.state.co.us/; http://geosurvey.state.co.us/

P122 Platteville Hydraulic Fracturing Infrastructure Survey

The extent of fracking infrastructure can only be seen below ground. Thousands of horizontally drilled boreholes sit around 2km below the surface - strangely ordered and aligned to the very same square mile grid

Sources: http://www.usgs.gov/; http://cogcc.state.co.us/; http://water.state.co.us/; http://geosurvey.state.co.us/

P123 Platteville Proposed Remediation Strategy

The seed dispersal units, supplied by communal clean growing schemes in the nearby towns, will be shifted eastwards in stages over time. The area of remediated land is determined by the changing local wind patterns throughout the year - the toxic crops are harvested and disposed of before future contaminated seed development.

P124-125 Remediation Strategy in Section

The section illustrates the timescale between seed supply, dispersal, harvest and controlled disposal. The charts below describe the quantities of seed required, how differing wind speeds and changing heights affect the distance of seed travel, and how much the concentration of landed seeds changes over flown distance. The final volumes of contaminated harvest can be dramatically reduced through composting and incineration, but a small volume of concentrated toxins will always remain.

Sources: http://forecast.weather.gov/; http://mesonet.agron.iastate.edu/; http://www.amjbot.org/; Colorado Agricultural Statistics Annual Report 2012 (http://www.colorado.gov/); http://www.hgca.com/)

P126 Prototype Seed Dispersal Unit: site components and collapsing procedure

P127 Prototype Seed Dispersal Unit: exploded axonometric and construction concept diagram

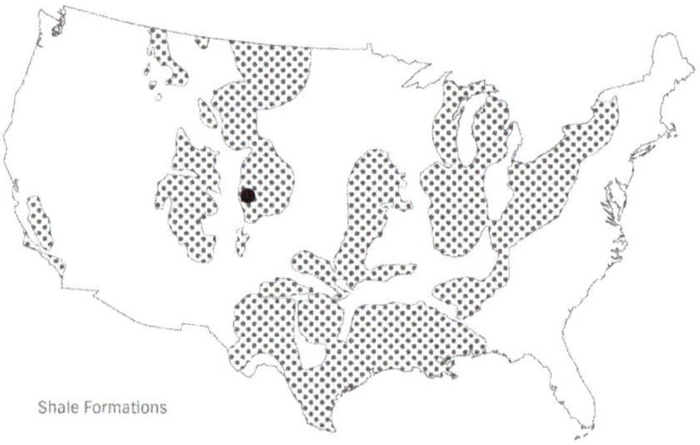

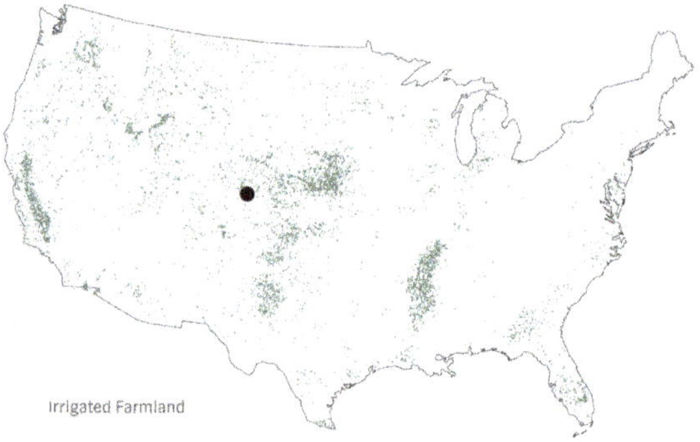

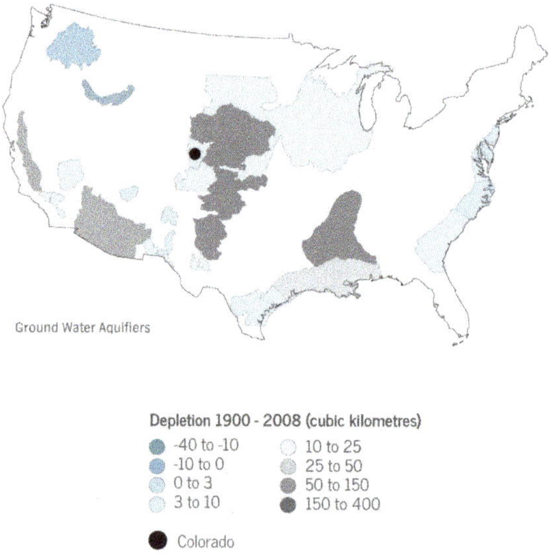

Known Fracking Ground Soil Contaminants

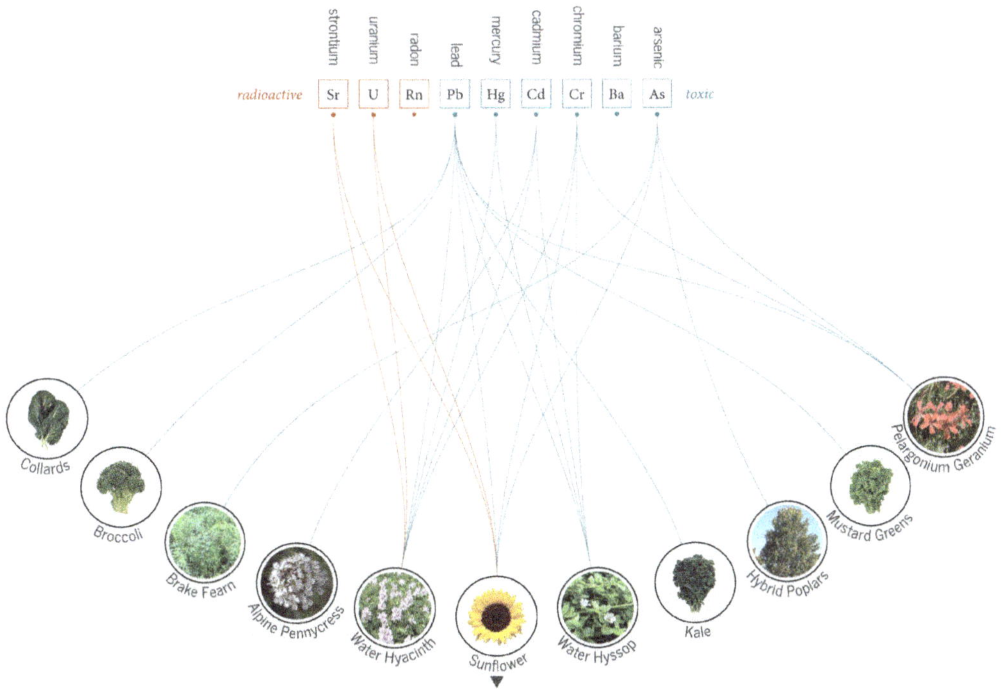
strontium
uranium
radon
lead
mercury
cadmium
chromium
barium
arsenic
radioactive
Sr
U
Rn
Pb
Hg
Cd
Cr
Ba
As
toxic
Collards
Broccoli
Brake Fearn
Alpine Pennycress
Water Hyacinth
Sunflower
Water Hyssop
Kale
Hybrid Poplars
Mustard Greens
Pelargonium Geranium

Bioremediating Plants Species

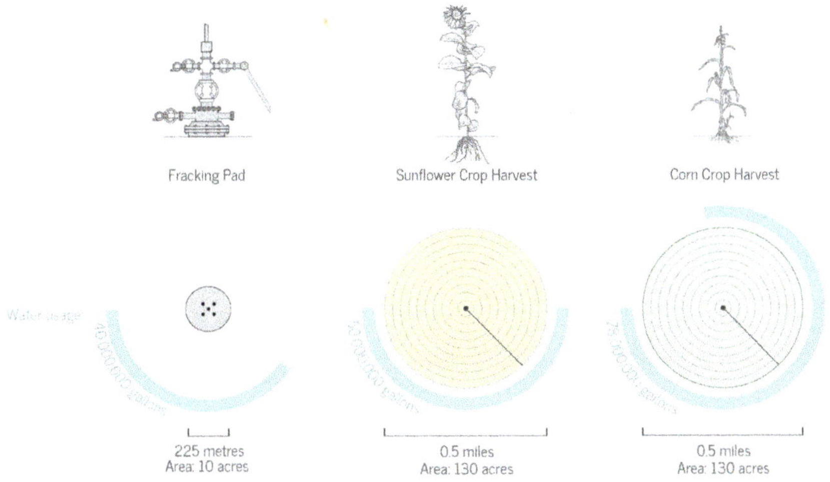
Fracking Pad
Sunflower Crop Harvest
Corn Crop Harvest
225 metres
Area: 10 acres
0.5 miles
Area: 130 acres
0.5 miles
Area: 130 acres

Platte Valley Crops + Water Consumption
Alfalfa
Grass hay/pasture
Dry Beans
Corn
Potatoes
Small Vedgetables
Sorghum, grain
Spring grains
Sugar beets
Sunflower
Wheat, winter
Colorado Crop Harvested Acres
Alfalfa
Grass hay/pasture
Dry Beans
Corn
Potatoes
Small Vedgetables
Sorghum, grain
Spring grains
Sugar beets
Sunflower
Wheat, winter

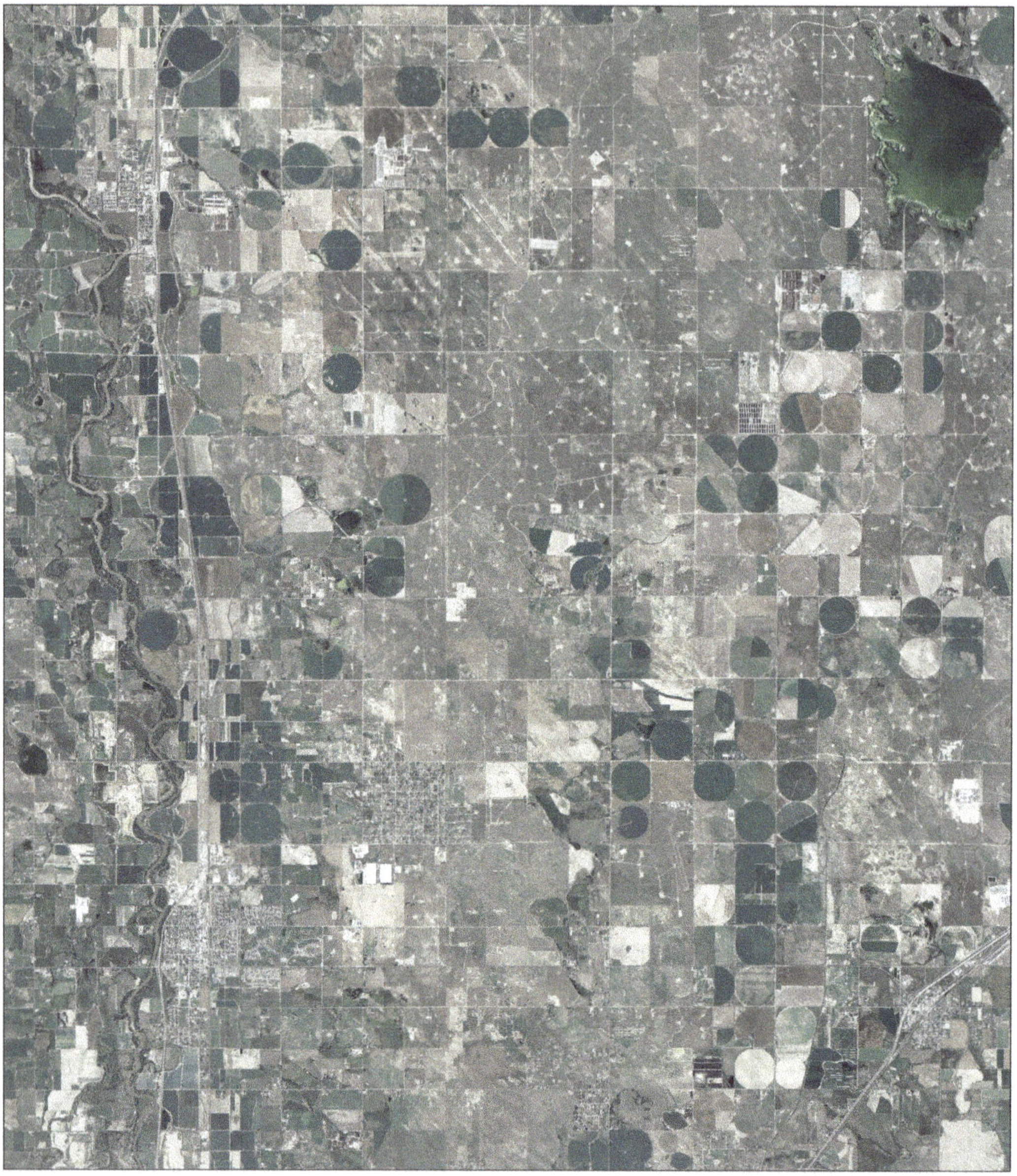

Milton Reservoir

Centre Pitvot Irrigation

Platteville

Hydraulic Fracturing Site

Flooding Contamination

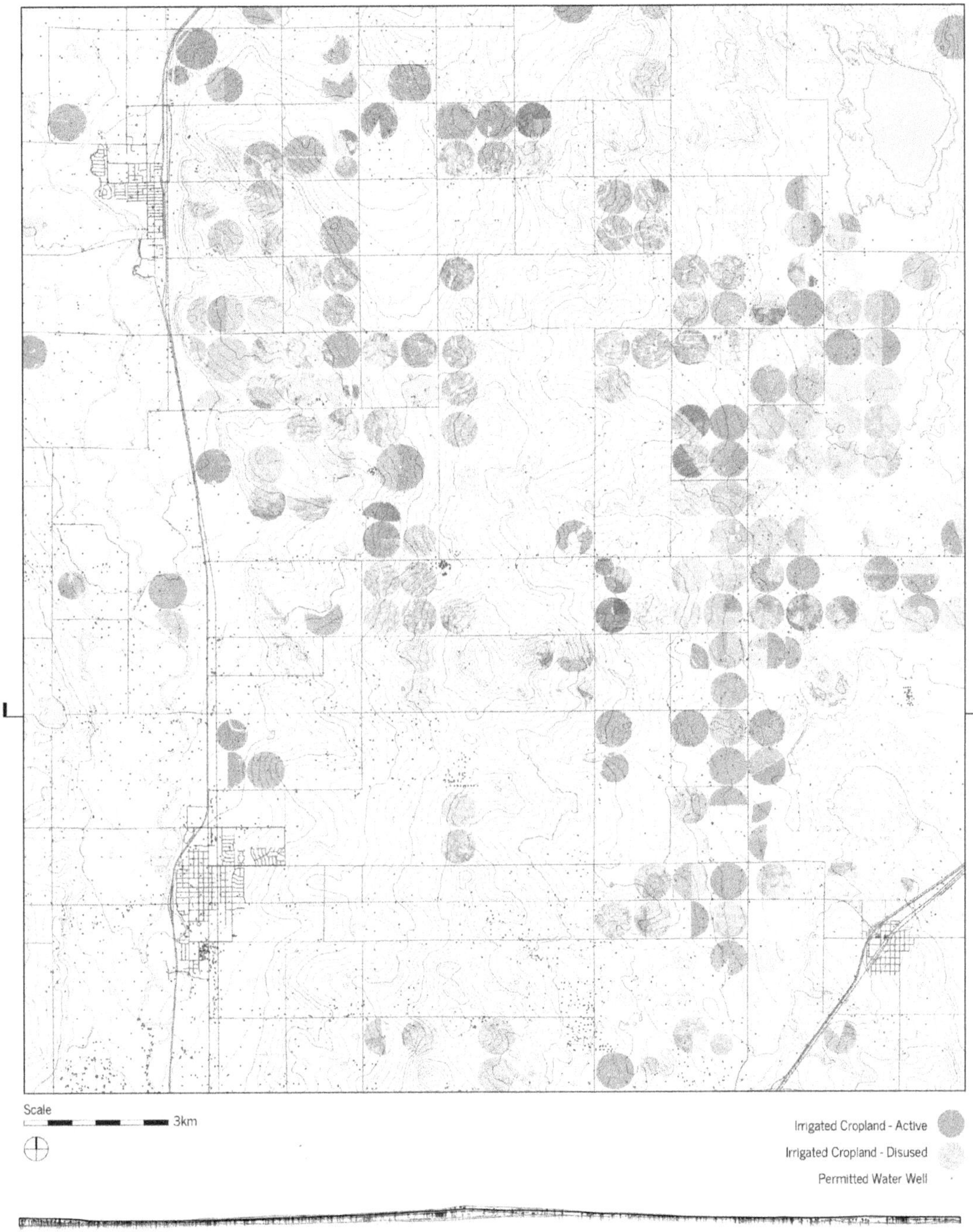
Scale
3km
Irrigated Cropland - Active
Irrigated Cropland - Disused
Permitted Water Well

Scale
3km
Fracking Well
Horizontal Well Direction
Historic / Abadoned / Permitted Oil and Gas Well Locations

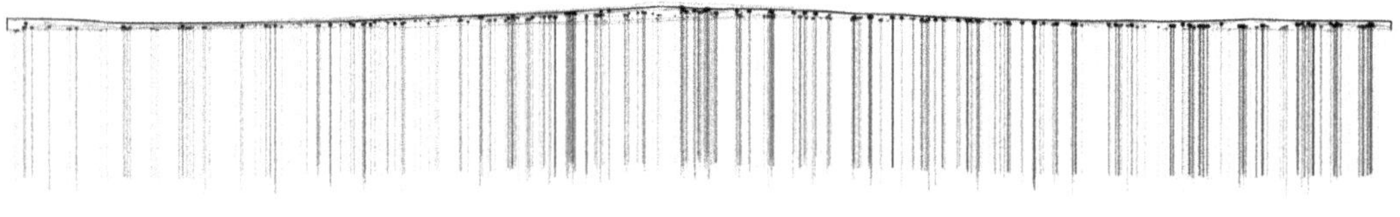

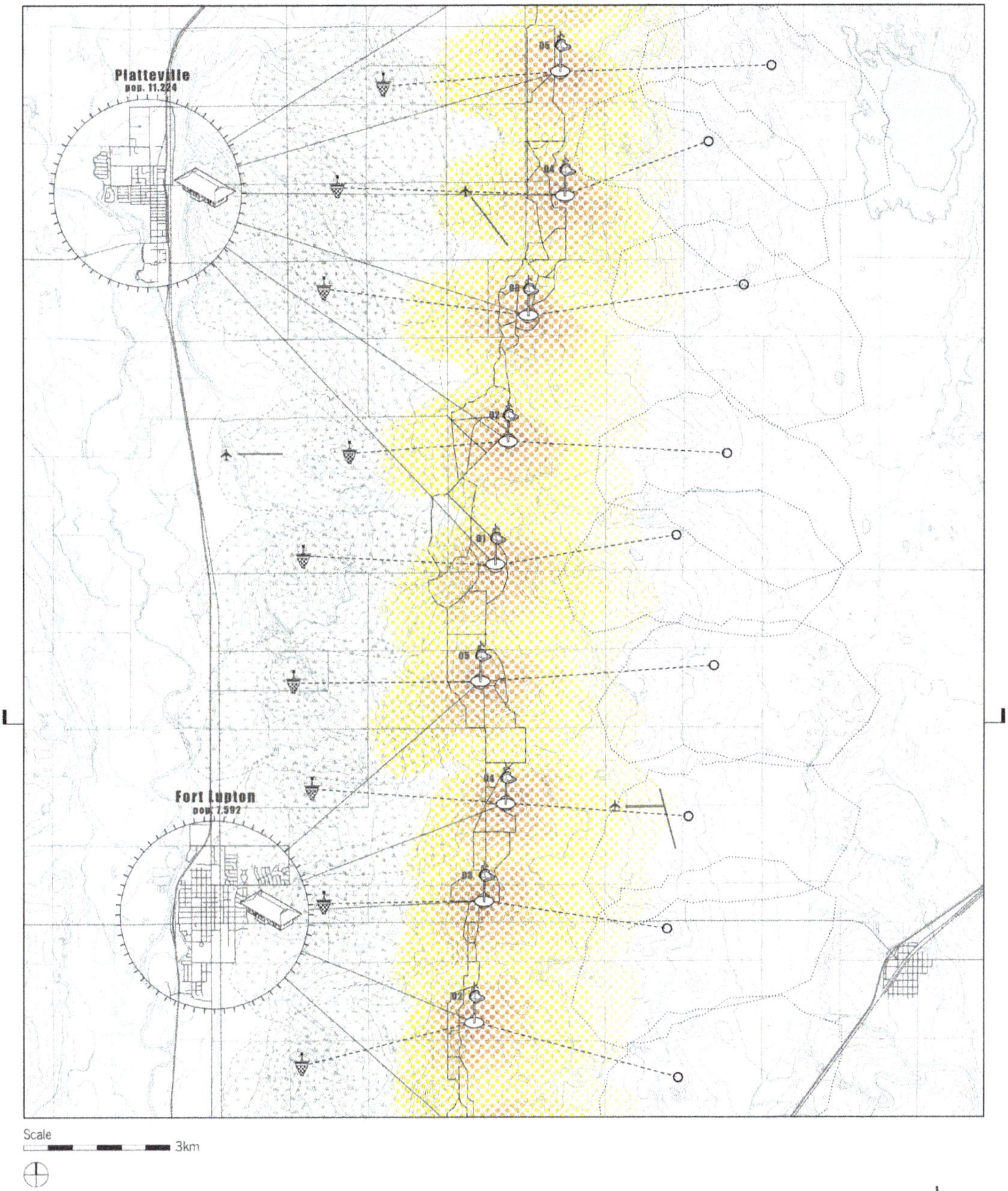
Platteville
pop. 11,224
Fort Lupton
pop. 7,592
Scale
3km

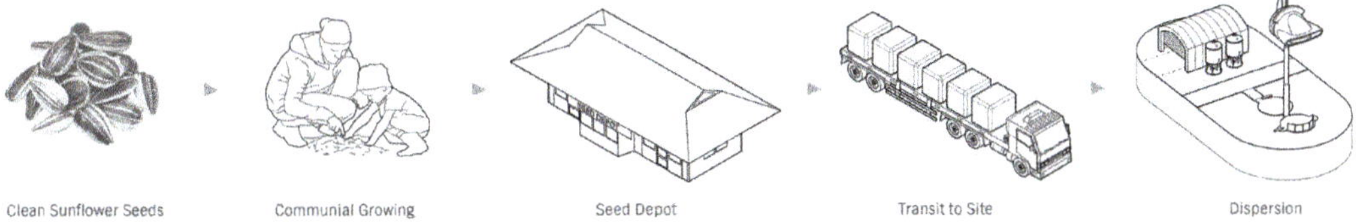
Clean Sunflower Seeds
Communial Growing
Seed Depot
Transit to Site
Dispersion

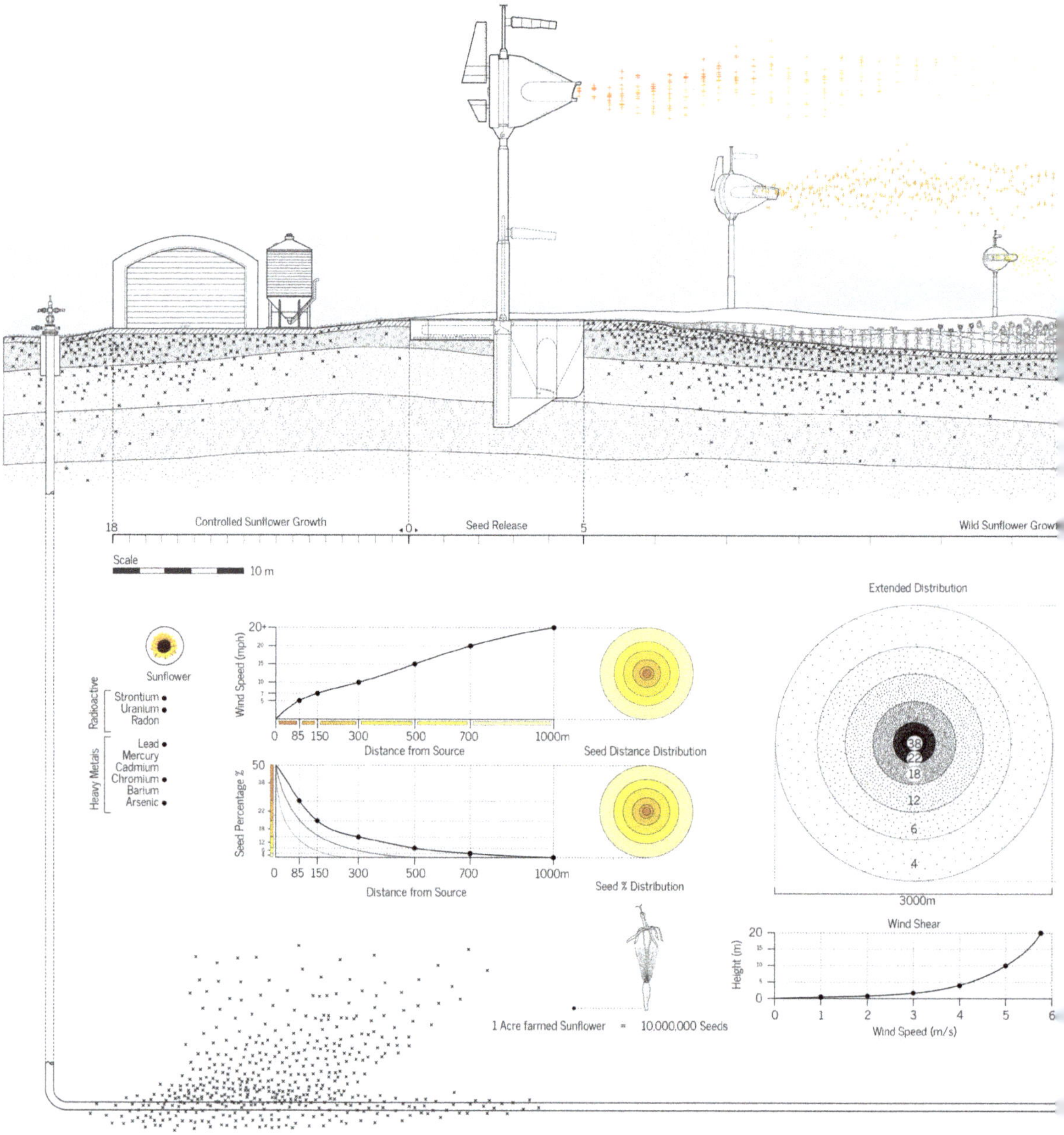

Controlled Sunflower Growth
Seed Release
Wild Sunflower Grow
18
0
5
Scale
10 m
Sunflower
Radioactive
Strontium
Uranium
Radon
Heavy Metals
Lead
Mercury
Cadmium
Chromium
Barium
Arsenic
Wind Speed (mph)
Distance from Source
0 85 150 300 500 700 1000m
Seed Distance Distribution
Seed Percentage %
Distance from Source
Seed % Distribution
Extended Distribution
38
22
18
12
6
4
3000m
Wind Shear
Height (m)
Wind Speed (m/s)
1 Acre farmed Sunflower = 10,000,000 Seeds

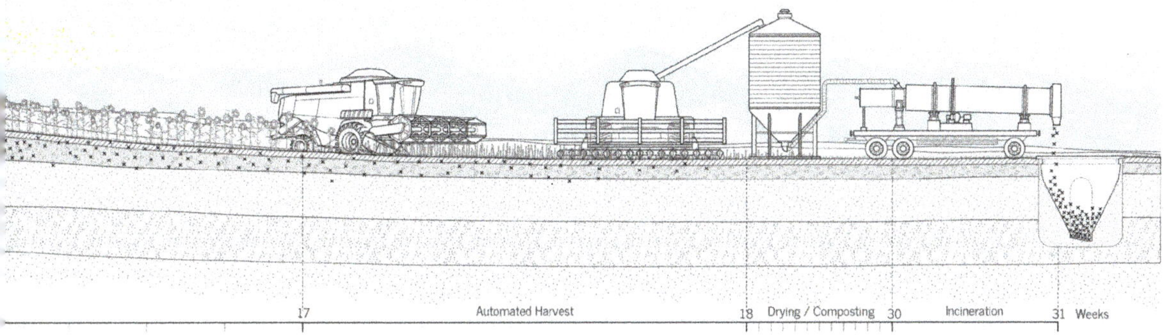

Greeley Wind Distribution

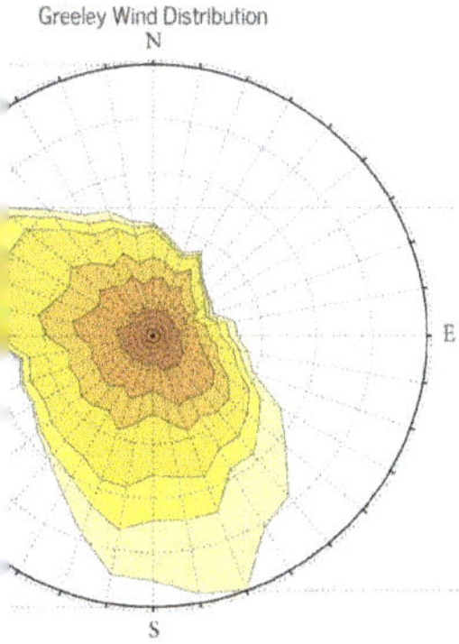

Automated Harvesting:
2483 acres
20 acres per hour
125 Hours

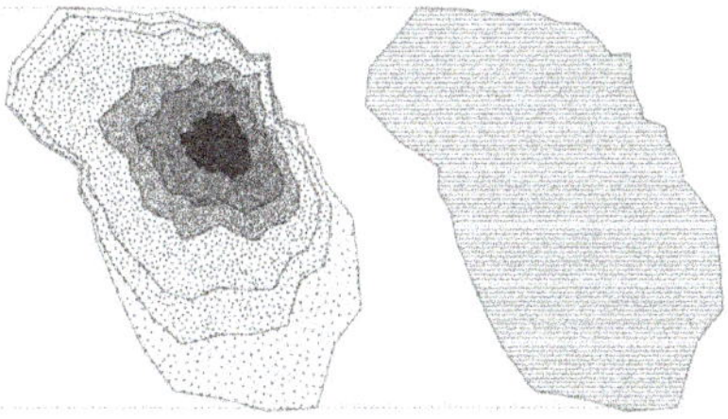

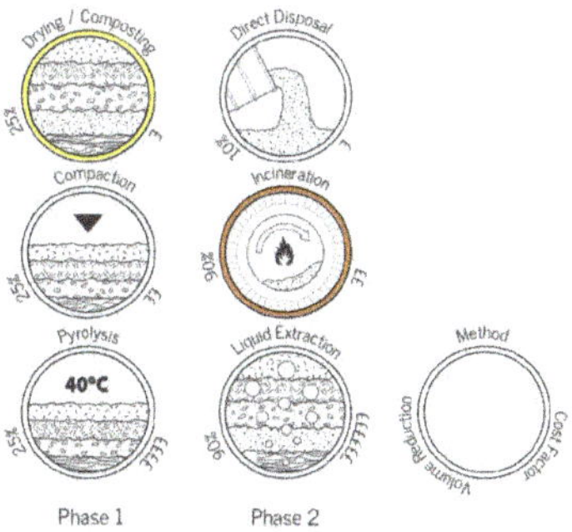

%	Plant Pop.	Plants/Acre	Yield/Acre (lbs)	Yield Total (lbs)
[illegible]	836,000	10,000	675	55,900
[illegible]	484,000	1,820	122	32,500
[illegible]	396,000	1,215	82	26,685
[illegible]	264,000	345	23	17,780
	132,000	306	20	8,865
	88,000	143	10	6,140
[illegible]	2,200,000			130,090 lbs /4185 bu

450 x (Plants/Acre x 0.00005) x 0.55

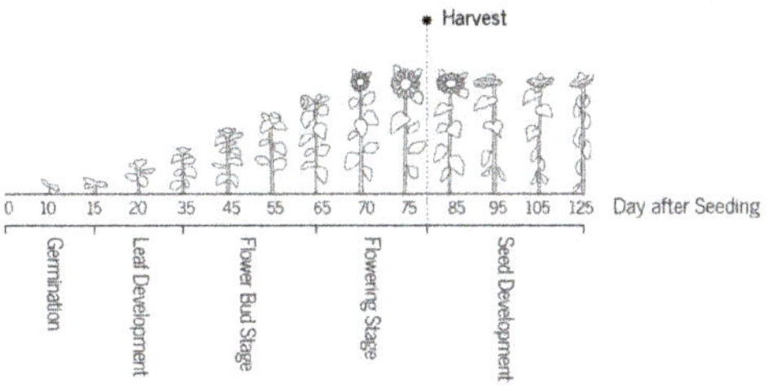

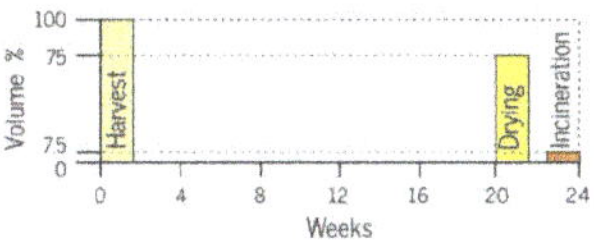

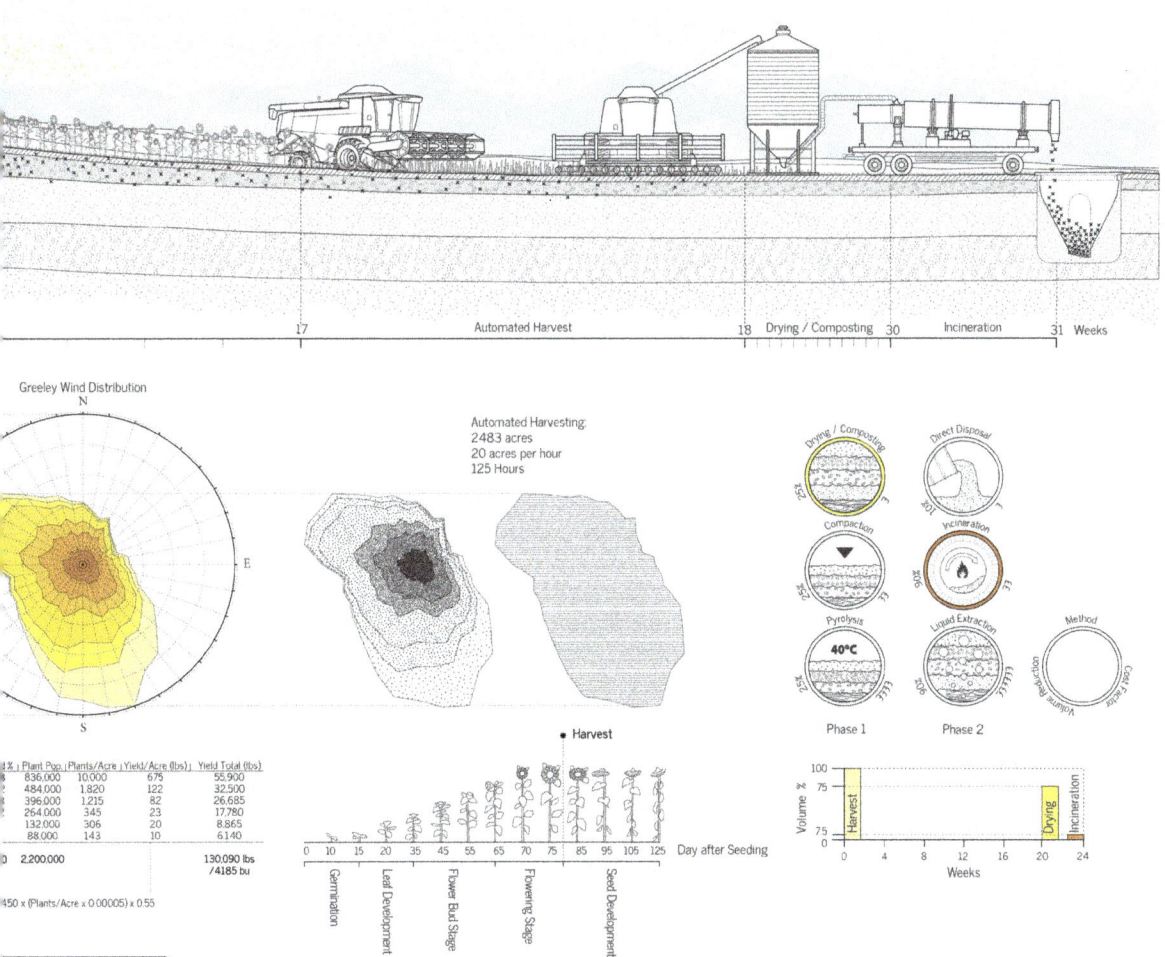

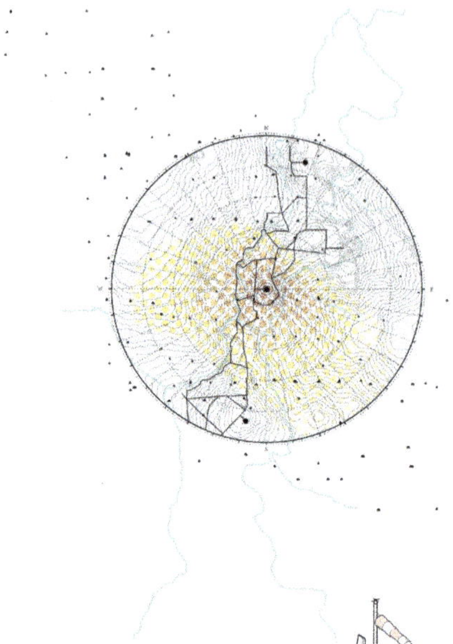

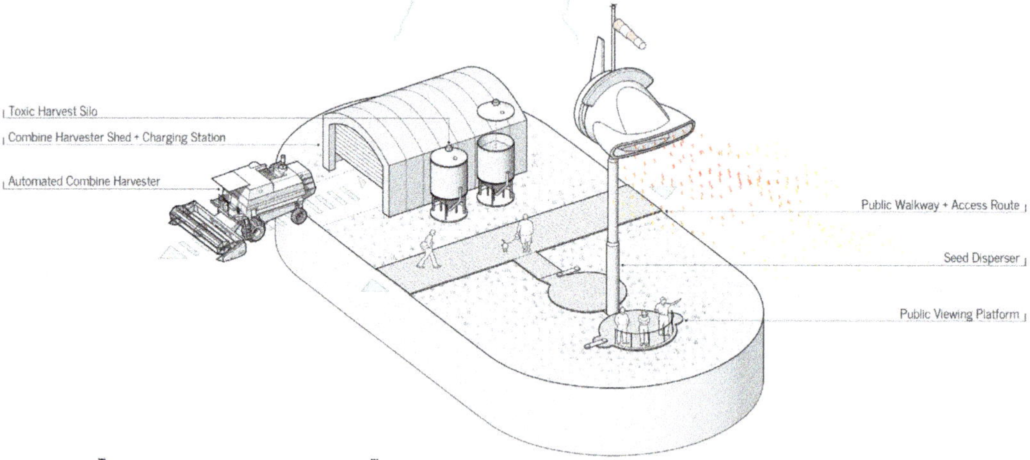

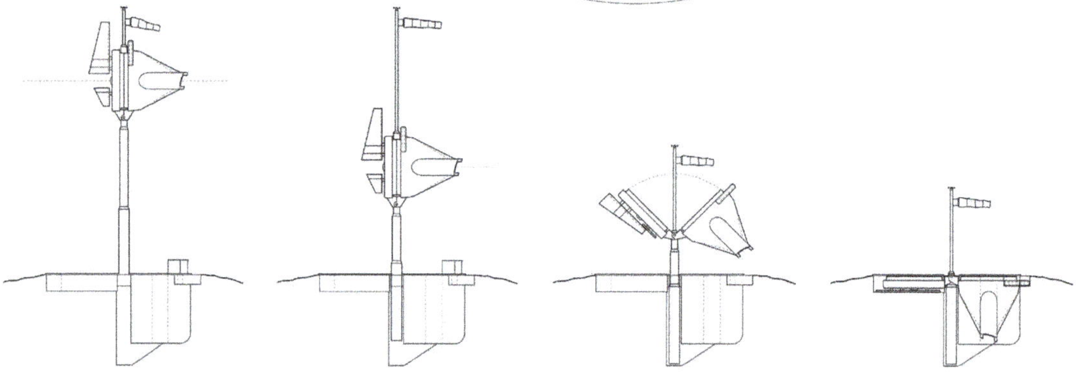

Low Winds Elevation High Winds Elevation Collapsing Procedure Hibernation

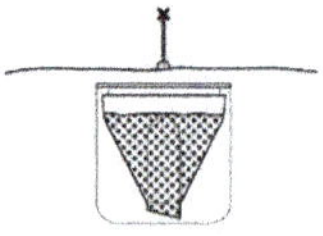

Future
Toxic Vault

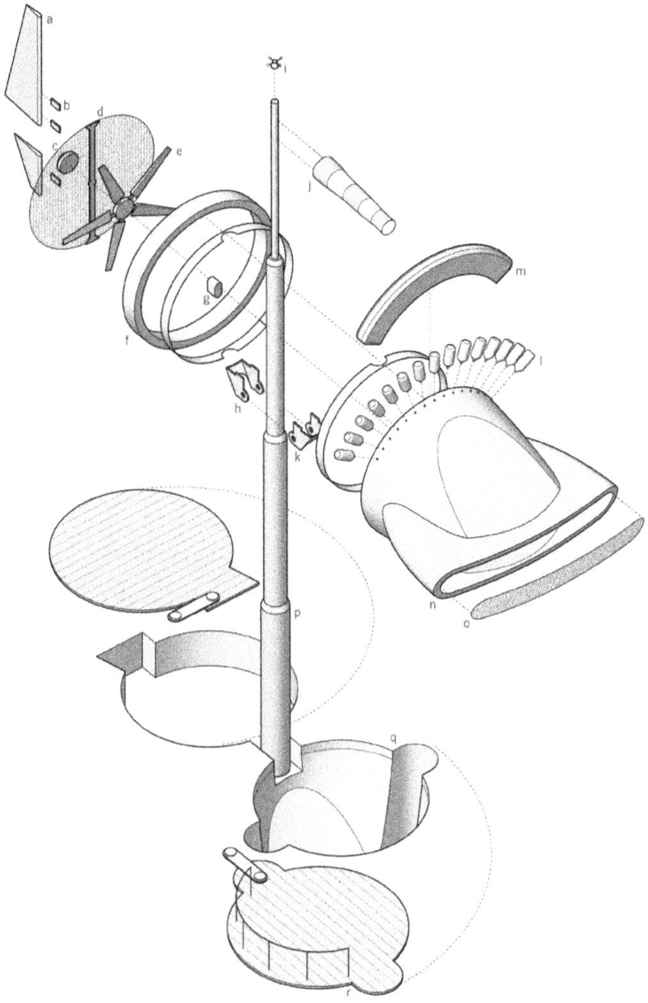

a - Tail Fin
b - Tail Fin Rotator
c - Turbine Cap
d - Brace + Bird Netting
e - Pin Wheel Rotors
f - Turbine Ring
g - Stator Junction Box
h - Turbine Hinge
i - Anemometer
j - Wind Sock
k - Seed Head Hinge
l - Seed Hoppers
m - Hopper Casing
n - Seed Head Air Disperser
o - Bird Netting
p - Telescopic Mast
q - Storage Vault
r - Public Viewing Platforms

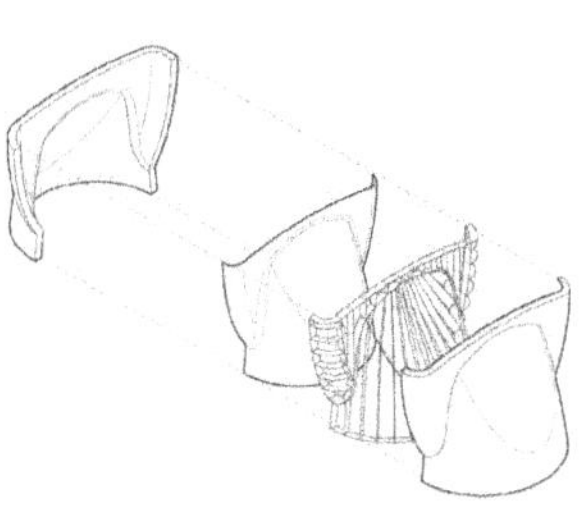

Seed Head Assembly

Construction

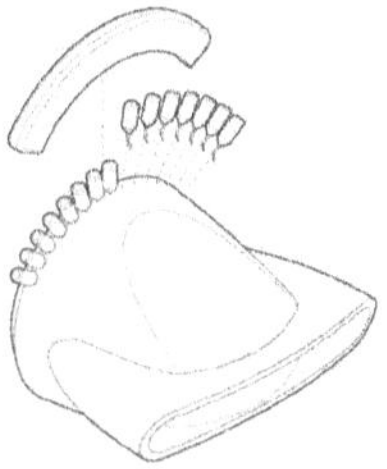

Seed Hoppers

Seed Head

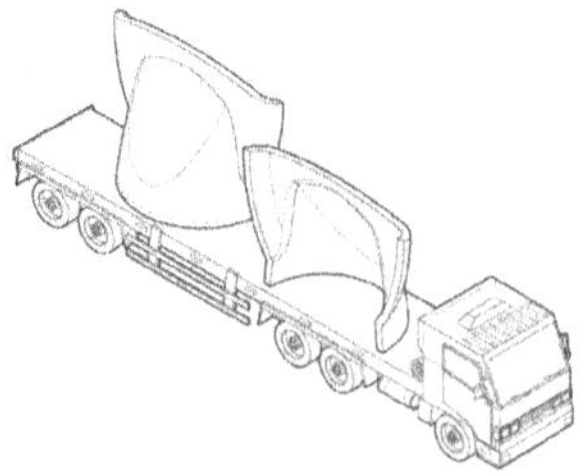

Transportation to Site

Delivery

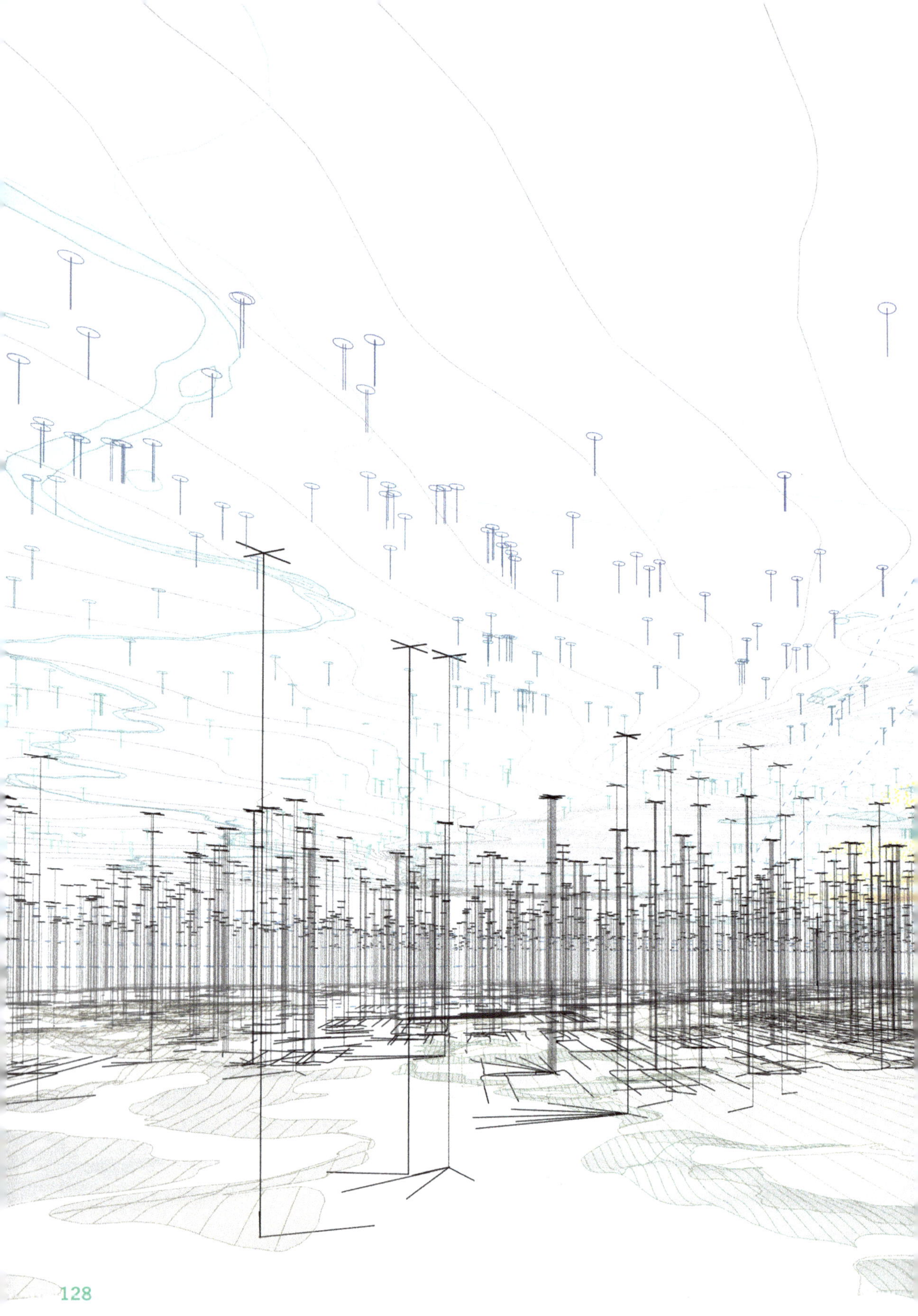

RESIDENTIAL RESPIRATOR

PHILIP HURRELL

How does one inhabit an invisibly toxic landscape?

Increasing smog and related health issues in rural and wilderness areas of America are drastically impacting peoples lives. (01) Clandestine violence is seeping and swirling across the American dream, revealing itself through debilitating symptoms and carcinogenic fumes. Unable to move due to increasing fracking activities, many people are living with the effects of air pollution from the recent explosion in domestic gas production. In response to this interrupted sense of place, Residential Respirator is a site specific prototype that creates a second skin around an existing home in Washington County, Pennsylvania. Beginning with the idea of a gas mask for a house, the proposal creates a filter layer against toxic air pollutants that is capable of breaking down the pollutants into CO^2 and water through the use of a specialised titanium dioxide treatment on it's skin. Using this new living condition, the underlying structure explores new opportunities for spatial experience by expanding the home between its current footprint and the buffer zone. A distributed network of air monitoring devices stimulate the buffer zone to deploy the skin and protect it's inhabitants as a response to the altered landscapes of fracked urbanism.

Notes

(01) ' New York Times - In Pinedale, Wyo., Residents Adjust to Air Pollution', http://www.nytimes.com/2011/03/10/ us/10smog.html

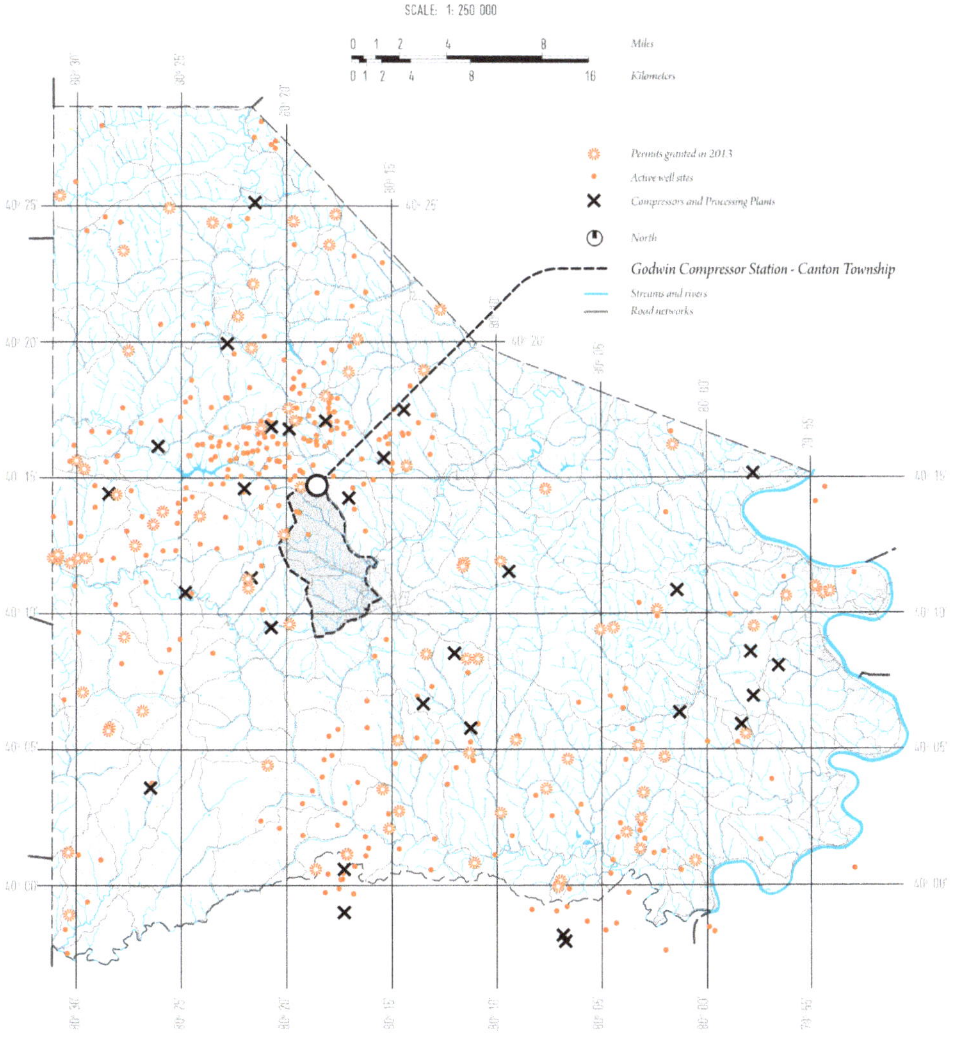
SCALE: 1: 250 000
0 1 2 4 8 Miles
0 1 2 4 8 16 Kilometers
Permits granted in 2013
Active well sites
Compressors and Processing Plants
North
Godwin Compressor Station - Canton Township
Streams and rivers
Road networks

P131 [top] Residential Respirator Problem Statement
[bottom] Fracking Infrastructure Study, Washington County, PA
Sources: http://www.dep.pa.gov/; http://www.fractracker.org/2013/06/pa-unconventional-production-data-aggregated/; http://stateimpact.npr.org/pennsylvania/drilling/

P133 [top] Gretna Community, Canton Township, Washington County, PA
Sources: https://abrahampacana.wordpress.com/2012/04/30/a-pagoda-in-pennsylvania/; https://endlessmountains.wordpress.com/page/15/; Google Maps, Google Streetview.
[bottom] Site topography and Intensity of Pollutant Sources
Sources: http://www.dep.pa.gov/; http://www.fractracker.org/2013/06/pa-unconventional-production-data-aggregated/

P134 Health Impacts: Agents and their Effects on the Human Body
Sources: McKenzie, L.M., Witter, R.Z., Newman, L.S. and Adgate, J.L. (2012). Human health risk assessment of air emissions from development of unconventional natural gas resources. Science of The Total Environment, 424, pp. 79–87. doi: 10.1016/j.scitotenv.2012.02.018; Colborn, T., Schultz, K., Herrick, L. and Kwiatkowski, C. (2013). An exploratory study of air quality near natural gas operations. Human and Ecological Risk Assessment: An International Journal, 20(1), pp. 86–105.

P135 Site and Wind Study
Sources: http://www.dep.pa.gov/; http://www.fractracker.org/2013/06/pa-unconventional-production-data-aggregated/

P136 Analysis of the occupation of a typical house over the course of the day

P137 Residential Respirator
[top] Roof Plan
[bottom left] Ground Floor Plan
[bottom right] First Floor Plan

P138 Residential Respirator: Prototype Trials and Development

P139 Residential Respirator: Components

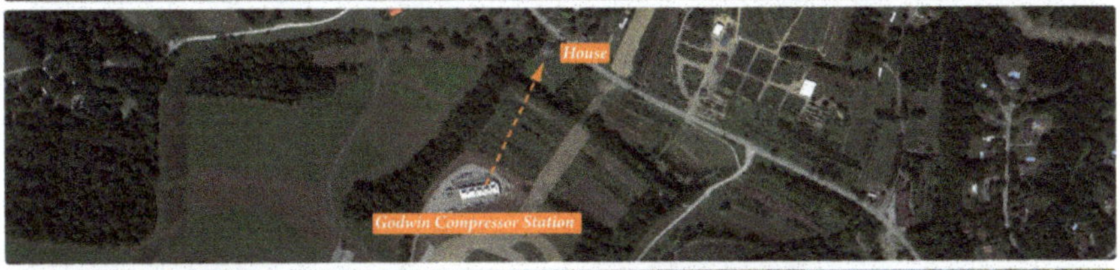
House
Godwin Compressor Station

Godwin Compressor Station
House

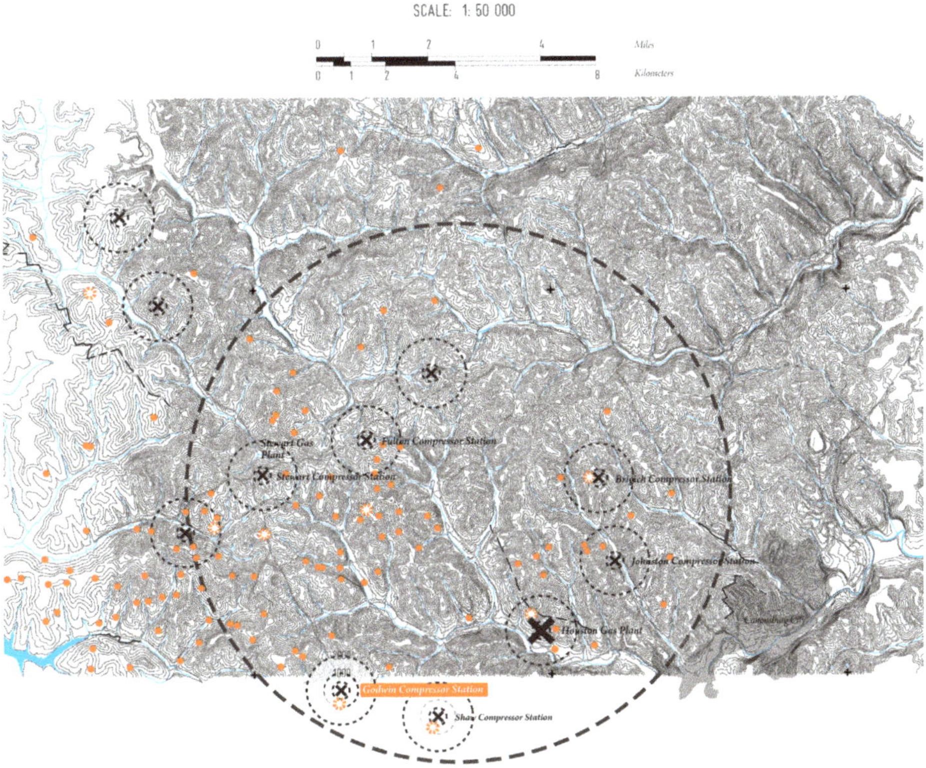
SCALE: 1: 50 000
0 1 2 4
Miles
0 1 2 4 8
Kilometers
Stewart Gas Plant
Stewart Compressor Station
Johnston Compressor Station
Houston Gas Plant
Godwin Compressor Station
Shaw Compressor Station

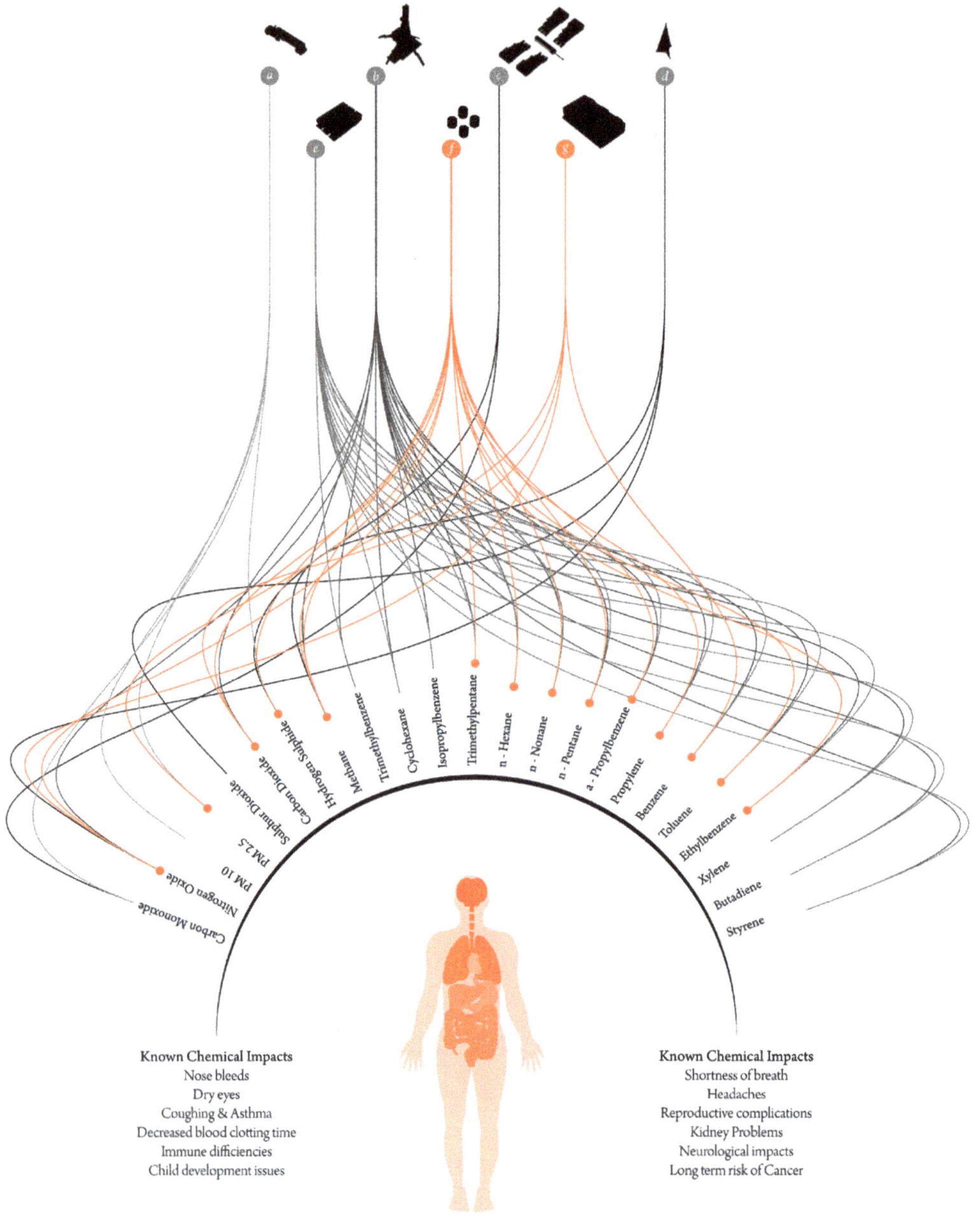
Carbon Monoxide
Nitrogen Oxide
PM 10
PM 2.5
Sulphur Dioxide
Carbon Dioxide
Hydrogen Sulphide
Methane
Trimethylbenzene
Cyclohexane
Isopropylbenzene
Trimethylpentane
n - Hexane
n - Nonane
n - Pentane
a - Propylbenzene
Propylene
Benzene
Toluene
Ethylbenzene
Xylene
Butadiene
Styrene
Known Chemical Impacts
Nose bleeds
Dry eyes
Coughing & Asthma
Decreased blood clotting time
Immune difficiencies
Child development issues
Known Chemical Impacts
Shortness of breath
Headaches
Reproductive complications
Kidney Problems
Neurological impacts
Long term risk of Cancer

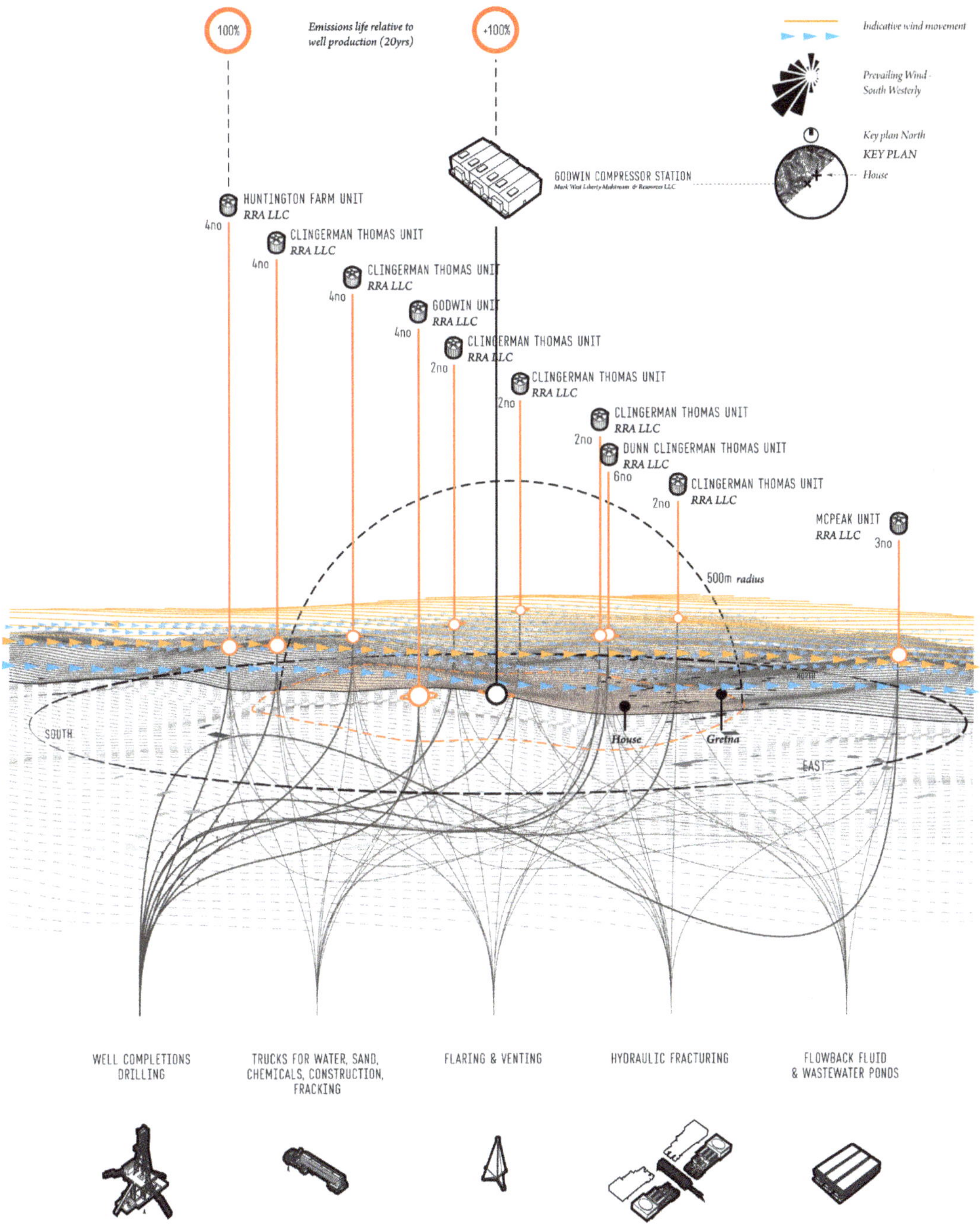
100%
Emissions life relative to well production (20yrs)
+100%
Indicative wind movement
Prevailing Wind - South Westerly
Key plan North
KEY PLAN
House
GODWIN COMPRESSOR STATION
HUNTINGTON FARM UNIT
RRA LLC
4no
CLINGERMAN THOMAS UNIT
RRA LLC
4no
CLINGERMAN THOMAS UNIT
RRA LLC
4no
GODWIN UNIT
RRA LLC
4no
CLINGERMAN THOMAS UNIT
RRA LLC
2no
CLINGERMAN THOMAS UNIT
RRA LLC
2no
CLINGERMAN THOMAS UNIT
RRA LLC
2no
DUNN CLINGERMAN THOMAS UNIT
RRA LLC
6no
CLINGERMAN THOMAS UNIT
RRA LLC
2no
MCPEAK UNIT
RRA LLC
3no
500m radius
SOUTH
House
Gretna
EAST
WELL COMPLETIONS DRILLING
TRUCKS FOR WATER, SAND, CHEMICALS, CONSTRUCTION, FRACKING
FLARING & VENTING
HYDRAULIC FRACTURING
FLOWBACK FLUID & WASTEWATER PONDS

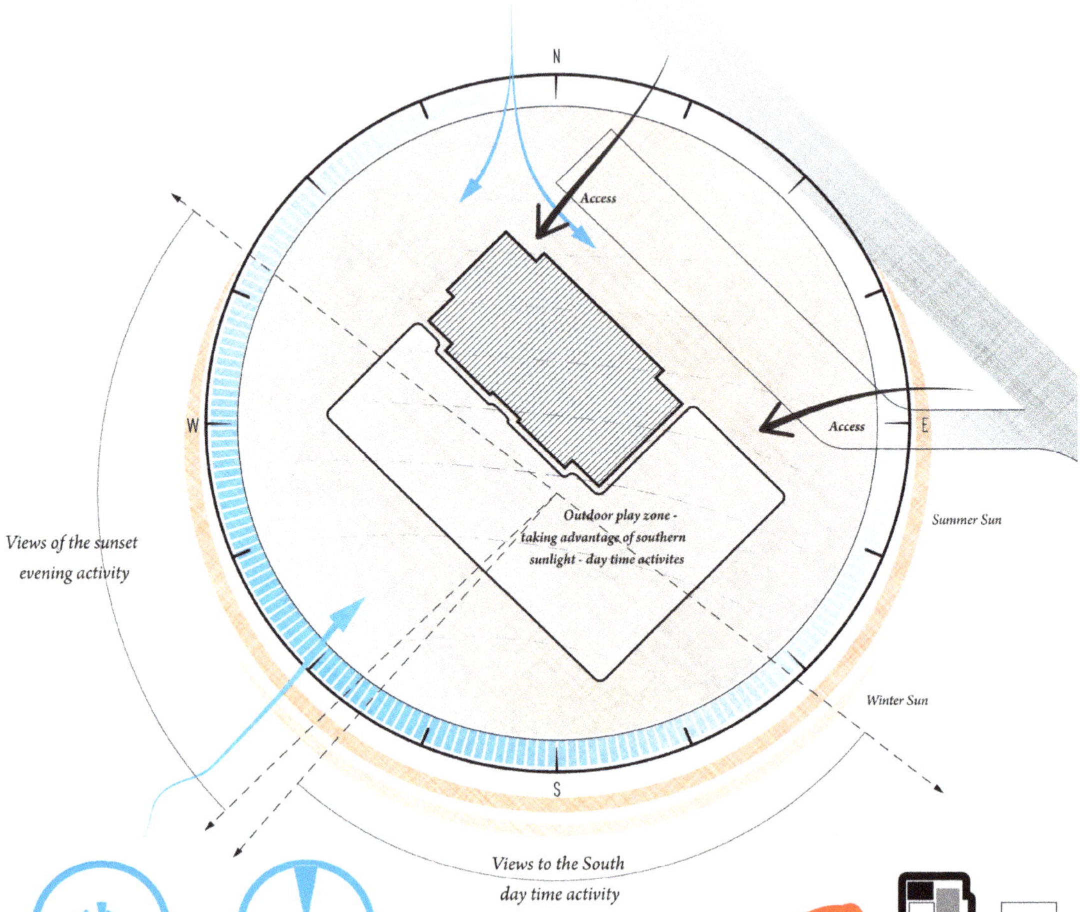

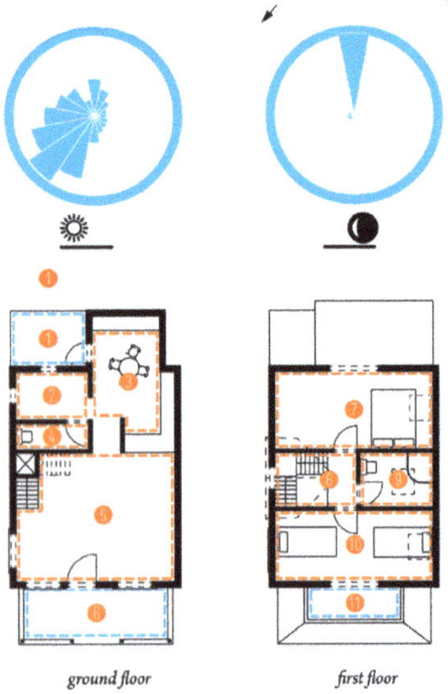

1. Outdoor & back patio (exposed)
2. Store
3. Kitchen & Dining
4. WC
5. Living room
6. Front Porch (exposed)
7. Master Bedroom
8. Landing & Stairwell
9. Bathroom
10. Bedroom
11. Balcony (exposed)

ACTIVITY LEVEL

North high low

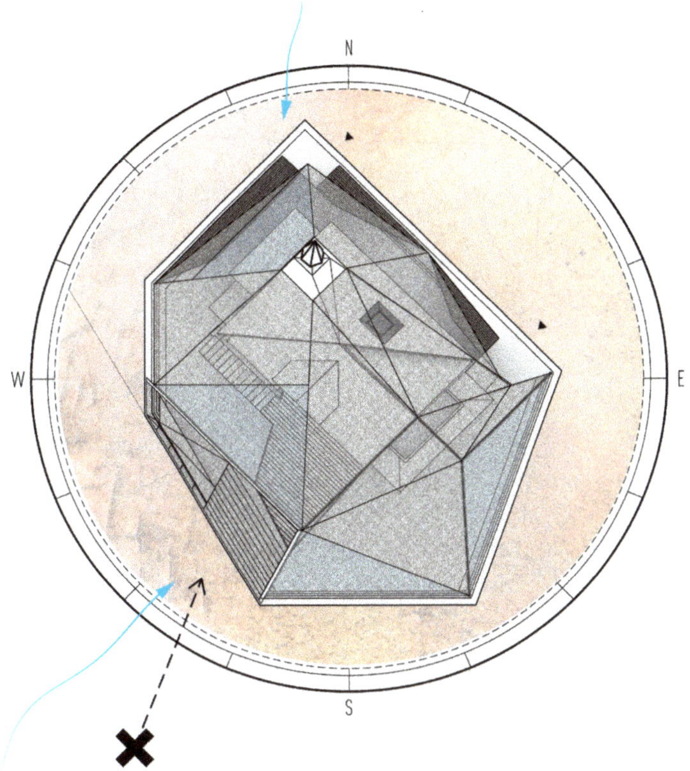

Key

1. Outdoor and back patio
2. Store
3. Kitchen and dining
4. WC
5. Living room
6. Frontporch
7. Master bedroom
8. Landing and stairwell
9. Bathroom
10. Bedroom
11. Balcony
12. Outdoor play zone

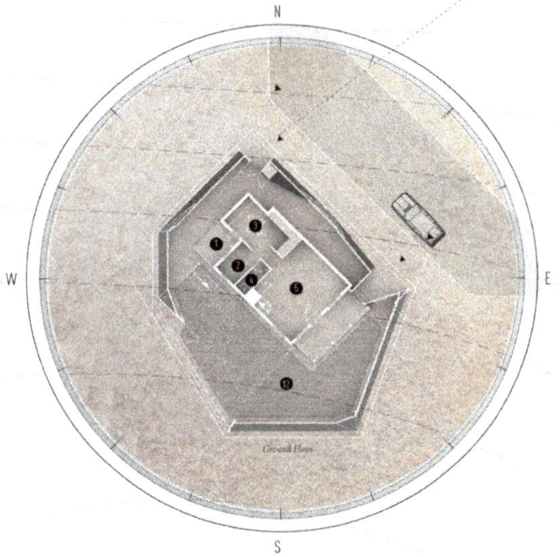

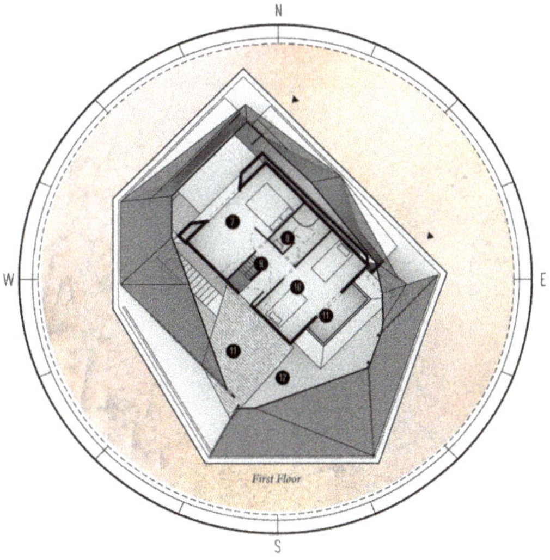

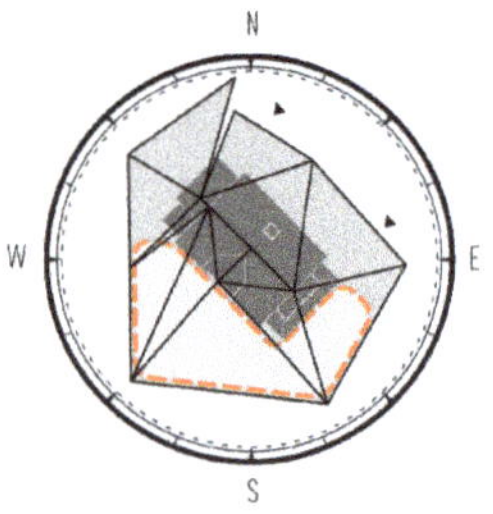

MASK PROTOTYPE
TRIAL 1

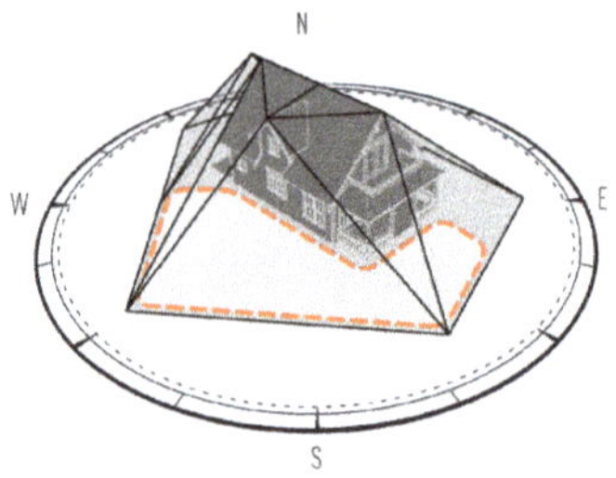

West Elevation

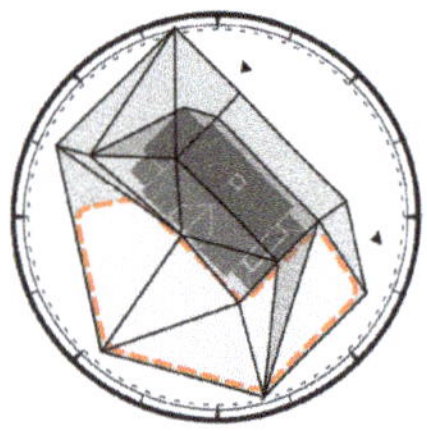

MASK PROTOTYPE
TRIAL 2

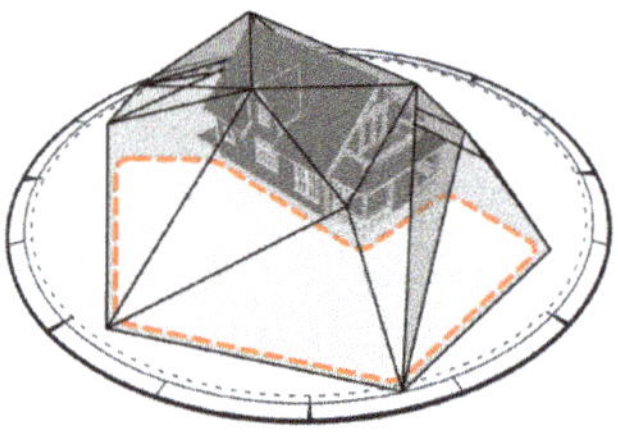

West Elevation

MASK PROTOTYPE FINAL
DEVELOPMENT BASE

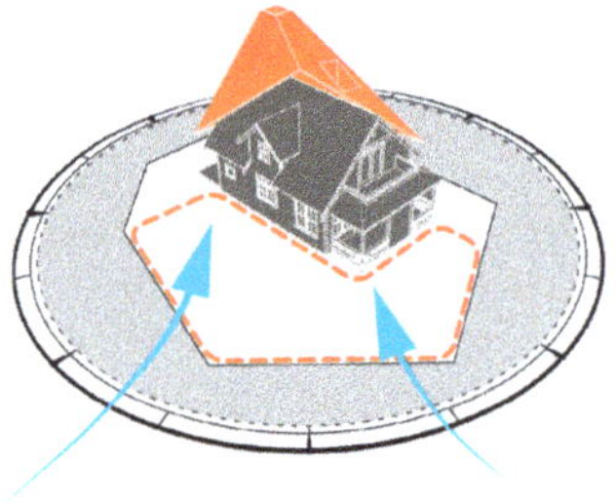

East Elevation

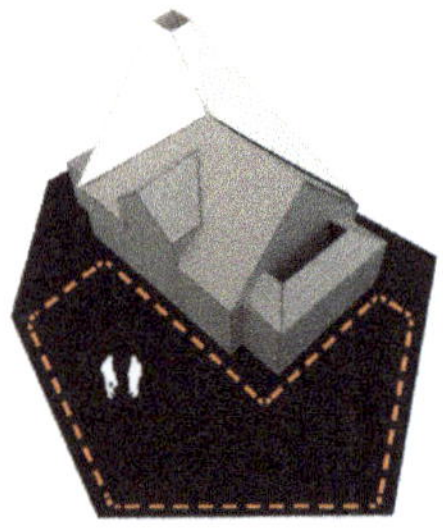

PHYSICAL MODEL
house scale and chimney

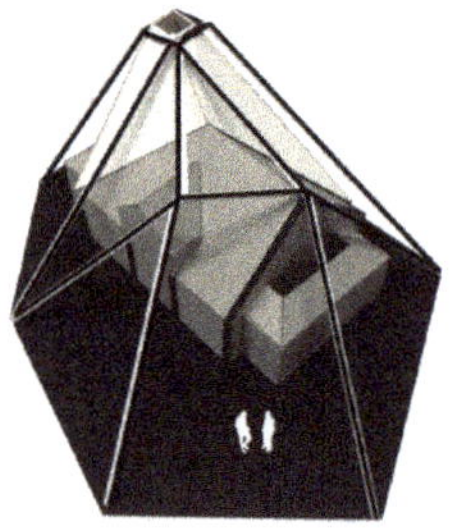

trial skin fold lines

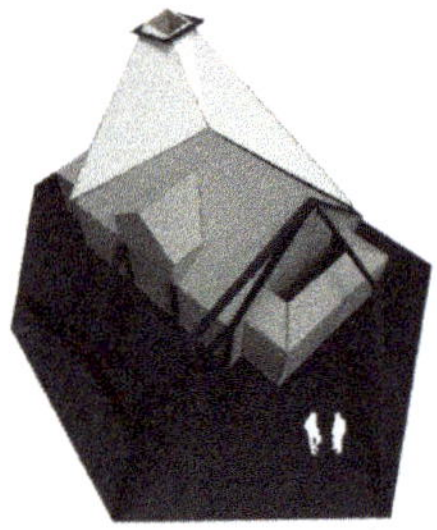

exploring substructure similar to a camping tent

HYPEROBJECTS / MICROPUBLICS

SEM 2 2013/14

INTRODUCTION

LINDSAY BREMNER AND ROBERTO BOTTAZZI

These days, powerful private interests over-determine politics and private matters prevail over public concerns. The fracking industry is a case in point. It is driven by powerful, profit seeking private interests, shrouded in secrecy, accountable to no one but its share-holders, yet having profound and far reaching public consequences for the earth and all who live on it. The State such an industry operates in is not a public realm, but an institution or set of institutions to administer and hold these private households together in nested units: the city, the nation, the United Nations etc. This model is profoundly out of touch with the reality of the Anthropocene, the time we live in. This is the time of what Timothy Morton calls "Hyperobjects," (01) the time of things that are massively distributed in time and space relative to humans. They involve profoundly different temporalities and scales than the ones we are used to and do not fit neatly into the nested model of contemporary politics.

In *Politics of Nature* (02) Bruno Latour proposes an alternative model for a public realm to address this state of affairs. He proposes that publics (plural) be mobilized around common matters of concern, or what he calls 'things.' Frei and Bohlen remind us in *MicroPublicPlaces* (03) that the word 'thing' originally meant an assembly or a courtroom where people gathered to discuss a matter of concern. A thing is not a material object, but a gathering one takes part in. Today, things are complicated networks of globally extending relations through which humans and non-humans are gathered together.

It was this inherent public-ness of things that students were asked to make explicit in the second design brief of 2013/14. They were asked to design a new public institution, a MicroPublicPlace that gathered humans and non-humans around a thing they had in common. Non-humans (the earth, global warming, polluted water etc.) were to be given a voice in this conversation, through science and / or aesthetic means. Students were required to design the constitution or set of rules for this public realm, to enable participants who did not agree to present their points of view; they were required to design a place where humans and non-humans could gather, both physically and virtually, to speak, deliberate and act in concert. This would include: a new building type to facilitate relationships

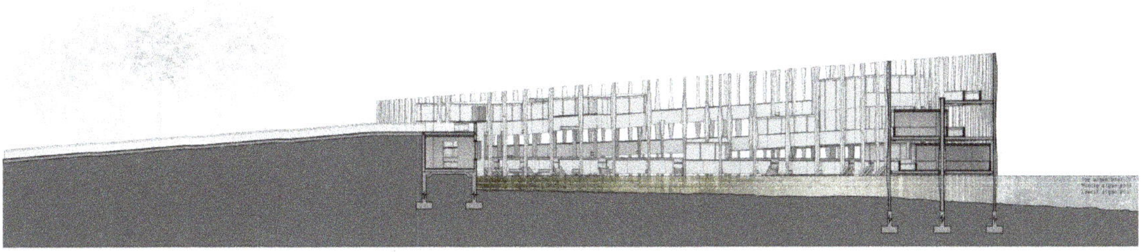

Algae Observation Centre

A facility to research, present and theatricise the process of eutrophication and algae bloom in Poole Harbour on the Dorset Coast. Members of the public curious enough to approach the building walk around the rim of a structural petri dish and observe the algae as it grows.

Alex Jaggs

amongst participants; links between the local and the global, human and non-human, human experience and data, science and aesthetics, the real and the virtual; and uses of technical and social media and computational design.

The site for this brief was the Isle of Purbeck on the Dorset Coast of England, which we visited in January 2014 and were guided around by local geologist Alan Holiday. This site offered opportunities for engagement with an extraordinary range of geological sites, paleontological remains and resource extraction histories such as Jurassic era fossils, Purbeck stone quarries, geothermal wells, the largest on-shore oil drilling site in Europe (Wytch Farm) and the Kimmeridge oil shales. The Dorset Council had approved a planning application in November 2013 for an exploratory shale gas well at California Quarry just outside Swanage. Students were asked to select one anthropocenic 'thing' on the Isle of Purbeck and to design a MicroPublicPlace to draw attention to its global extent.

This book presents four responses to this brief. Andrew Baker-Falkner project, 'Hydrological Slice,' was a Micropublic Water Treatment Plant to filter and purify produced water and organise new human relations with it, should fracking commence at California Farm near Swanage in Dorset. John Cook, drawing on his first semester's work, proposed a 'Micropublic Seed Bank' in a disused quarry on the sea-facing cliffs on the Isle of Purbeck, as part of an existing global network of seed vaults for collecting, storing and distributing bio-remediating plant seeds. Claire Holton's 'Micropublic Powerhouse' reactivated disused geothermal wells around Dorset to supply geothermal electricity to serve emergency facilities during electricity blackouts. This responded to the severe storms in the UK in December 2013 that had resulted in electricity blackouts around the country. She focused on the design of one such facility in the village of Wareham in Dorset, which was conceptualised as a 'Micropublicteahouse' that used geothermal energy to grow, dry and serve tea. Michael O'Hanlon's project 'Micro Seismic Pavilion' explored the use and application of micro-seismic data to create sonic experiences.

Notes

(01) T. Morton, *Hyperobjects, Philosophy and Ecology after the End of the World*, Minneapolis, University of Minnesota Press, 2013.

(02) B. Latour, *Politics of Nature, How to bring the Science into Democracy*, Cambridge, MA., Harvard University Press, 2004.

(03) H. Frei and M. Bohlen, *MicroPublicPlaces, Situated Technologies Pamphlet no. 6*, New York, Architectural League, 2010, http://www.situatedtechnologies.net/?q=node/104 (accessed 10 July 2015).

HYDROLOGICAL SLICE

MICROPUBLIC WATER TREATMENT PLANT

ANDREW BAKER FALKNER

The controversial gas extraction method, fracking, sits between politics and the environment. Its use of approximately 18Ml of freshwater per well per year raises questions of the value of water as a resource in an age where there is increased pressure to maximise new energy production methods and technologies. This project deals with these dual aspects of environmental/human concern, by filtering and purifying produced water, while also encouraging people to reflect on their own and others use of water through a direct and indirect experience of the water and the landscape it inhabits. The Micropublic Water Treatment Plant is a centralized produced water treatment plant which is placed along ordered hydrological slices in the landscape, regulated by natural, softer, elements such as water, speed of movement and views. It brings together the public and large multinational companies into the same sphere, curated by a resource vital to both, water. It aims to encourage discussion and raise issues connected to the increasing consumption of water. By placing the plant in an area whose natural environment already attracts people, it also simultaneously acts as a machine for enhancing the experience of the landscape.

P145 The Hydrological Slice

P146 [top] Exploded Isometric of the Hydrological Slice
[bottom] Analysis of site lines from the Hydrological Sice

P147 [top left] Rights of Way, Isle of Purbeck
The Isle of Purbeck attracts a large number of tourists each year. As a result, there are a number of paths and rights of way running across the landscape, and a number concentrated south of Swanage overlooking the coast where California Quarry is located. The intervention utilized these existing elements and paths as clues for design.
Source: http://www.rowmaps.com/kmls/DT/
[top rght] Site Location, California Quarry
California Quarry is owned and run by Suttle Stone Quarries, a local company specializing in materials local to the area, such as Purbeck Stone. In 2014, they obtained planning permission to drill an exploratory borehole to test and evaluate the Portland - Wight reservoir for commercial extraction of oil and gas. It was expected that unconventional gas would be found in the same locality, which would require hydraulic fracturing,. The site chosen was an existing footpath that runs north-south near to California Quarry, with an existing road to the north and the South West Coastal Path to the south. This runs along the south coast of England for 630 miles, passing through Exmoor, Cornwall, Devon and Dorset. As a key tourist attraction it draws a lot of walkers, from the more casual walkers to serious ramblers, especially in the summertime. Integrating these publics at key points along the Hydrological Slice was central to the proposal.
[bottom] California Quarry
Photographs: Lindsay Bremner

P148 Study of east-west linear landscape elements such as walls, roads, paths etc.

P149 Study of north-south inear landscape elements such as walls, roads, paths etc.

P150 Walk along the line of the Hydrological Slice

P151 Landscape Analysis of the Hydrological Slice

P152 Plan of the Hydrological Slice

P153 Water Monitoring Devices along the Hydrological Slice
The water is monitored at five points in relation to the different stages of filtration. These become moments along the walk to pause and engage with the processes underway.

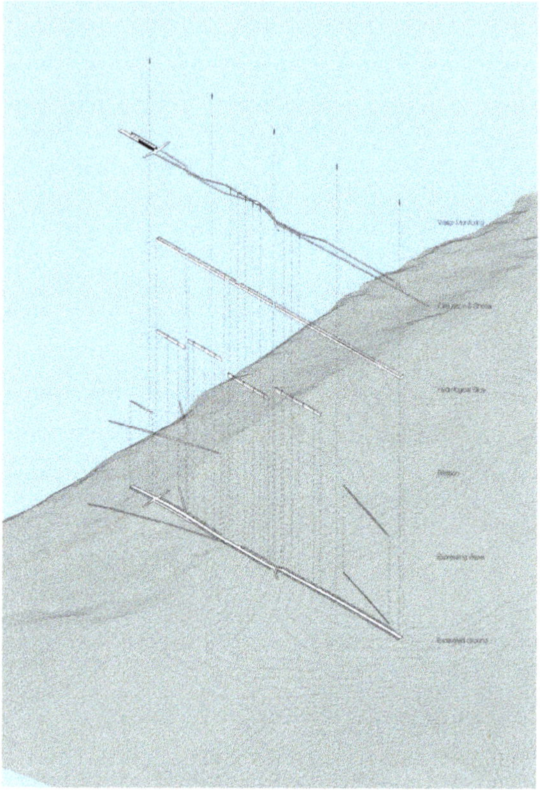

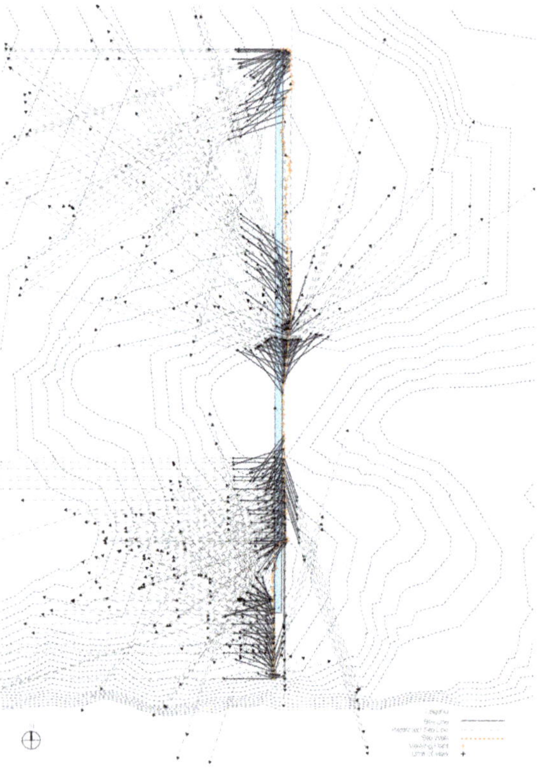

Southward Quarry
Site
Eastern Quarry
California Quarry
South West Coast Path

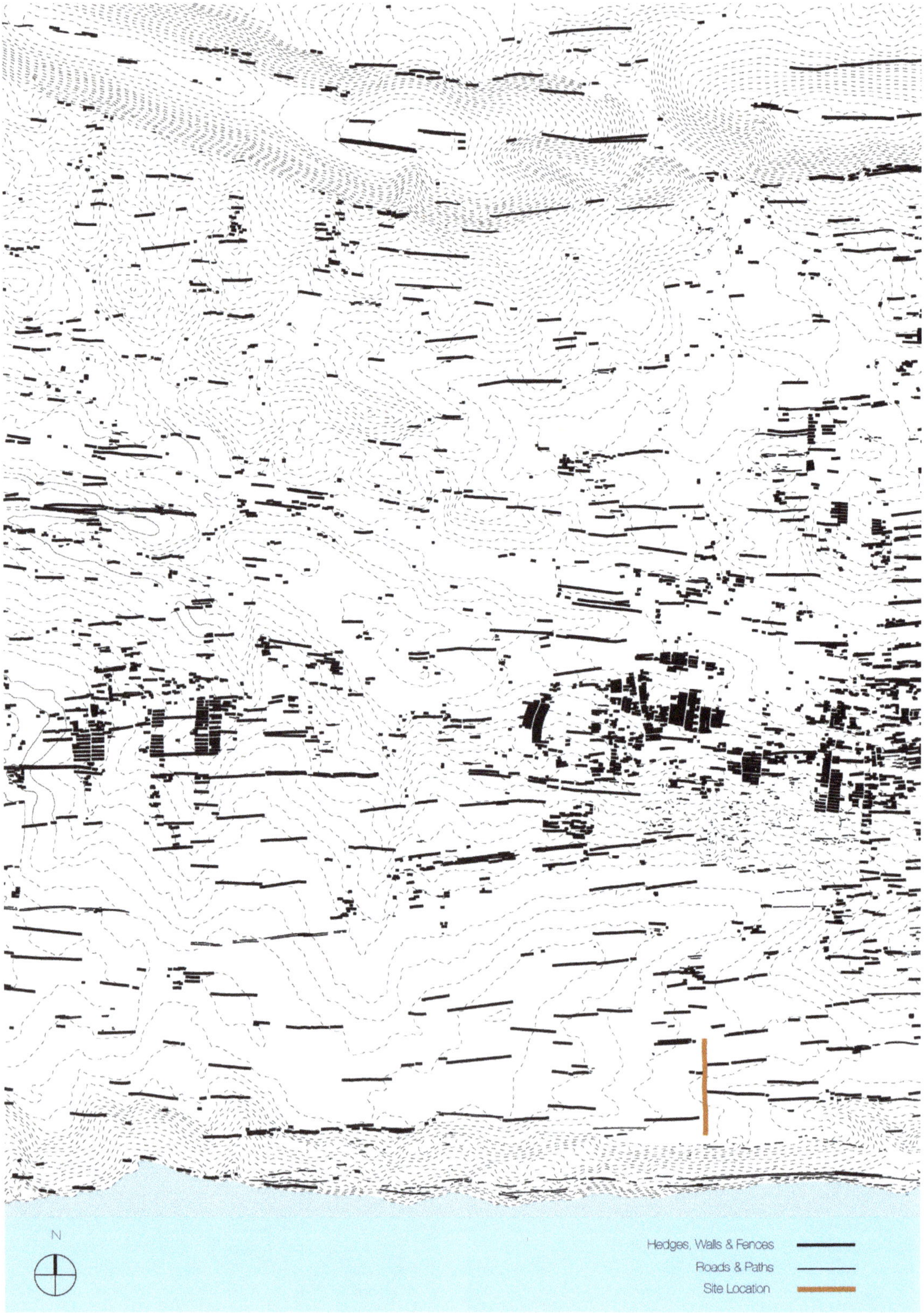
N
Hedges, Walls & Fences
Roads & Paths
Site Location

N
Hedges, Walls & Fences
Roads & Paths
Site Location

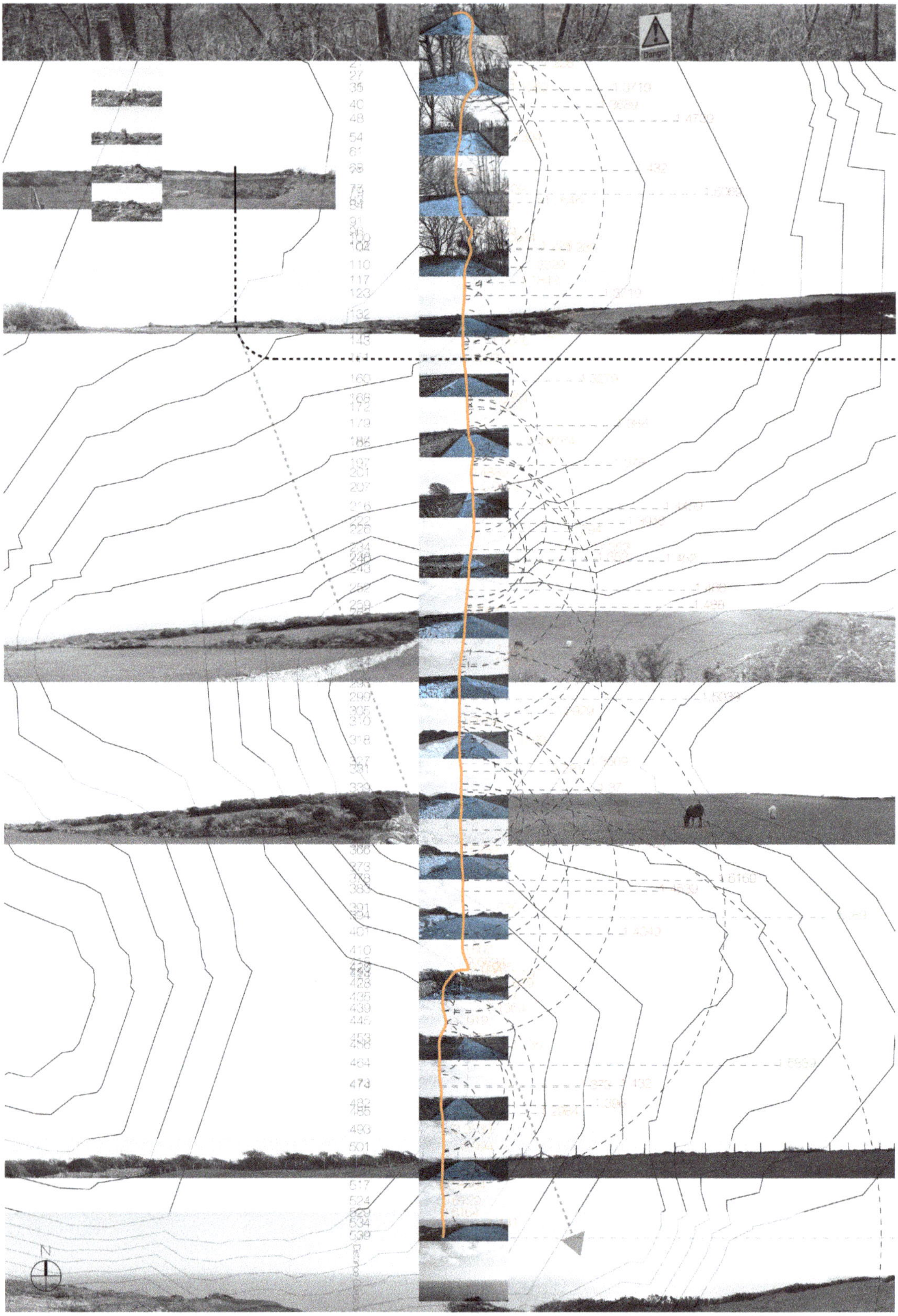

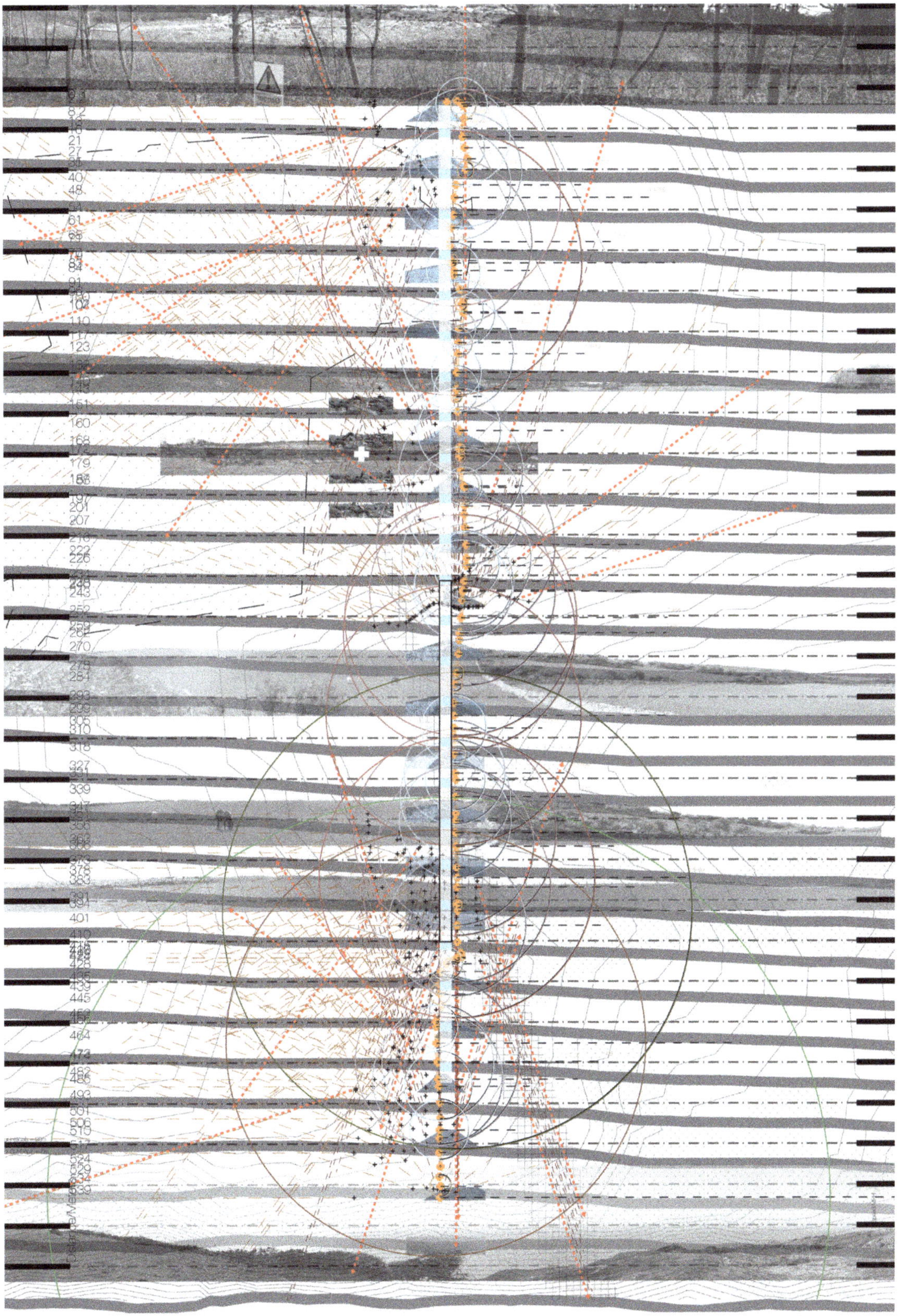

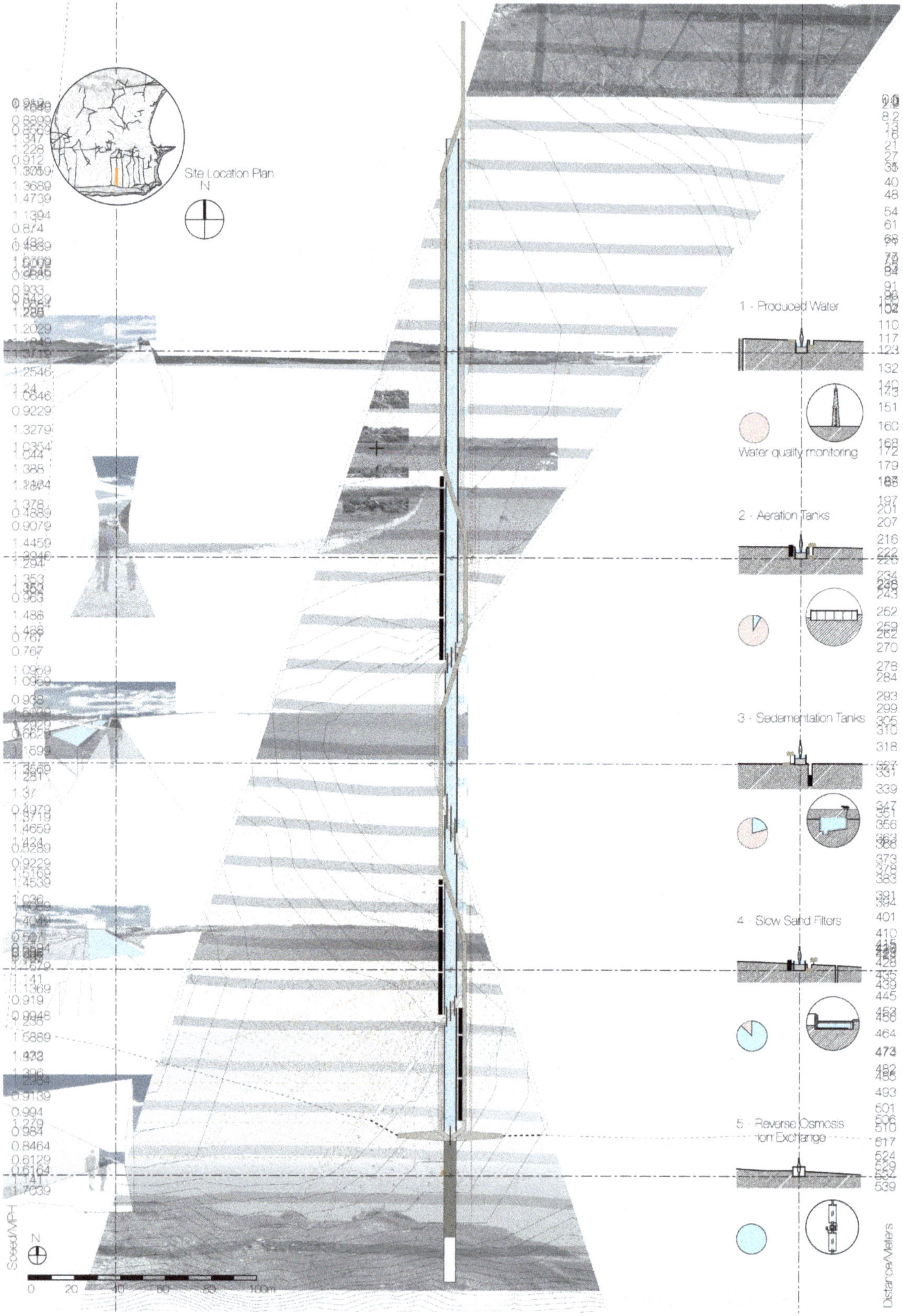

Site Location Plan
N
1 - Produced Water
Water quality monitoring
2 - Aeration Tanks
3 - Sedementation Tanks
4 - Slow Sand Filters
5 - Reverse Osmosis
Ion Exchange
Speed/MPH
Distance/Meters
0 20 40 60 80 100m

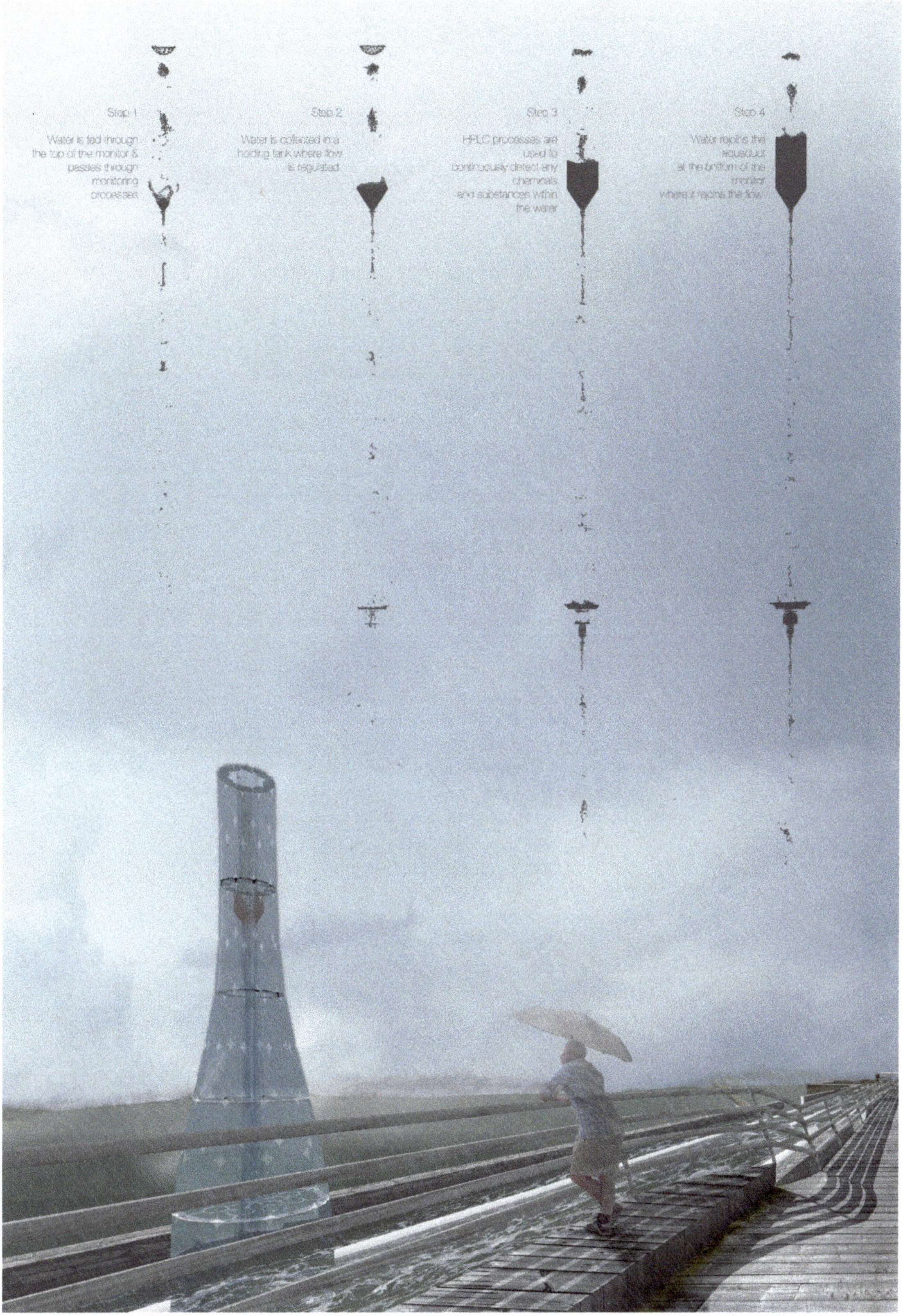
Step 1
Water is fed through the top of the monitor & passes through monitoring processes
Step 2
Water is collected in a holding tank where flow is regulated
Step 3
HPLC processes are used to continuously detect any chemicals and substances within the water
Step 4
Water rejoins the aqueduct at the bottom of the monitor where it rejoins the flow.

MICROPUBLIC SEED BANK

JOHN COOK

Continuing the investigations conducted for the previous semester's proposal, Remediated Landscape, during the second semester, I looked further back along the time-line of the proposed remediation strategy - at the collection, storage and distribution of phytoremediating plant seeds. A seed bank was proposed on the Isle of Purbeck in Dorset, to be buried within an abandoned quarry, the Winspit Cave. This would form part of a global network of seed vaults. The seed bank is formed by excavating into the old quarry, with space being created through subtraction rather than addition. The dimensions of mining machinery heads and detonation arrays determine the widths and proportions of the carved out spaces and corridors. A plant nursery, lab facilities, virtual seedbank and the temperature controlled seed vaults themselves are either carved out from the rock or installed as prefabricated components. The security and stable conditions offered by this environment are vital in preserving the seeds at a constant -180C. The public awere invited to observe the processes of seed preservation, highlighting the species, sources, quantities and economics of the operation, and the role the strategy played in the context of wide scale destructive mineral extraction. As visitors journey through the seed bank, the cavernous environment becomes an experiential time-line in itself- the rock walls are both formed from and reveal the fossilised remains of previously living flora, now billions of years old. Through its unique location and publically observable procedures, the Isle of Purbeck Micropublic Seed Bank aimed to communicate wider geological processes, deep time and the consequences of the anthropocene.

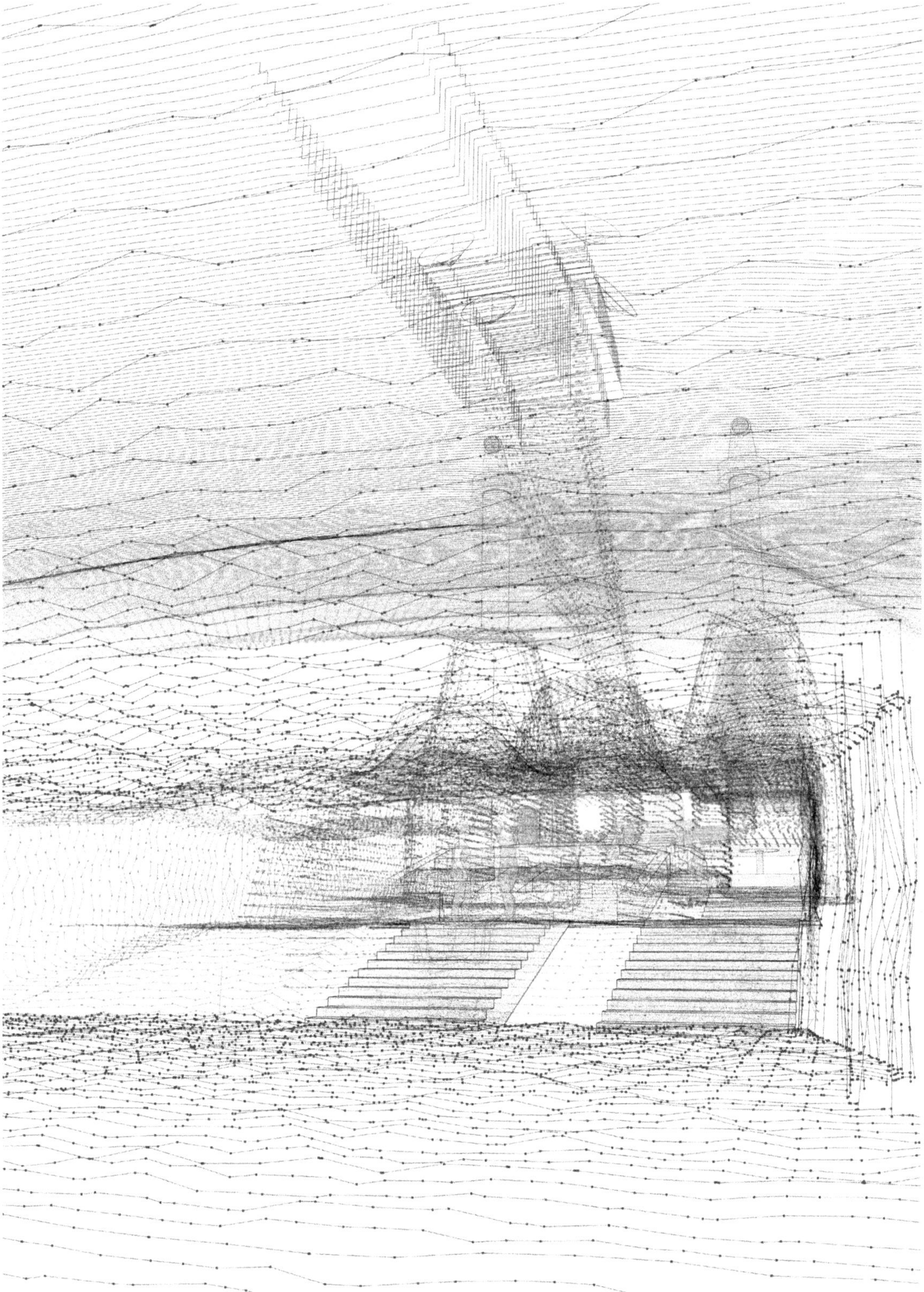

P155 Interior Perspective

P157 Global Production of Phytoremediating Plant Seeds
This chart traces the source of a number of key phytoremediating crop varieties, their economic worth, and the toxins they are able to absorb.
Source: http:// www.earthstat.org

P158 Global Network Map of Seed Banks
There are thousands of operational seed banks around the world -some are as small as a home-freezer, others the size of a warehouse. Their size is determined by the number of 'accessions' (seed samples) they hold - the largest being the Svalbard Global Seed Bank in Norway. Seedbanks merely preserve and protect their collection, the seeds remain in the ownership of the country that deposited them.
Sources: http://www.bgci.org/map/php; http://www.agrowebcee.net/sunflower/sunflower-research-network/; http://www.slideshare.net/DagEndresen/2009-01-29-cac-global-information-systems; http://www.seedsnatcher.com/2010/10/seed-banks-around-world.html; http://www.agprofessional.com/news/Update-on-the-worlds-15-largest-seed-banks-217990631.html

P159 Global List of Seed Banks
Sources: http://www.bgci.org/map/php; http://www.agrowebcee.net/sunflower/sunflower-research-network/; http://www.slideshare.net/DagEndresen/2009-01-29-cac-global-information-systems; http://www.seedsnatcher.com/2010/10/seed-banks-around-world.html; http://www.agprofessional.com/news/Update-on-the-worlds-15-largest-seed-banks-217990631.html

P160 Photographs of Winspit Caves
[top right] http://www.strollingguides.co.uk/
Others: John Cook

P161 Winspit Caves in plan and section,
The Winspit coastal free-stone quarry operated until 1940 producing stone for major buildings in London. Stone pillars were left to keep the cave roof stable, and excavated rock was craned down onto waiting boats below. During the World War II it was used a site for air and naval defences.

P162 Micropublic Seed Bank, subsurface level plan
The public enter through the cave's mouth, slowly journeying upwards through the various excavated caverns, until finally exiting onto the hilltop above.

P163 Micropublic Seed Bank, surface level plan
The seedbank is only evidenced above ground through its air vents and skylight antennae.

P164 Micropublic Seed Bank, excavation and construction procedures
The architecture of the seed bank is dictated by the sizes, processes and order of quarrying techniques. Space is created through 'corridor mining' and the use of explosives to create doorways, troughs or ceiling vaults.

P165 Micropublic Seed Bank, construction procedures and components
The seed bank's specialised components such as lab facilities and seed vaults themselves must be prefabricated and installed on site. Just how the quarried stone left, the components are delivered by boat and moved into place on carved out 'rutways'.

P166 Micropublic Seed Bank, Terraced Plant Nursery + Solar Time
The plant nursery uses harvested rainwater to grow samples of plants periodically to test a seed collection's viability. The living plants are vitalized to grow in the dark conditions by a remote skylight system, which autonomously tracks the path of the sun high above the surface.

P167 Micropublic Seed Bank, Virtual Seed Vault
The virtual seed vault provides an interactive display of live-data to do with stock levels, sources and seed varieties, for both the Purbeck Seed Vault itself and throughout the global seed bank network.

Shale Deposits

Recoverable Shale gas Reserves (Trillions Cubic Feet)

China	1275
United States	862
Argentina	774
Mexico	681
South Africa	485
£/MMBtu (million british thermal units) :	3

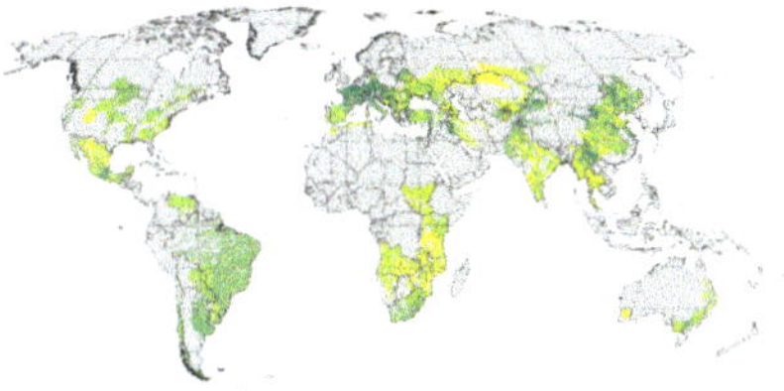

Sunflower Yield — 3 t/ha

Sunflower
Helianthus Annuus
52 known species

Sunflower Production - 2007 (Million Metric Tonnes)

Russian Federation	5.671
Ukraine	4.174
Argentina	3.497
India	1.463
United States	1.318
£/lb :	6.75

Chemicals Absorbed:

- Ni nickel
- Sr strontium
- U uranium
- Pb lead
- Cr chromium
- As arsenic
- Cd cadmium
- Cs cesium
- Cu copper
- Mn manganese
- Zn zinc

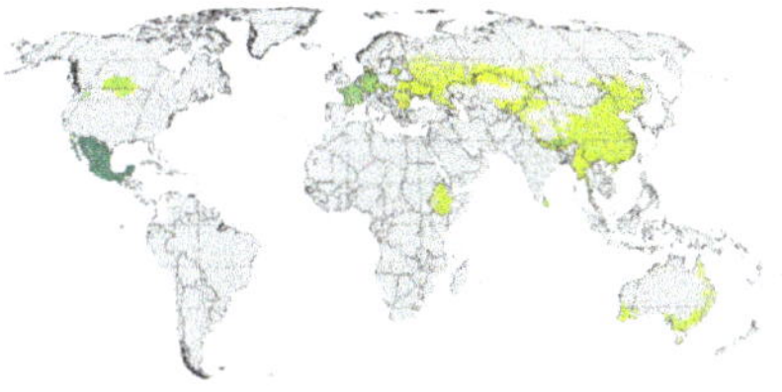

Mustard Greens Yield — 3 t/ha

Mustard Greens
Brassica *Juncea*
40 known species

Mustard Seed Production - 2007 (Million Metric Tonnes)

China	204
India	152
Indonesia	69
Viet Nam	44
Thailand	38
£/lb :	3.6

- Sr strontium
- Se selenium
- Pb lead
- Cr chromium
- Ag silver
- Cd cadmium
- Cs cesium
- Cu copper
- Mn manganese
- Zn zinc

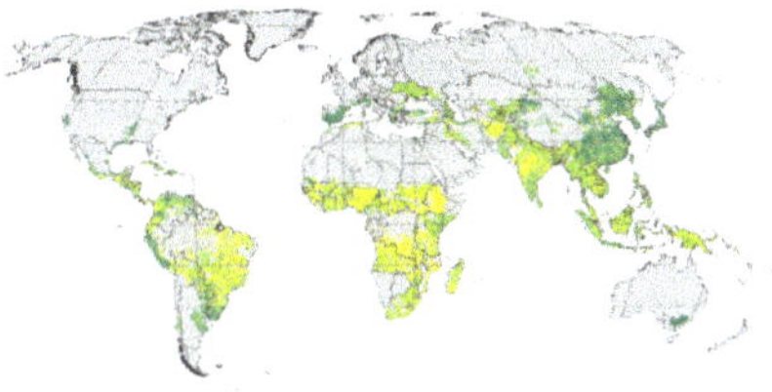

Rice Yield — 8 t/ha

Rice
Oryza Sativa
23 known species

Rice Production - 2014 (Million Metric Tonnes)

Nepal	132.660
Canada	123.400
Myanmar	55.000
China	16.500
United States	14.436
£/lb :	12

- Cr chromium
- As arsenic

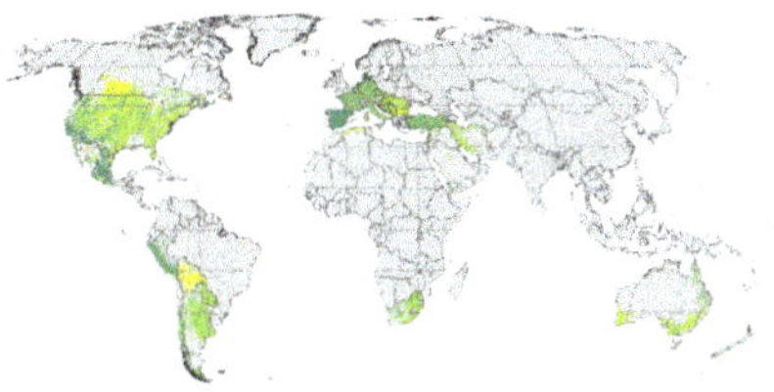

Alfalfa Yield — 60 t/ha

Alfalfa
Medicago Sativa
87 known species

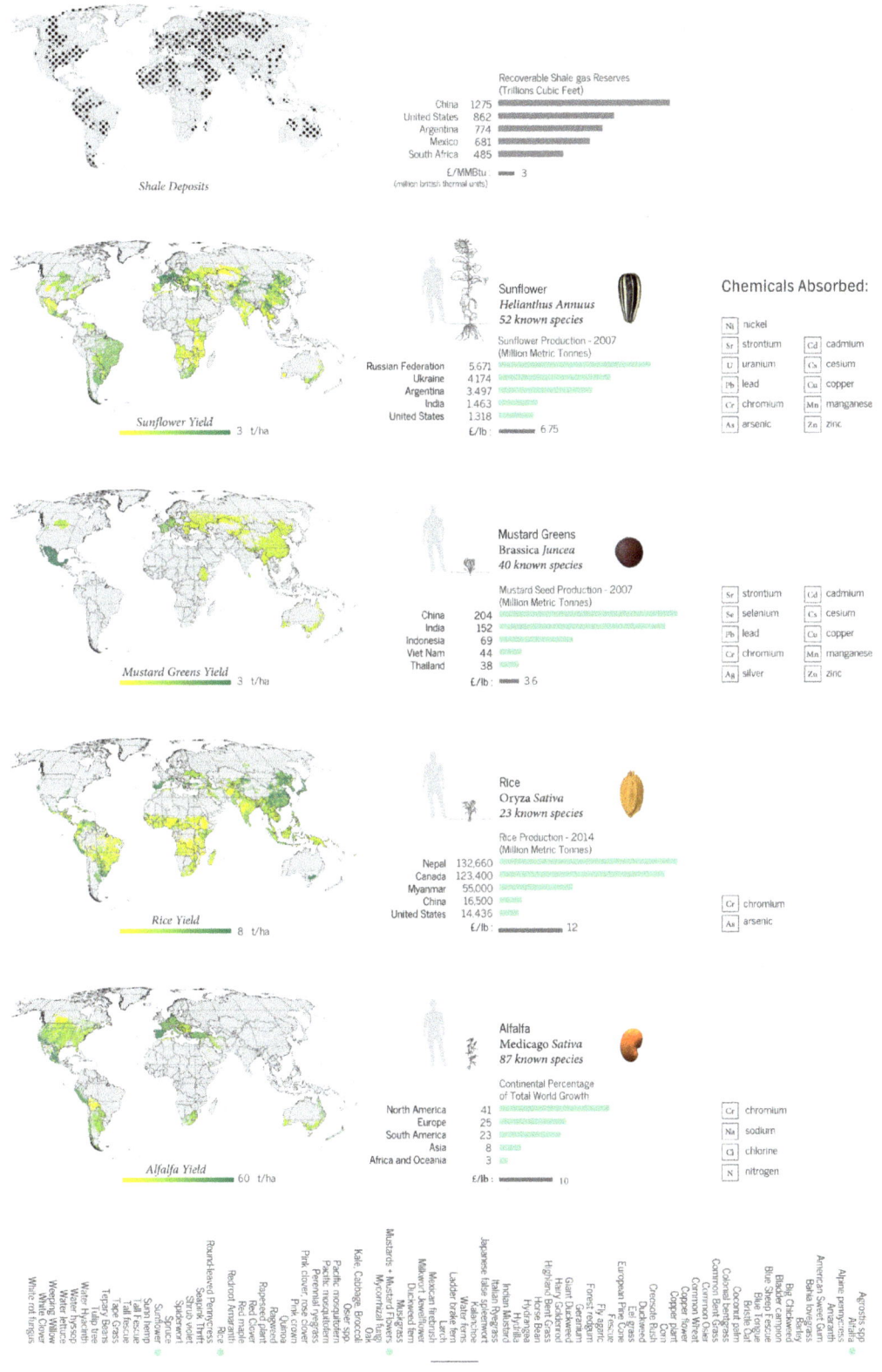

Continental Percentage of Total World Growth

North America	41
Europe	25
South America	23
Asia	8
Africa and Oceania	3
£/lb :	10

- Cr chromium
- Na sodium
- Cl chlorine
- N nitrogen

Hyeraccumulators

Agrostis spp
Alfalfa
Alpine pennycress
Amaranth
American Sweet Gum
Bahia lovegrass
Barley
Big Chickweed
Bladder campion
Blue Sheep Fescue
Blue Tongue
Bristle Oat
Coconut palm
Colonial bentgrass
Common Bent Grass
Common Osier
Common Wheat
Copper flower
Copper plant
Corn
Creosote Bush
Duckweed
Eel grass
European Pine Cone
Fescue
Fly agaric
Forest redgum
Geranium
Giant Duckweed
Hairy Goldenrod
Highland Bent Grass
Horse Bean
Hydrangea
Hydrilla
Indian Mustard
Italian Ryegrass
Japanese false spleenwort
Kalanchoe
Water ferns
Ladder brake fern
Larch
Mexican firebrush
Milkwort Jewelflower
Duckweed fern
Muskgrass
Mustards + Mustard Flowers
Mycorrhizal fungi
Oak
Kale, Cabbage, Broccoli
Osier spp.
Pacific mosquitofern
Pacific mosquitofern
Perennial ryegrass
Pink clover, rose clover
Pink crown
Quinoa
Ragweed
Rapeseed plant
Red Clover
Red maple
Redroot Amaranth
Rice
Round-leaved Pennycress
Seapink Thrift
Shrub violet
Spiderwort
Spruce
Sunflower
Sunn hemp
Tall Fescue
Tall fescue
Tape Grass
Tepary Beans
Tulip tree
Water Hyacinth
Water hyssop
Water lettuce
Weeping Willow
White Clover
White rot fungus

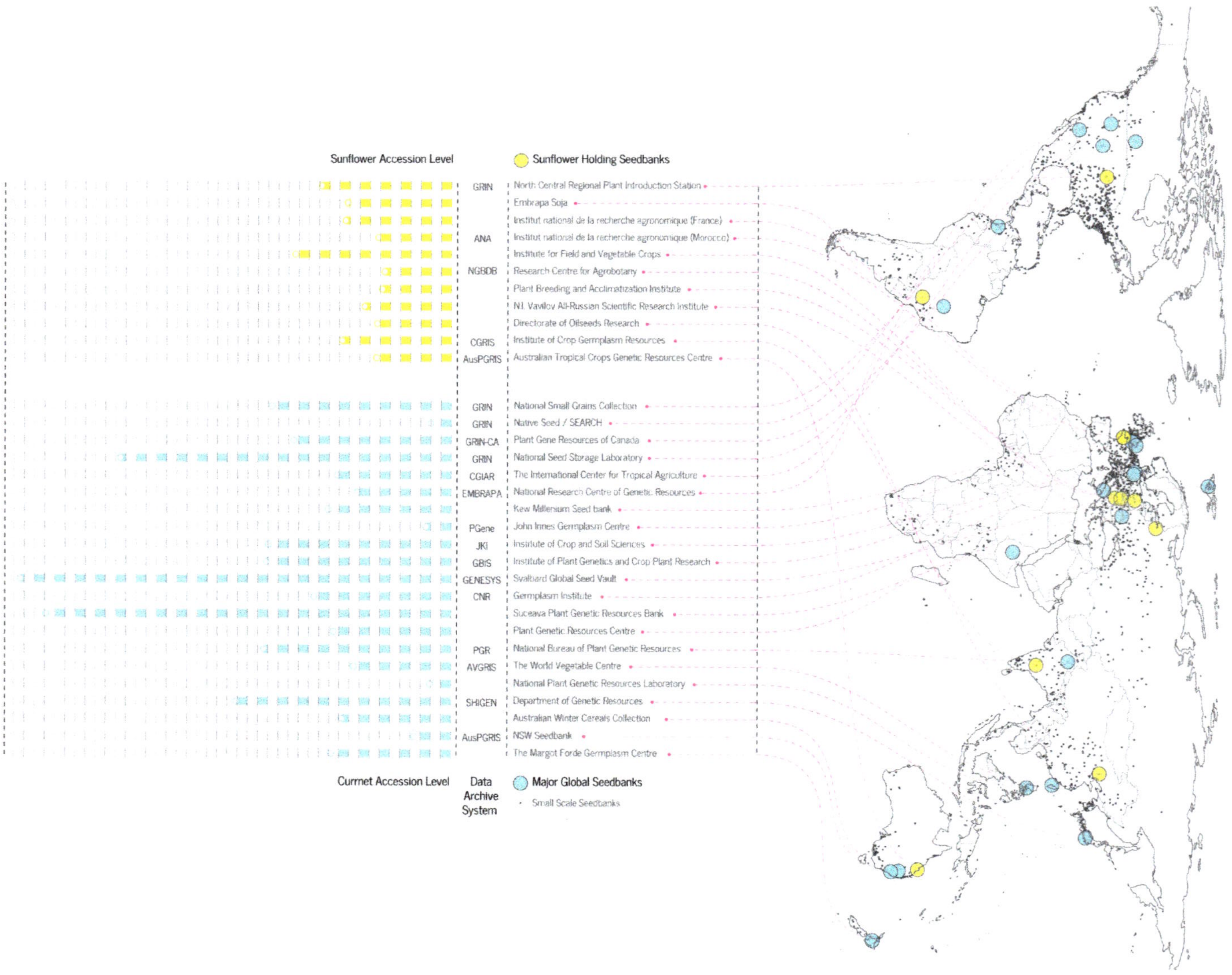
Sunflower Accession Level
Sunflower Holding Seedbanks
GRIN
North Central Regional Plant Introduction Station
Embrapa Soja
Institut national de la recherche agronomique (France)
ANA
Institut national de la recherche agronomique (Morocco)
Institute for Field and Vegetable Crops
NGBDB
Research Centre for Agrobotany
Plant Breeding and Acclimatization Institute
N.I. Vavilov All-Russian Scientific Research Institute
Directorate of Oilseeds Research
CGRIS
Institute of Crop Germplasm Resources
AusPGRIS
Australian Tropical Crops Genetic Resources Centre
GRIN
National Small Grains Collection
GRIN
Native Seed / SEARCH
GRIN-CA
Plant Gene Resources of Canada
GRIN
National Seed Storage Laboratory
CGIAR
The International Center for Tropical Agriculture
EMBRAPA
National Research Centre of Genetic Resources
Kew Millenium Seed bank
PGene
John Innes Germplasm Centre
JKI
Institute of Crop and Soil Sciences
GBIS
Institute of Plant Genetics and Crop Plant Research
GENESYS
Svalbard Global Seed Vault
CNR
Germplasm Institute
Suceava Plant Genetic Resources Bank
Plant Genetic Resources Centre
PGR
National Bureau of Plant Genetic Resources
AVGRIS
The World Vegetable Centre
National Plant Genetic Resources Laboratory
SHIGEN
Department of Genetic Resources
Australian Winter Cereals Collection
AusPGRIS
NSW Seedbank
The Margot Forde Germplasm Centre
Currnet Accession Level
Data Archive System
Major Global Seedbanks
Small Scale Seedbanks

Germplasm Institute

Institut national de la recherche agronomique (France)

Institut national de la recherche agronomique (Morocco)

Institute of Crop and Soil Sciences

Institute of Crop Germplasm Resources

Institute of Plant Genetics and Crop Plant Research

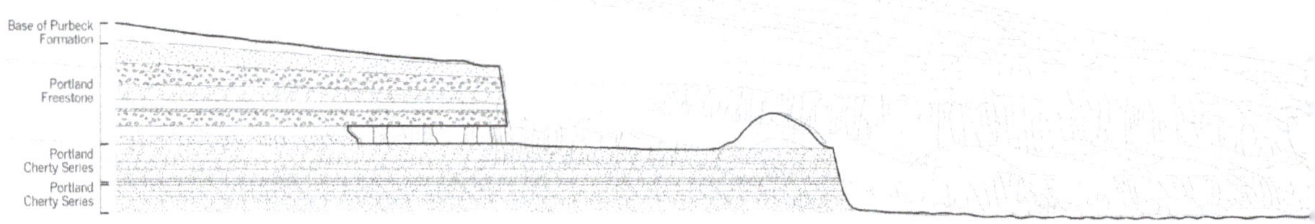
Base of Purbeck Formation
Portland Freestone
Portland Cherty Series
Portland Cherty Series

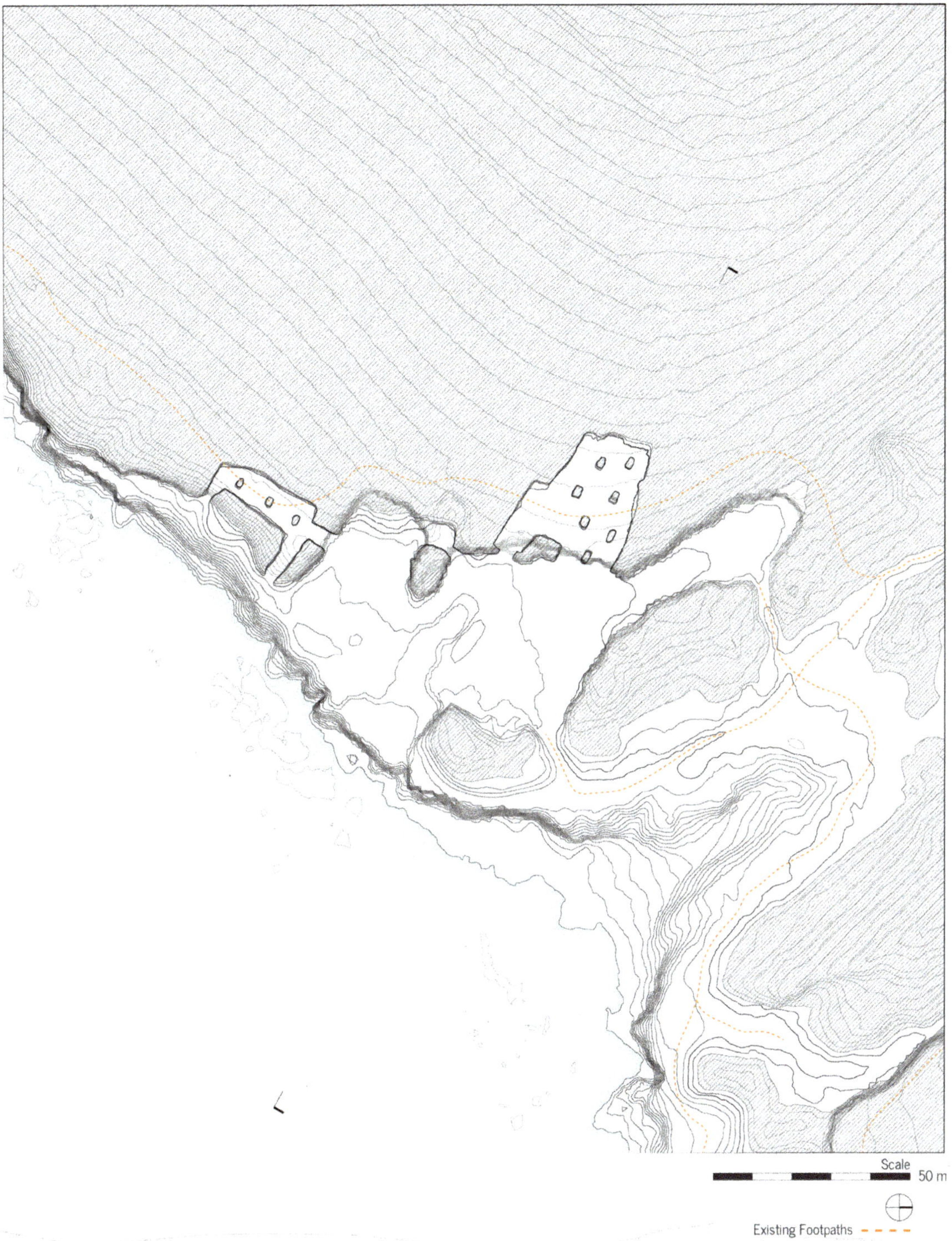
Scale
50 m
Existing Footpaths

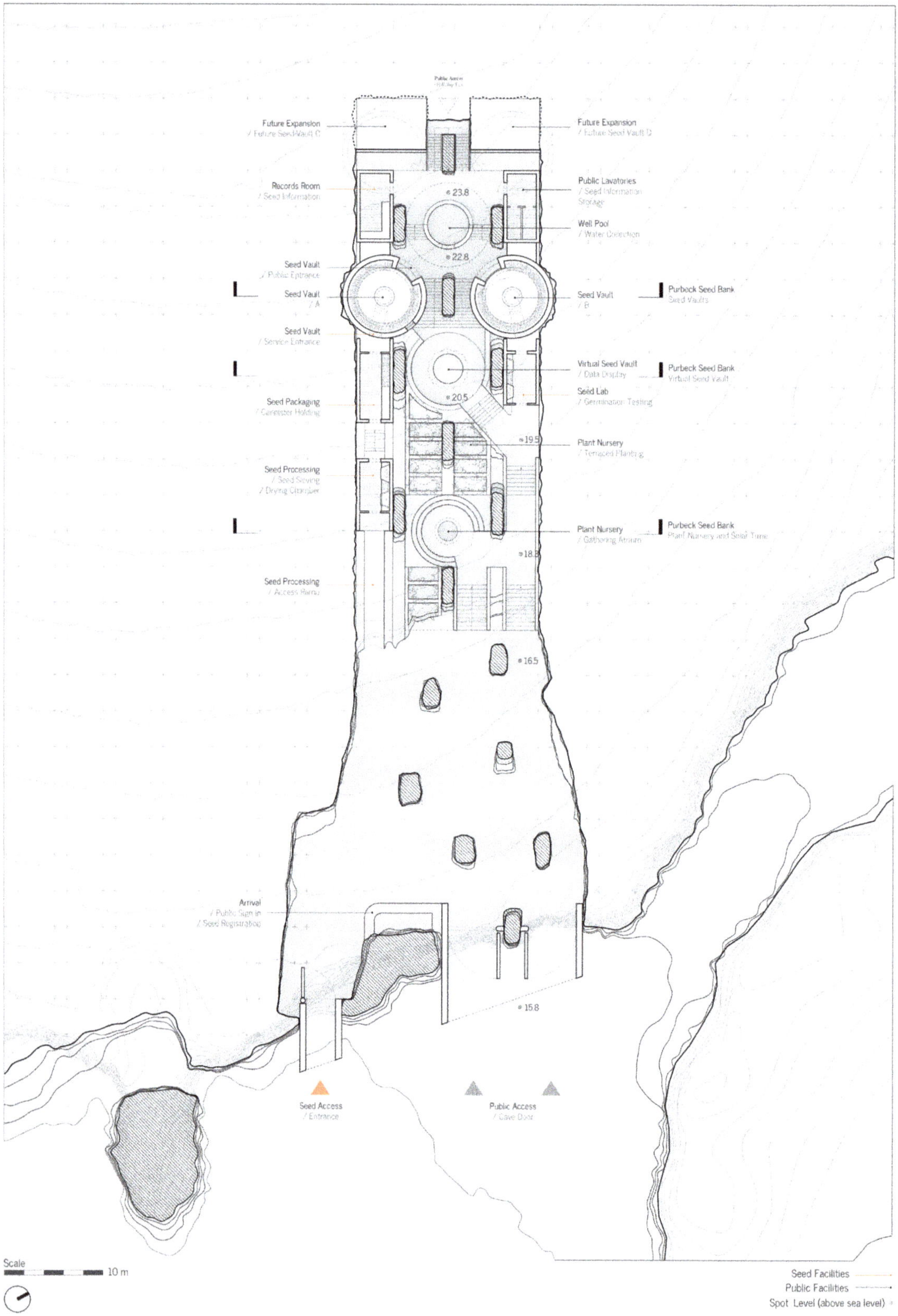
Future Expansion
/ Future Seed Vault C
Future Expansion
/ Future Seed Vault D
Records Room
/ Seed Information
Public Lavatories
/ Seed Information Storage
23.8
Well Pool
/ Water Collection
22.8
Seed Vault
/ Public Entrance
Seed Vault
/ A
Seed Vault
/ B
Purbeck Seed Bank
Seed Vaults
Seed Vault
/ Service Entrance
Virtual Seed Vault
/ Data Display
Purbeck Seed Bank
Virtual Seed Vault
Seed Lab
/ Germination Testing
20.5
Seed Packaging
/ Cannister Holding
19.5
Plant Nursery
/ Terraced Planting
Seed Processing
/ Seed Sieving
/ Drying Chamber
Plant Nursery
/ Gathering Atrium
Purbeck Seed Bank
Plant Nursery and Solar Time
18.3
Seed Processing
/ Access Ramp
16.5
Arrival
/ Public Sign In
/ Seed Registration
15.8
Seed Access
/ Entrance
Public Access
/ Cave Door
Scale
10 m
Seed Facilities
Public Facilities
Spot Level (above sea level)

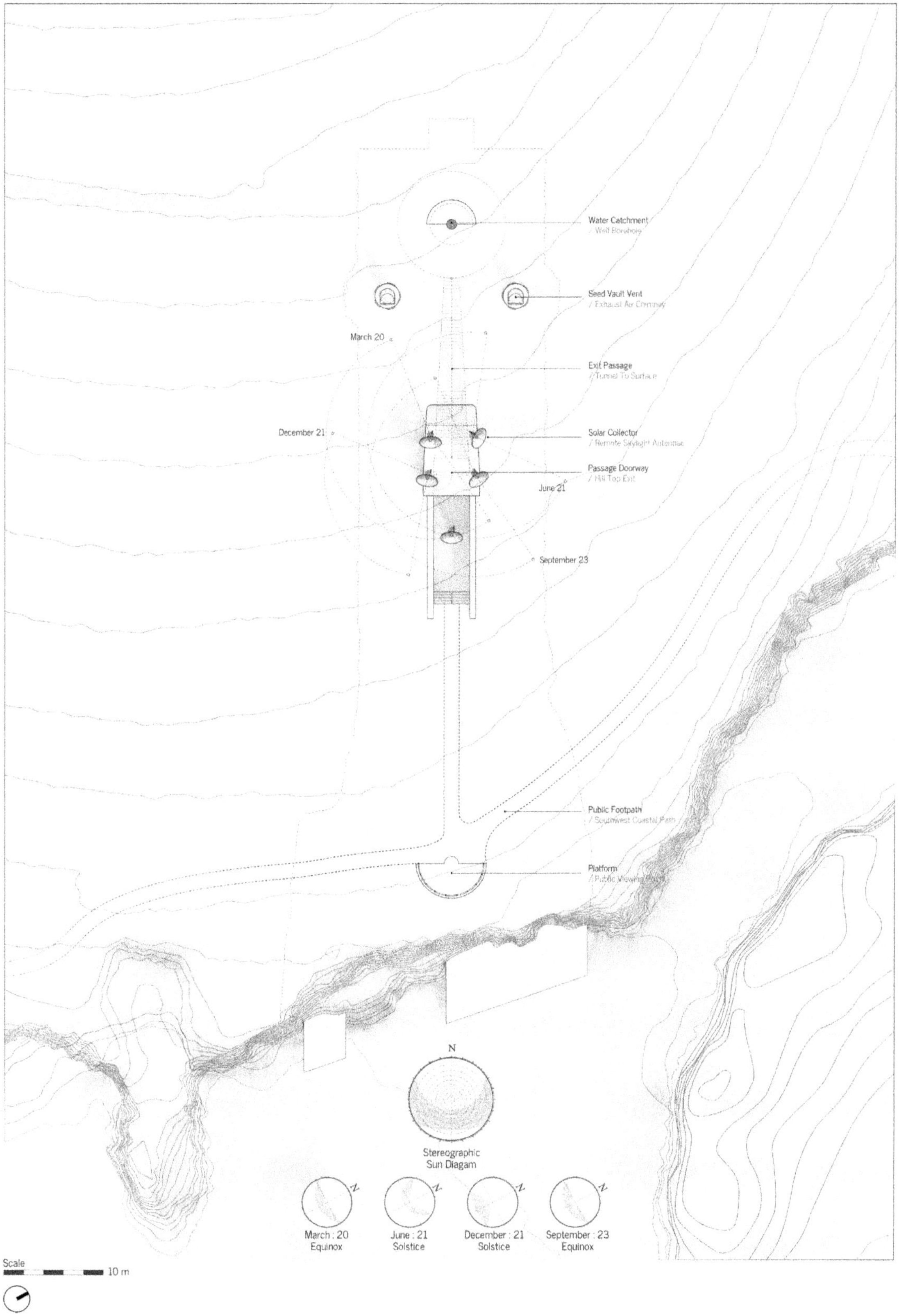
Water Catchment
/ Well Borehole
Seed Vault Vent
/ Exhaust Air Chimney
March 20
Exit Passage
/ Tunnel To Surface
December 21
Solar Collector
/ Remote Skylight Antennae
Passage Doorway
/ Hill Top Exit
June 21
September 23
Public Footpath
/ Southwest Coastal Path
Platform
/ Public Viewing
N
Stereographic
Sun Diagam
March : 20
Equinox
June : 21
Solstice
December : 21
Solstice
September : 23
Equinox
Scale
10 m

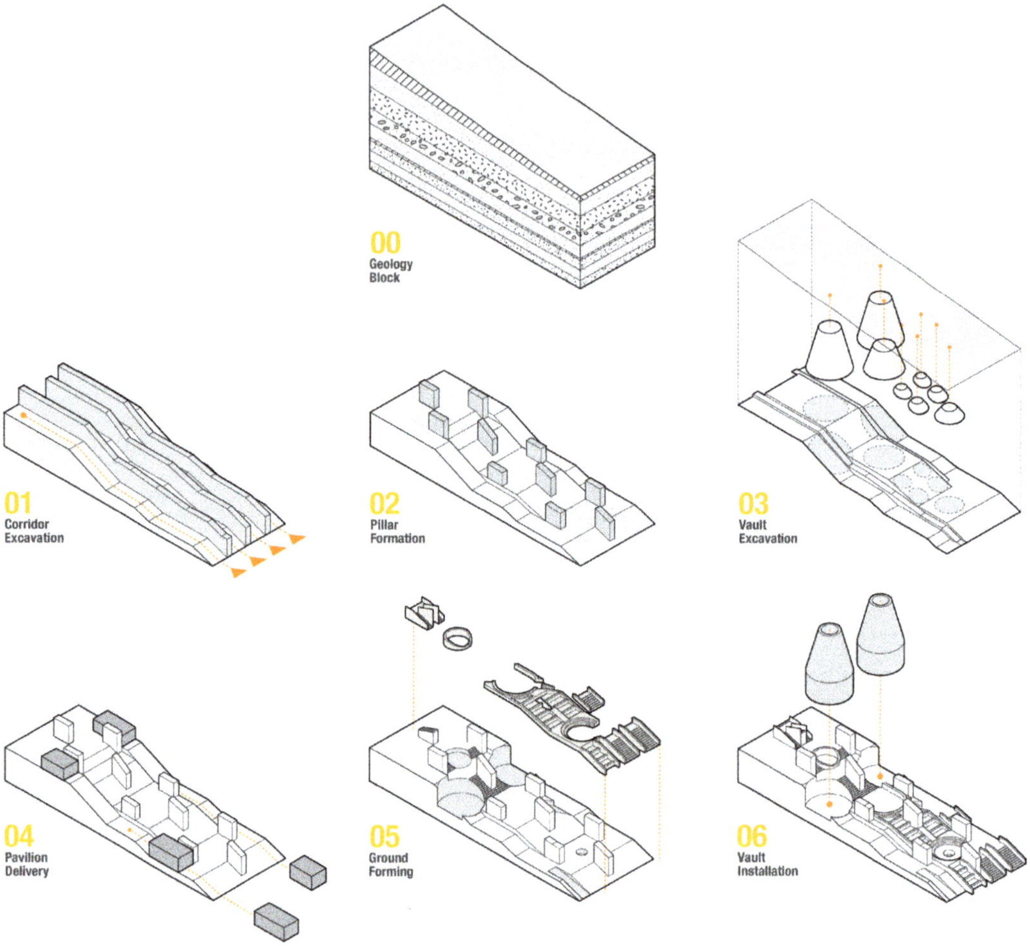

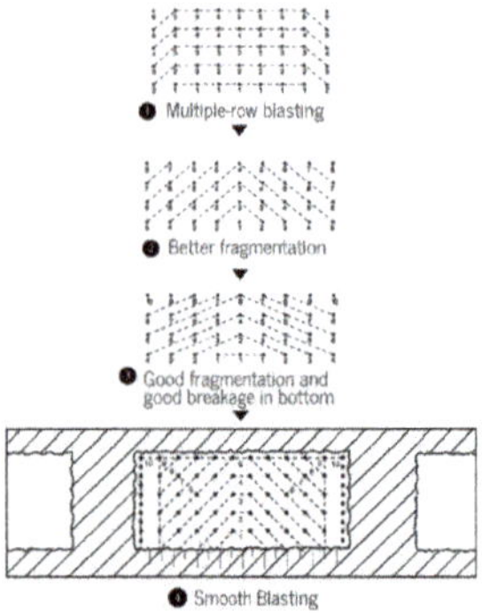

Pillar Forming
Detonator Arrangement and ingnition Sequence

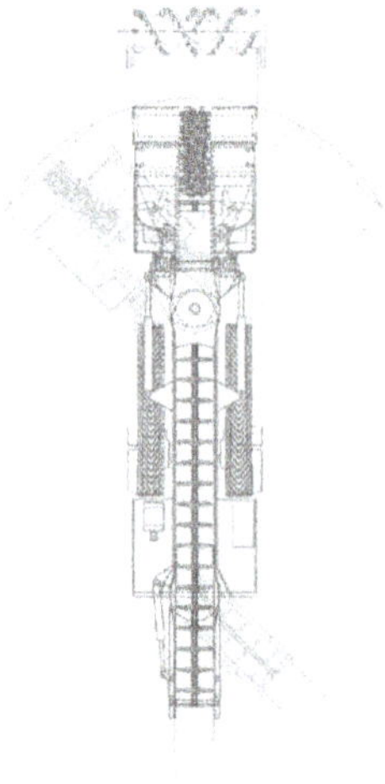

Corridor Mining
Continous Mining Machinery

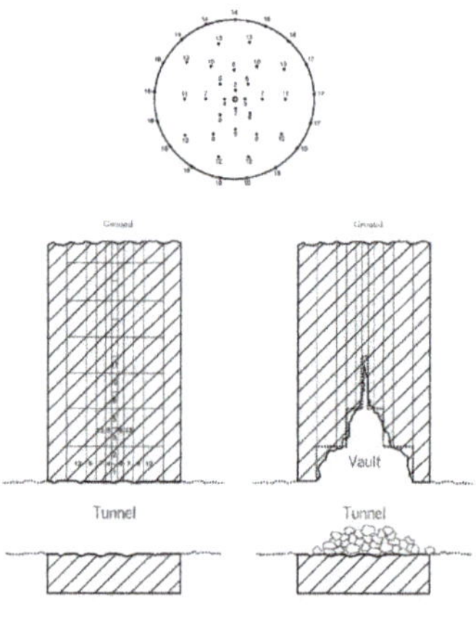

Vertical Excavation
Detonator Arrangement and ingnition Sequence

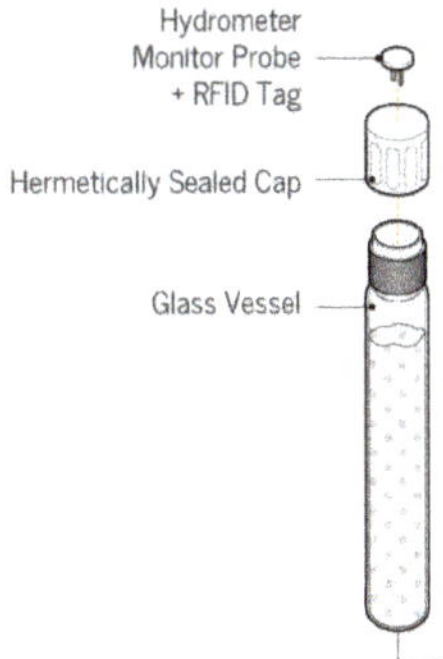

Seed Canister

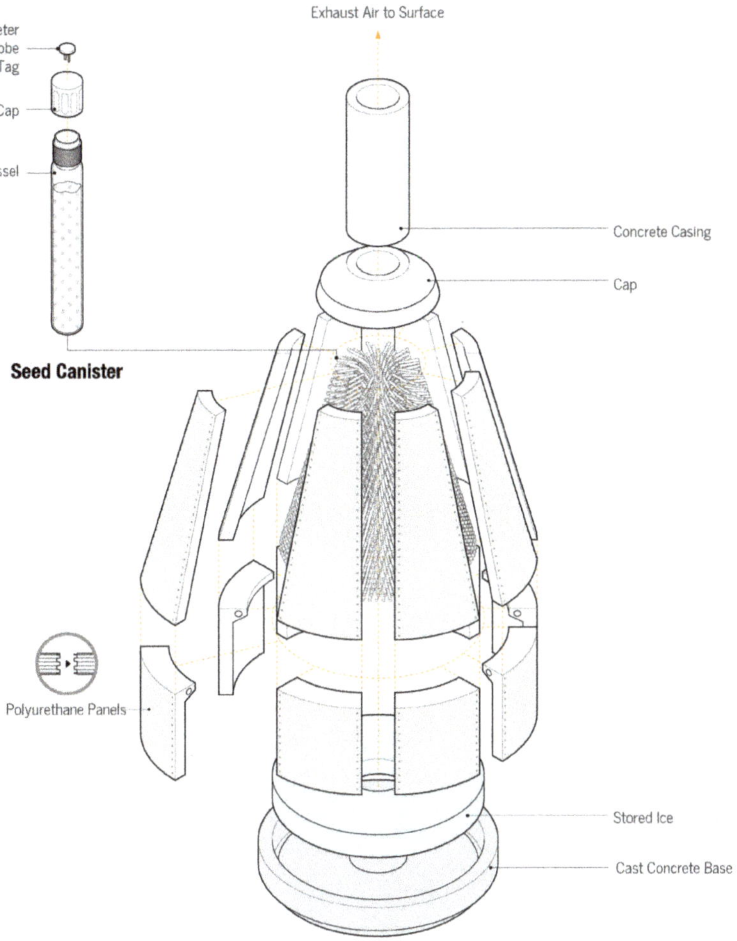

Seed Vault

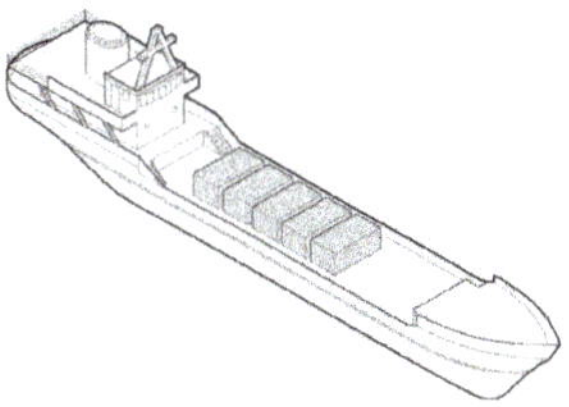

Delivery
Shipping Container Dimensions and Transportation Procedures

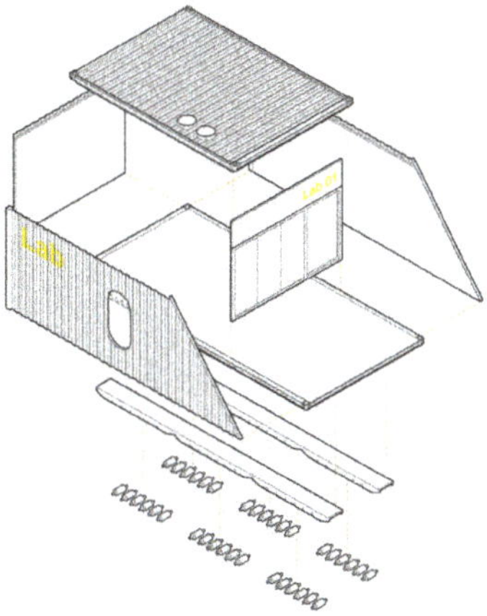

Construction
Prefabricated Construction Elements

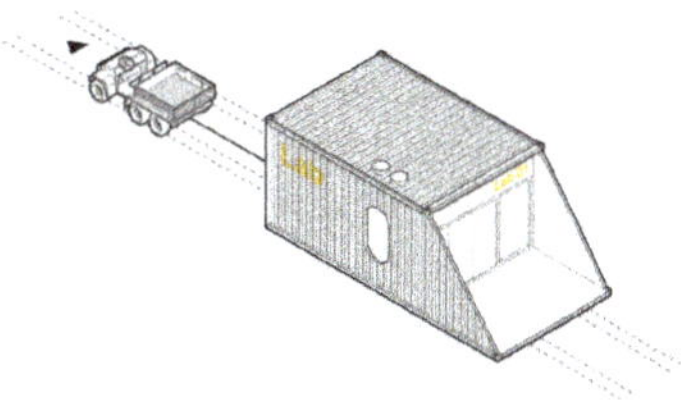

Mobility
Transportation and Flexibility On-Site

Pavilions

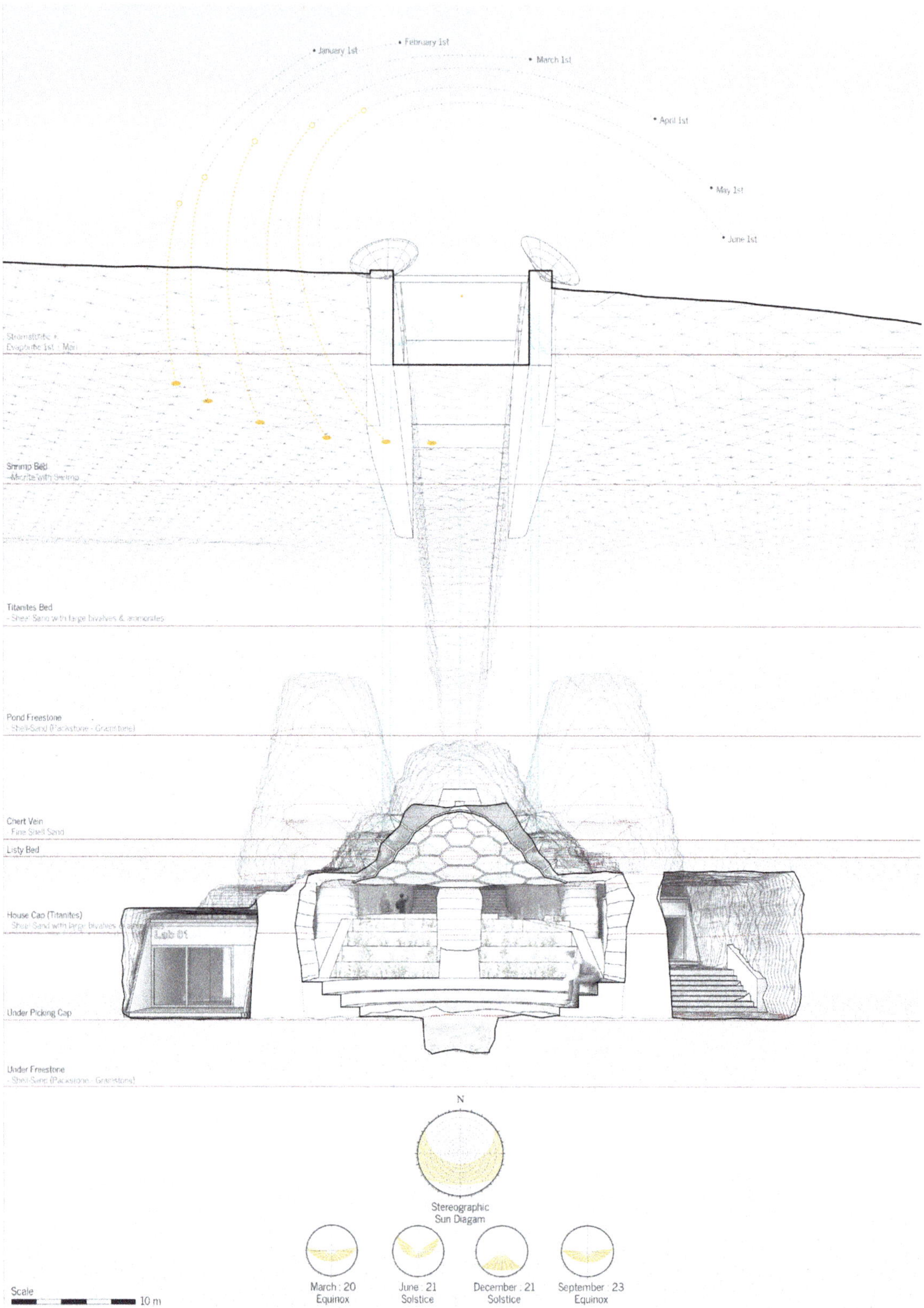
January 1st
February 1st
March 1st
April 1st
May 1st
June 1st
Shrimp Bed
Titanites Bed
Pond Freestone
Chert Vein
- Fine Shell Sand
Listy Bed
House Cap (Titanites)
Lab 01
Under Picking Cap
Under Freestone
N
Stereographic
Sun Diagam
March : 20
Equinox
June : 21
Solstice
December : 21
Solstice
September : 23
Equinox
Scale
10 m

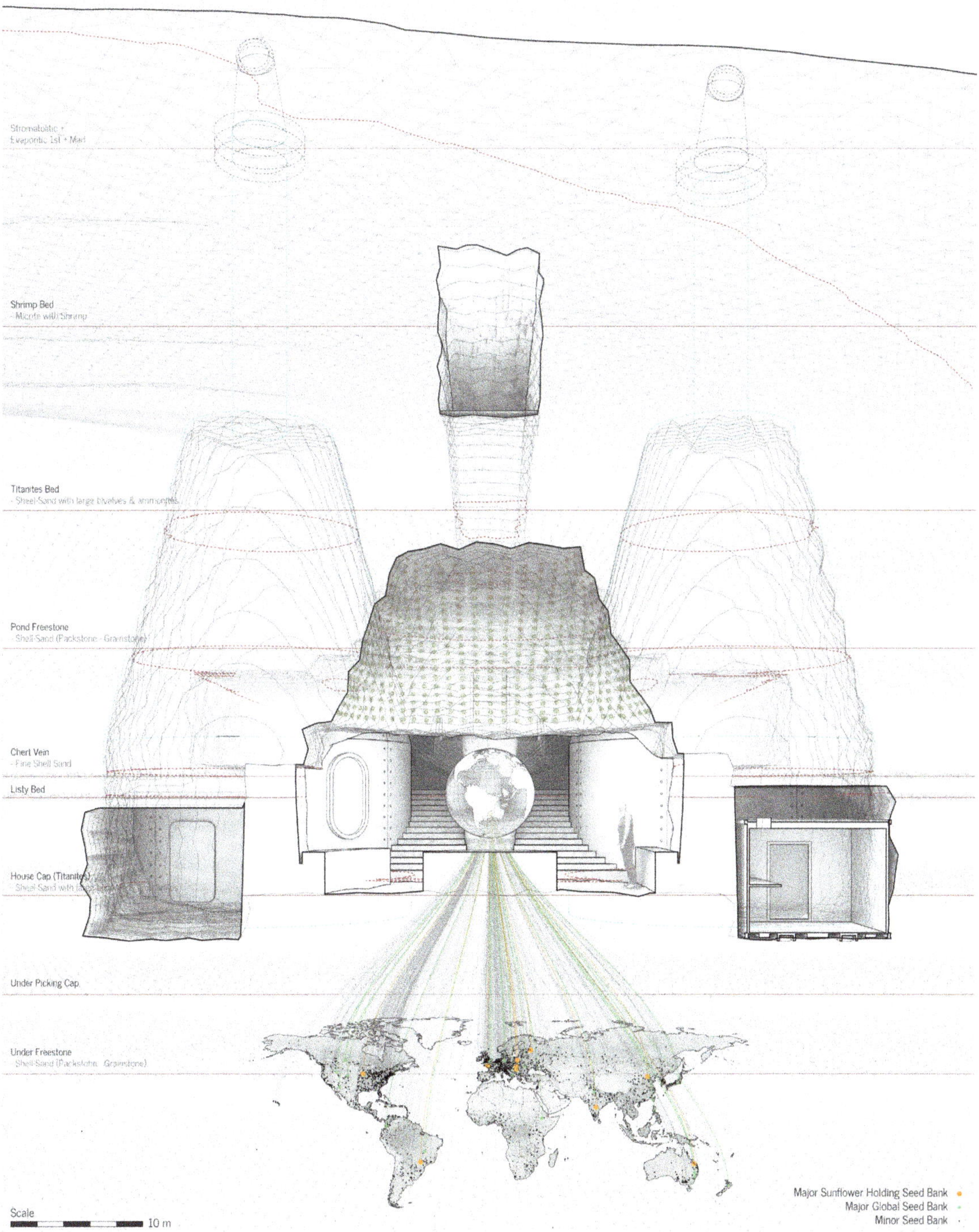
Stromatolitic +
Evaporitic 1st + Marl
Shrimp Bed
- Micrite with Shrimp
Titanites Bed
- Sheel-Sand with large bivalves & ammonites
Pond Freestone
- Shell-Sand (Packstone - Grainstone)
Chert Vein
- Fine Shell Sand
Listy Bed
House Cap (Titanites)
Under Picking Cap
Under Freestone
- Shell-Sand (Packstone - Grainstone)
Major Sunflower Holding Seed Bank
Major Global Seed Bank
Minor Seed Bank
Scale
10 m

MICROPUBLIC POWERHOUSE

CLAIRE HOLTON

The UK's energy infrastructure is becoming increasingly fragile and less reliable as unpredictable weather patterns impact it. As a direct response to the 2013 Christmas blackouts, the project proposed a prototypical MicroPublicPowerhouse (MPP) employing off-grid geothermal heat and Electricity to supply the village of Wareham, Dorset in times of crisis. The project was sited on the site of existing public toilets positioned above a 1,300 m borehole, which was drilled by BP in the 1960's. This was chosen for its central village location and its geological properties providing 70 c + heat. The project aimed to bring the public closer to its energy source. Gathering human and non-human actors around a thing of 'the earths heat' creating a public resource for discussion and learning away from private control. Within this micropublicplace, the earth's heat was given a voice through a micro public tearoom and public washrooms. Here the heat was used to supply geothermal electricity, to cultivate, dry and serve the tea, and provide washrooms with hot geothermal water. The public tearoom and public washrooms served day-to-day rituals, while the center as a whole provided one-off refuge facilities during extended blackouts. If a blackout occurred, the MPP could fuel surrounding public buildings (such as the community center and the church) with heat and light to provide extra shelter. It could also provide tea, hot food and washing facilities for those in need. The MPP presented itself as a prototype for future public energy in and around the Isle of Purbeck, where it could be replicated, geology permitting, to deliver off-grid heat and electricity.

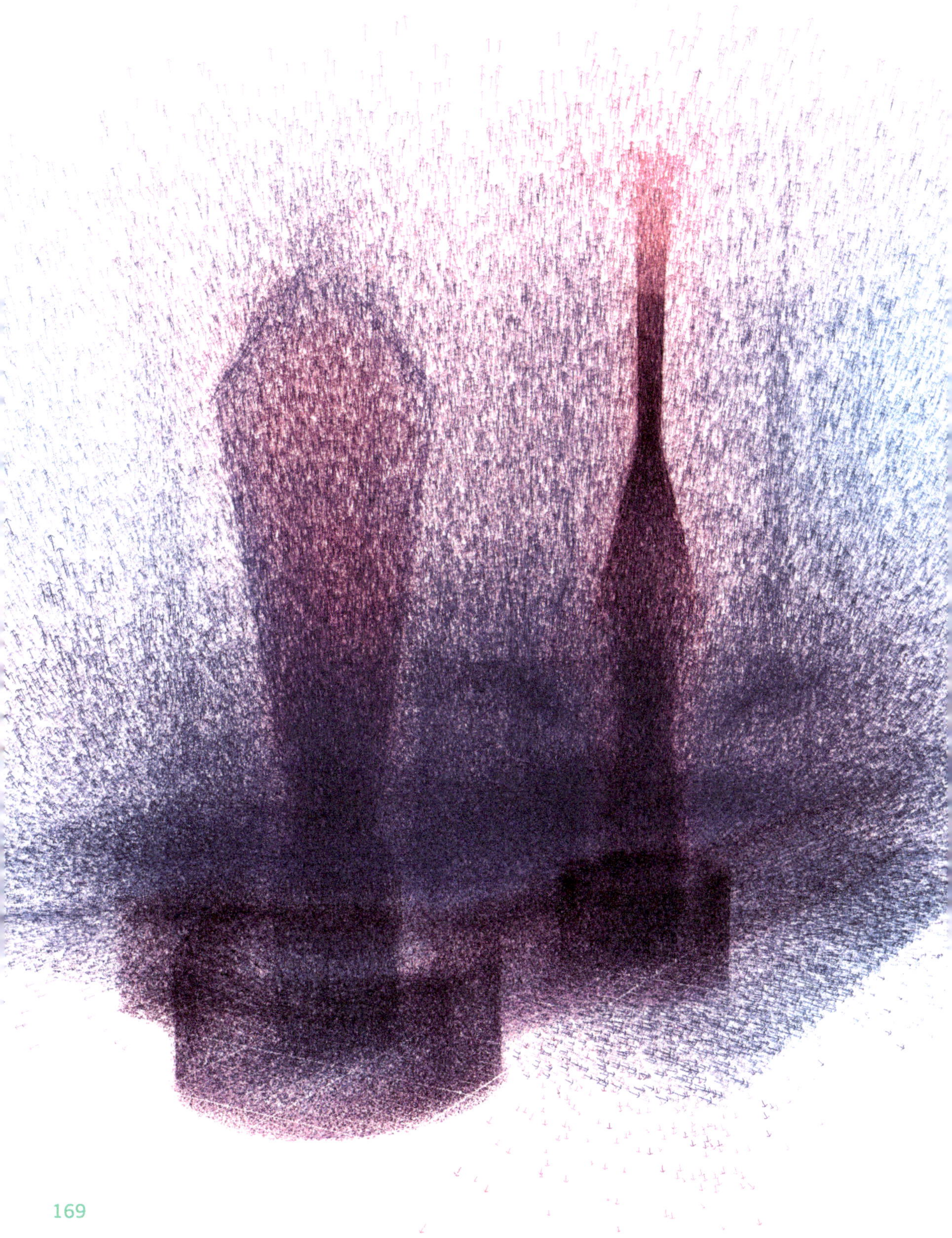

IMAGE CAPTIONS

P169 Heat Map, Micropublic Powerhouse

P171 The Isle of Purbeck showing existing geothermal boreholes
Sources: https://www.gov.uk/government/organisations/department-of-energy-climate-change/; http://www.nerc.ac.uk

P172 [top] Site Plan, Micropublic Powerhouse
[bottom] The site from across the River Frome
Photograph: Google Streetview
The Powerhouse is located on the site of an abandoned public toilet in the heart of the village of Wareham on the Isle of Purbeck. The public toilets were sited above a capped 1,300m borehole drilled by BP when prospecting for oil in the 1960's.
Source: http://www.nerc.ac.uk

P173 The Micropublic PowerHouse in a snow storm

P174 [top left] Basement Plan
[top right] Ground Floor Plan
[bottom] The site from the quay square
Photograph: Google Streetview

P175 [top left] First Floor Plan
[top right] Fifth Floor Plan

P176-177 Section B-B

P178-179 Geothermal Energy Strategy, Isle of Purbeck

P180-181 Micropublic Powerhouse during a Blackout

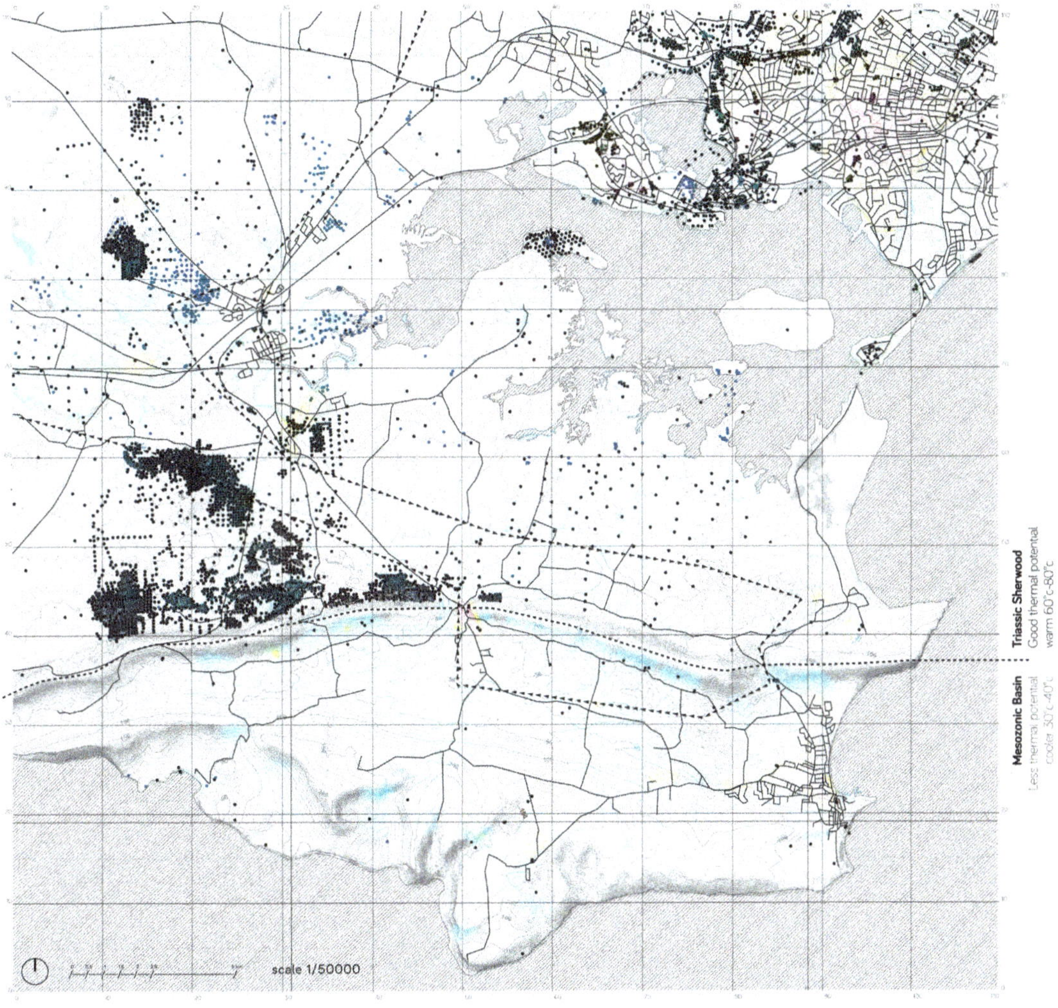
scale 1/50000
Triassic Sherwood
Good thermal potential
warm 60°c-80°c
Mesozonic Basin
Less thermal potential
cooler 30°c-40°c

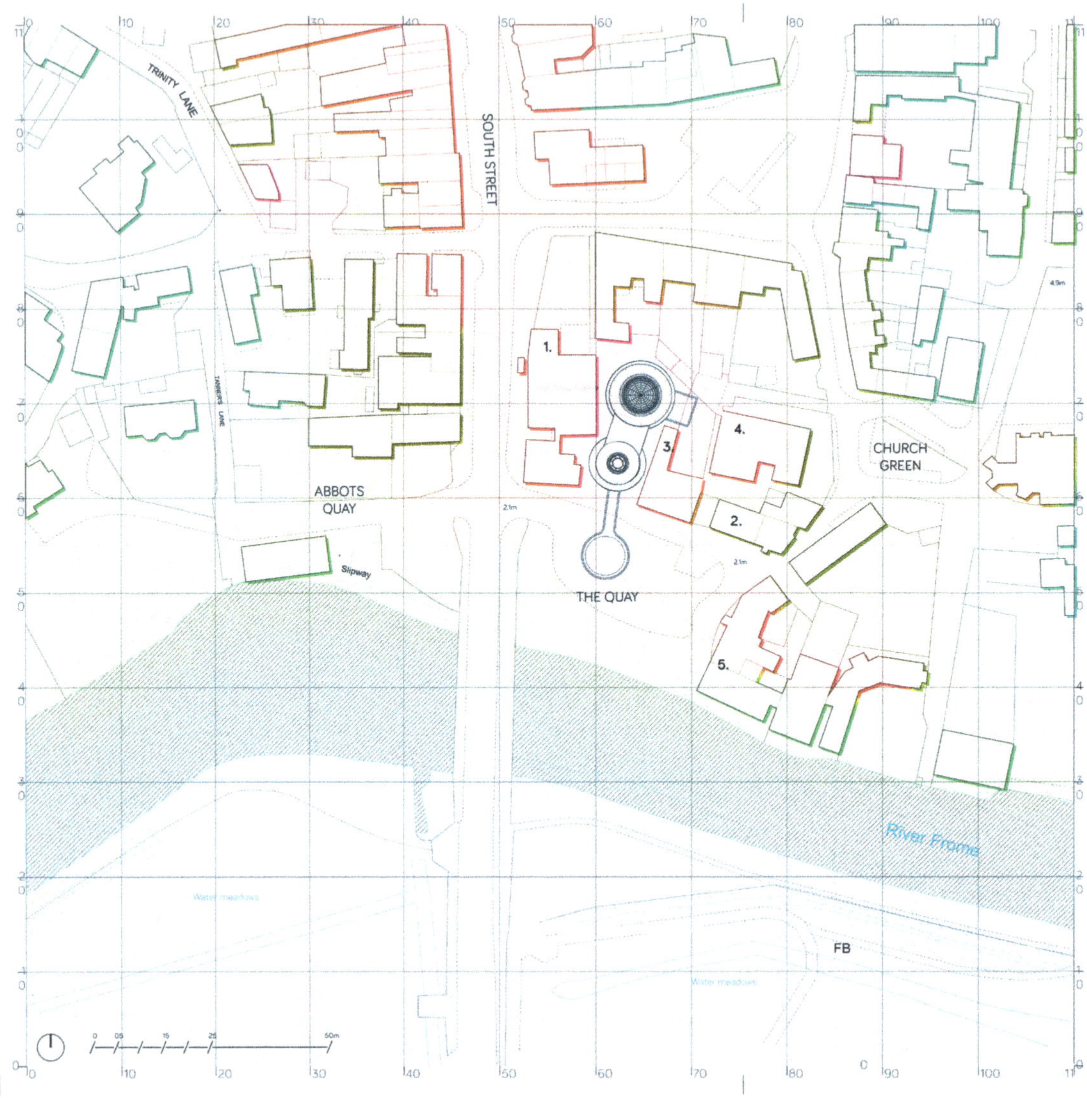
TRINITY LANE
SOUTH STREET
ABBOTS QUAY
THE QUAY
CHURCH GREEN
Slipway
River Frome
Water meadows
FB
1.
2.
3.
4.
5.

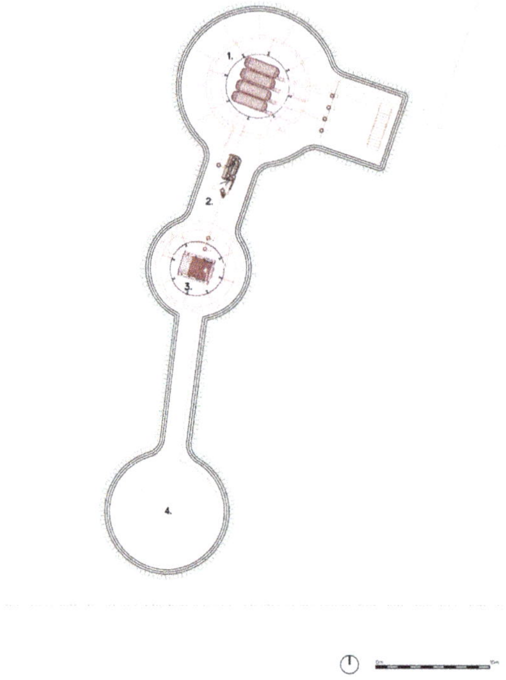

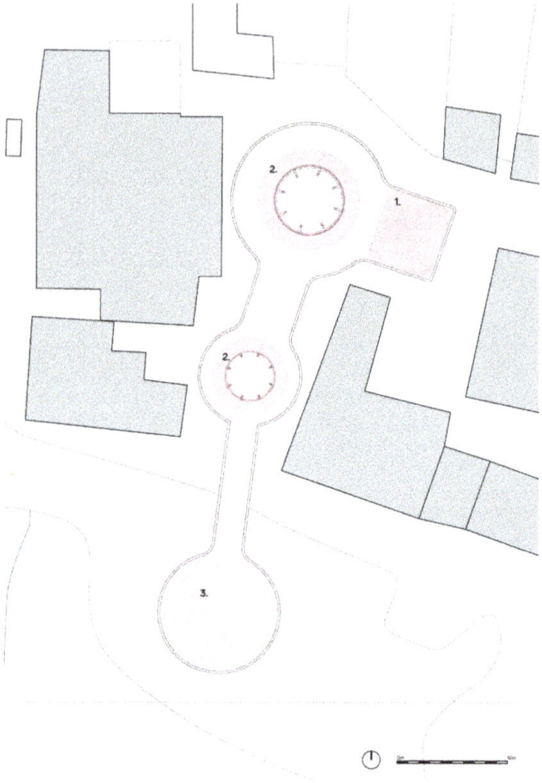

Basement Plan

1 Heat exchange; hot geothermal water converted into hot water for wash rooms (44C), laundry (30C) slow cookers and tea making (75C)

2 Geothermal Pump

3 Raser technology generator, uses 70C to turn working liquid into steam at 4C

4 Tea sorting space, storage and laundry

Ground Floor Plan

1 Large removable hatch to move geothermal machinery and provide ventilation into the basement and stair access

2 Vents around the chimney allow for ventilation plus a view down into the basement

3 Large vent delivers warm tea infused air into the quay's square. This will create a micro social climate attracting plant growth and provide the market with a source of heat during winter

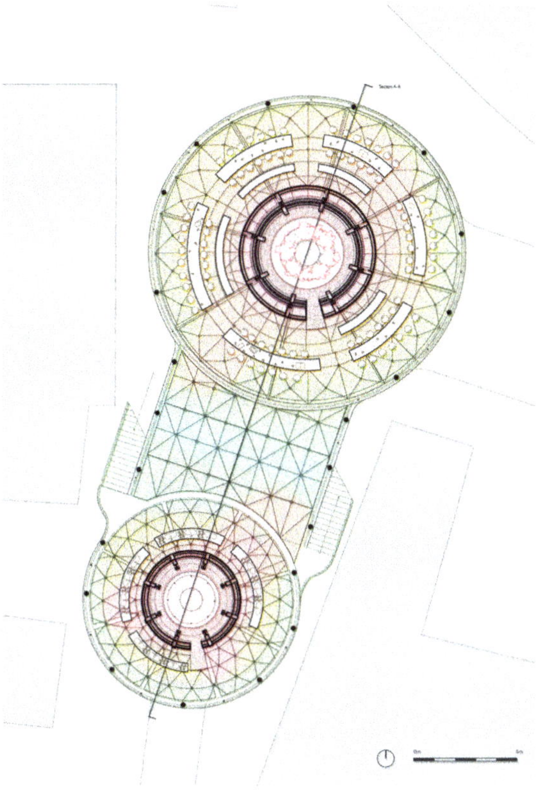

First Floor Plan

Public Tea-Rooms

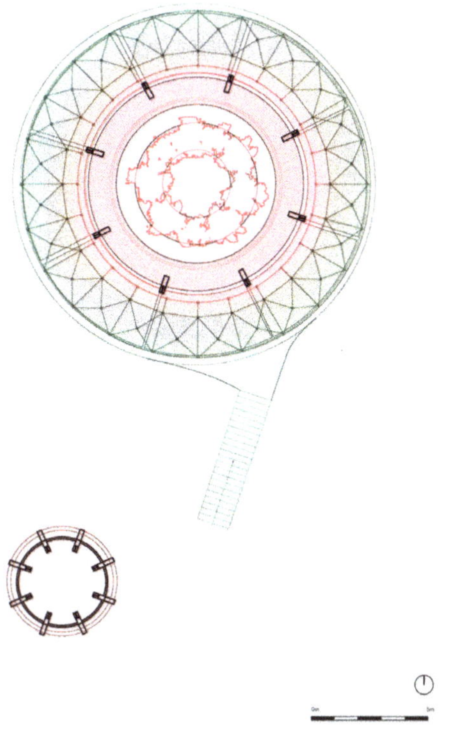

Fifth Floor Plan

Viewing deck / Stargazing

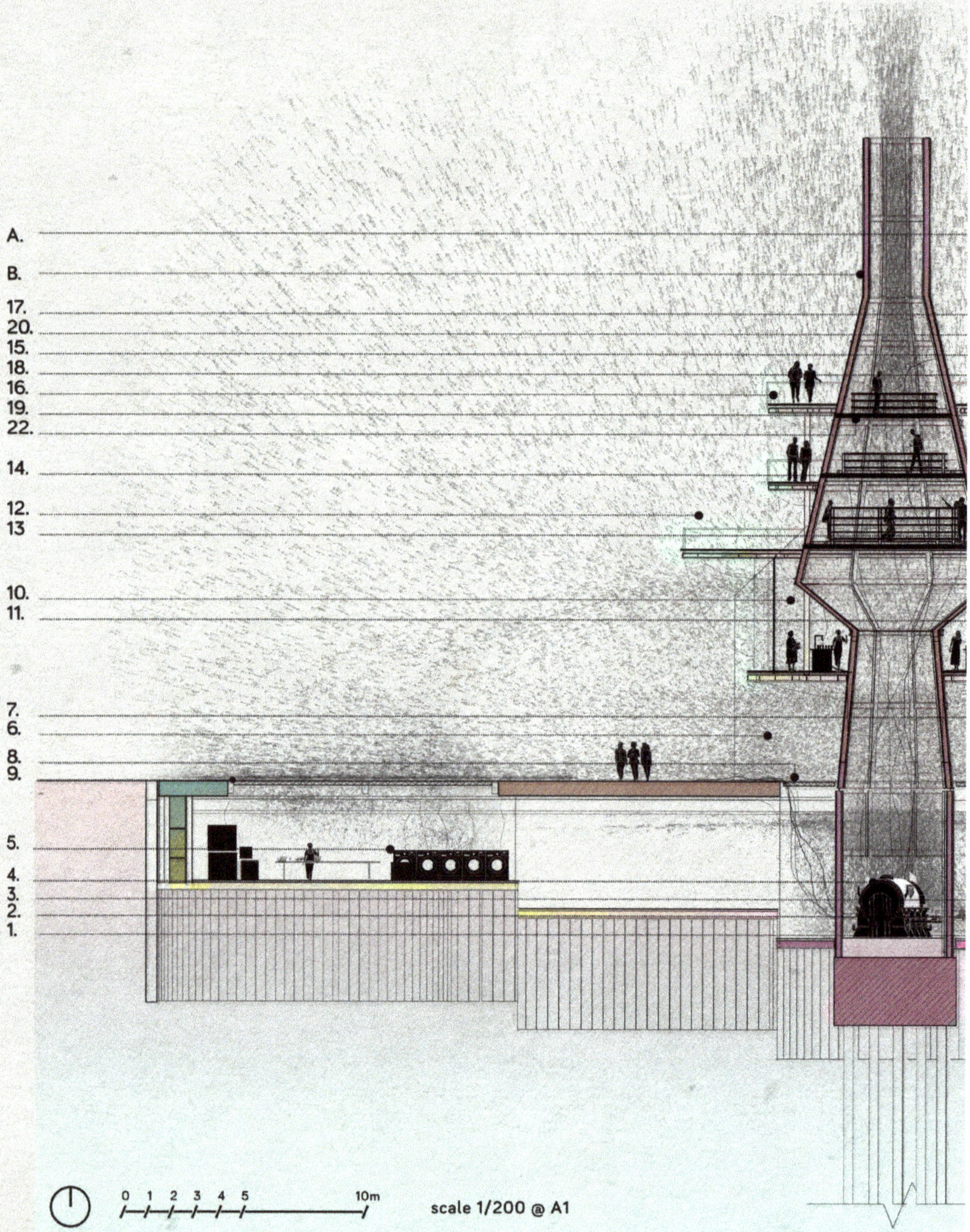

Basement (-1) Private
1. Heat exchange unit creates usable hot water from the geothermal system
2. Borehole pump - electricity supplied from generator
3. Well head
4. Raser technology generator
5. Laundry, tea sorting area and storgage facilities

Ground Level (0) Public
6. Circulation route to community center
7. Access to the MPP
8. Ventilation grates and views down to the basement floor
9. Ventilation and views down to the tea sorting area, creates a warm social space in the quay

First Platforms (+1) Public
10. MicroPublicTearooms (Serving area)
-Tea served at 70°c
11. MicroPublicTearooms (Drinking area)

Second Platform (+2) Public
12. MicroPublic Toilets
13. Viewing Platform

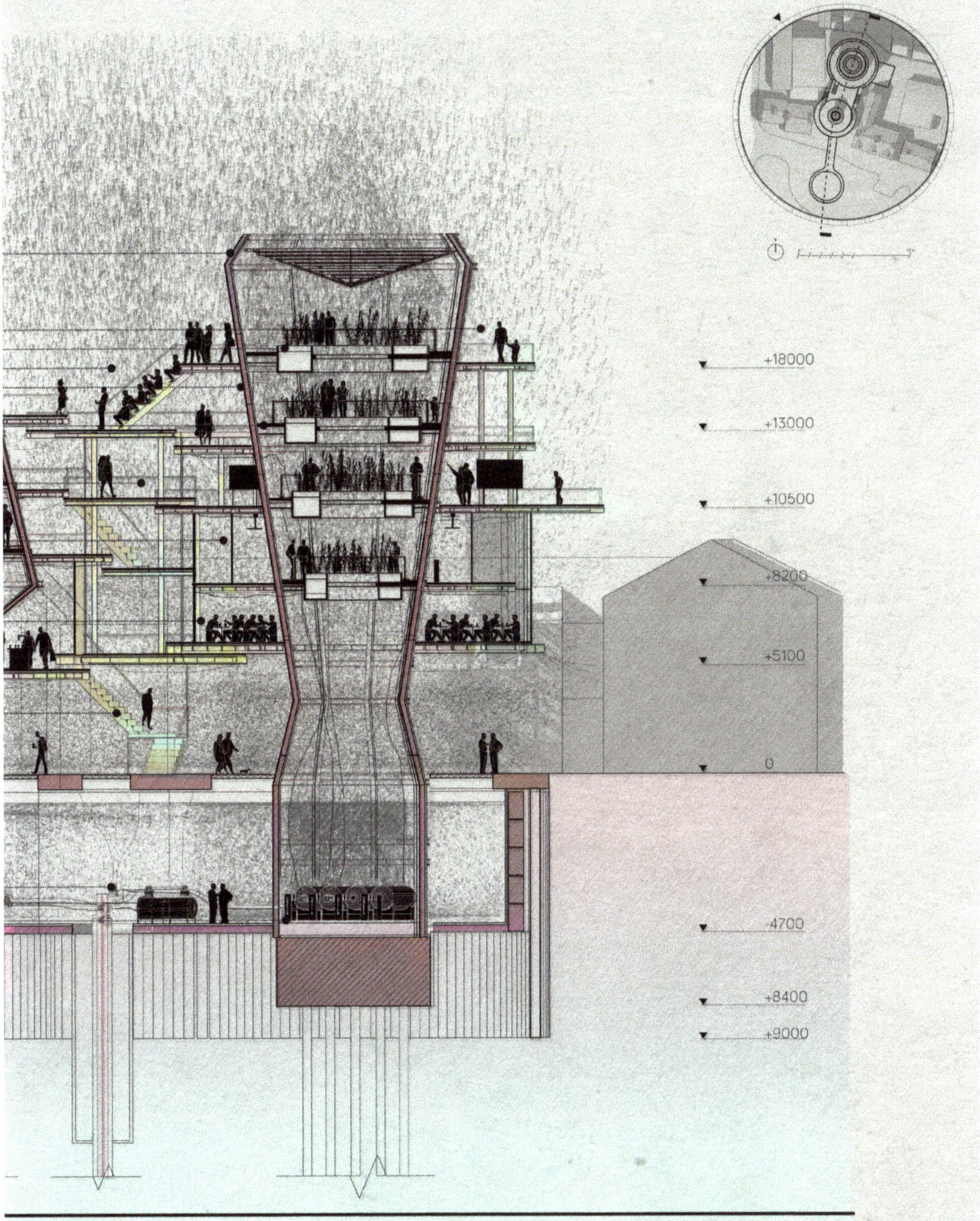

Third Platform (+1) Public
14. Gathering space and exhibiting space

Forth Platform (+2) Public
15. Gathering space and debate platform
16. Viewing platform
17. Stargazing deck

Tea Growing Chimney (A)
18. Access Portals
19. Perforated Walkway Public / Private
20. Tea plant cultivation platforms, plants grow in a **hot, humid** environment supplied by the basement level waste heat and humidity.

Tea Drying Chimney (B)
21. Access Portals
22. Perforated Walkway Public / Private
23. Porous drying racks, wither the leaves in a **dry, hot** atmosphere.

[ ] Enclosed to the elements

Prototypical -
MicroPublicPowerhouse
Prototype **replicated** to surrounding villages - off grid electricity and heat t
vulnerable rural populations + providing a public service from the wasted g
1.
2.
3.
3.
4.
2.
Village Hall
3.
Community Centre
HEAT SCALE
Warm
Cool
CREECH 2.
50.637, -2.140
Public Lumber Drying
Geothermal Chimney use : Dry lumber for fuel
Public building connect heat and electricity : Village Hall
3. CORFE
50.642, -2.062
Public Soup Kitchen
Geothermal Chimney use : Grow Vegetables
Public building connect heat and electricity : Library, Community Centre

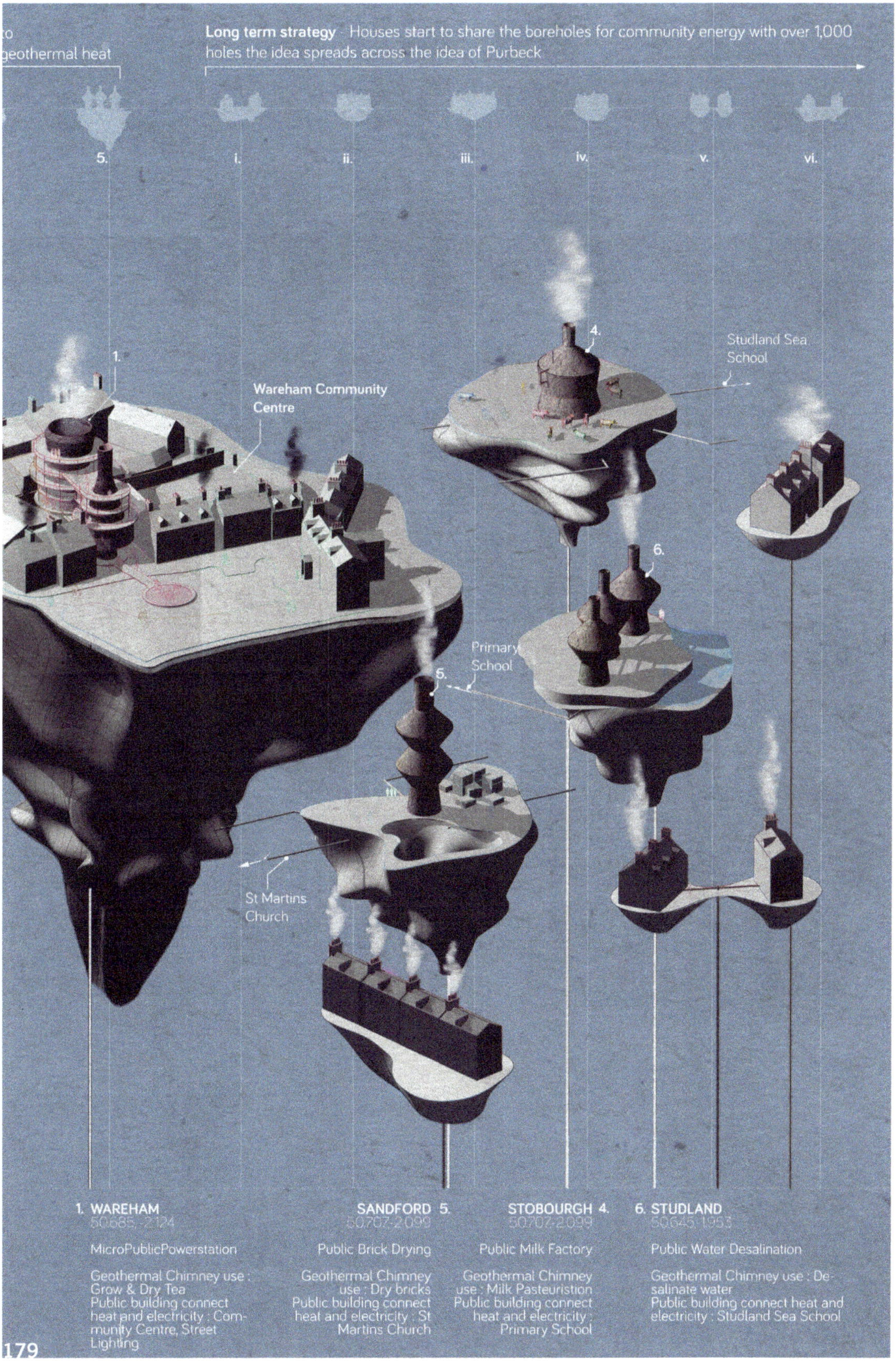

to
geothermal heat
Long term strategy - Houses start to share the boreholes for community energy with over 1,000 holes the idea spreads across the idea of Purbeck
5.
i.
ii.
iii.
iv.
v.
vi.
1.
Wareham Community Centre
4.
Studland Sea School
6.
Primary School
5.
St Martins Church
1. WAREHAM
50.685, -2.124
MicroPublicPowerstation
Geothermal Chimney use : Grow & Dry Tea
Public building connect heat and electricity : Community Centre, Street Lighting
SANDFORD 5.
50.707-2.099
Public Brick Drying
Geothermal Chimney use : Dry bricks
Public building connect heat and electricity : St Martins Church
STOBOURGH 4.
50.707-2.099
Public Milk Factory
Geothermal Chimney use : Milk Pasteuristion
Public building connect heat and electricity : Primary School
6. STUDLAND
50.645 1.953
Public Water Desalination
Geothermal Chimney use : Desalinate water
Public building connect heat and electricity : Studland Sea School

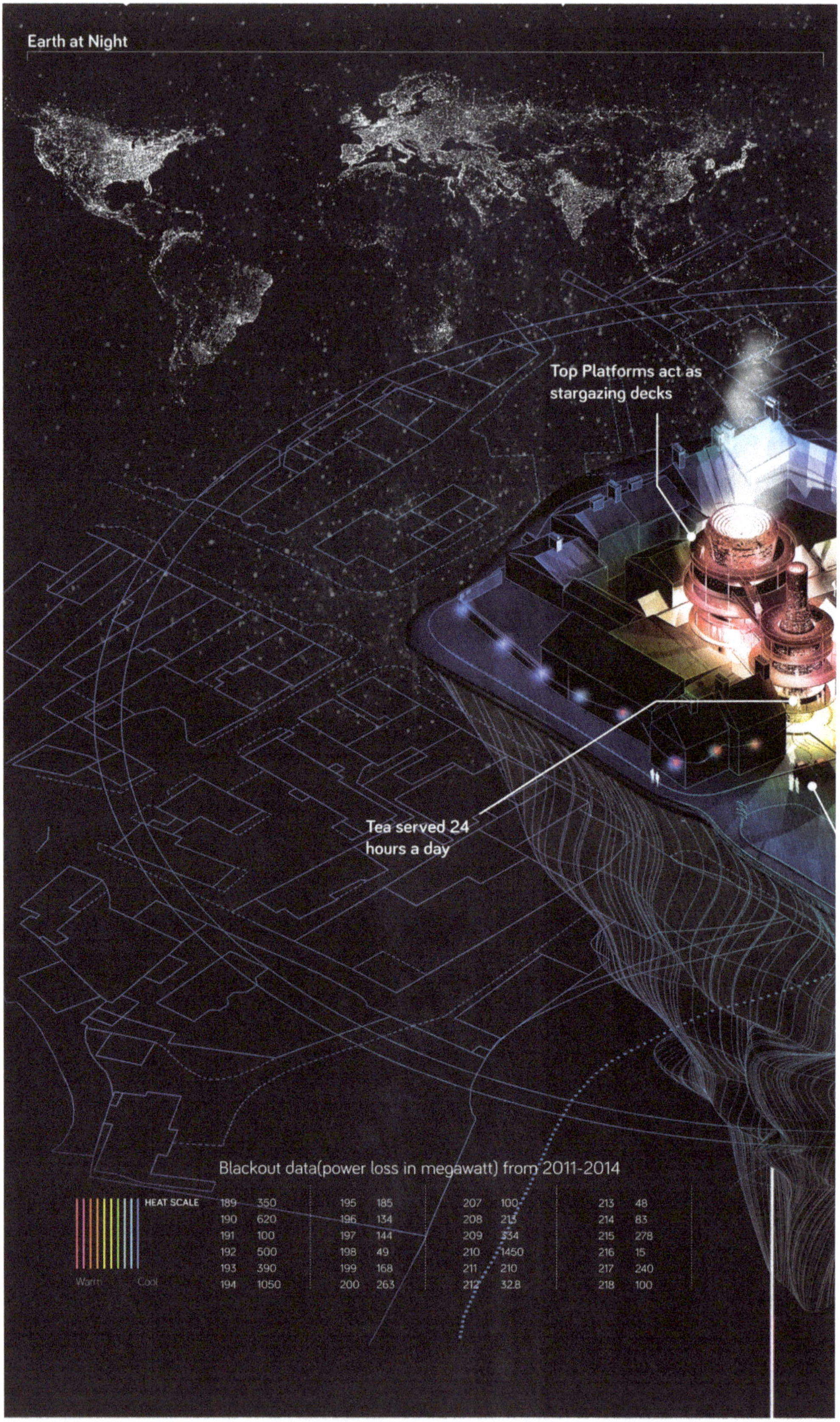

189	350	195	185	207	100	213	48
190	620	196	134	208	213	214	83
191	100	197	144	209	334	215	278
192	500	198	49	210	1450	216	15
193	390	199	168	211	210	217	240
194	1050	200	263	212	32.8	218	100

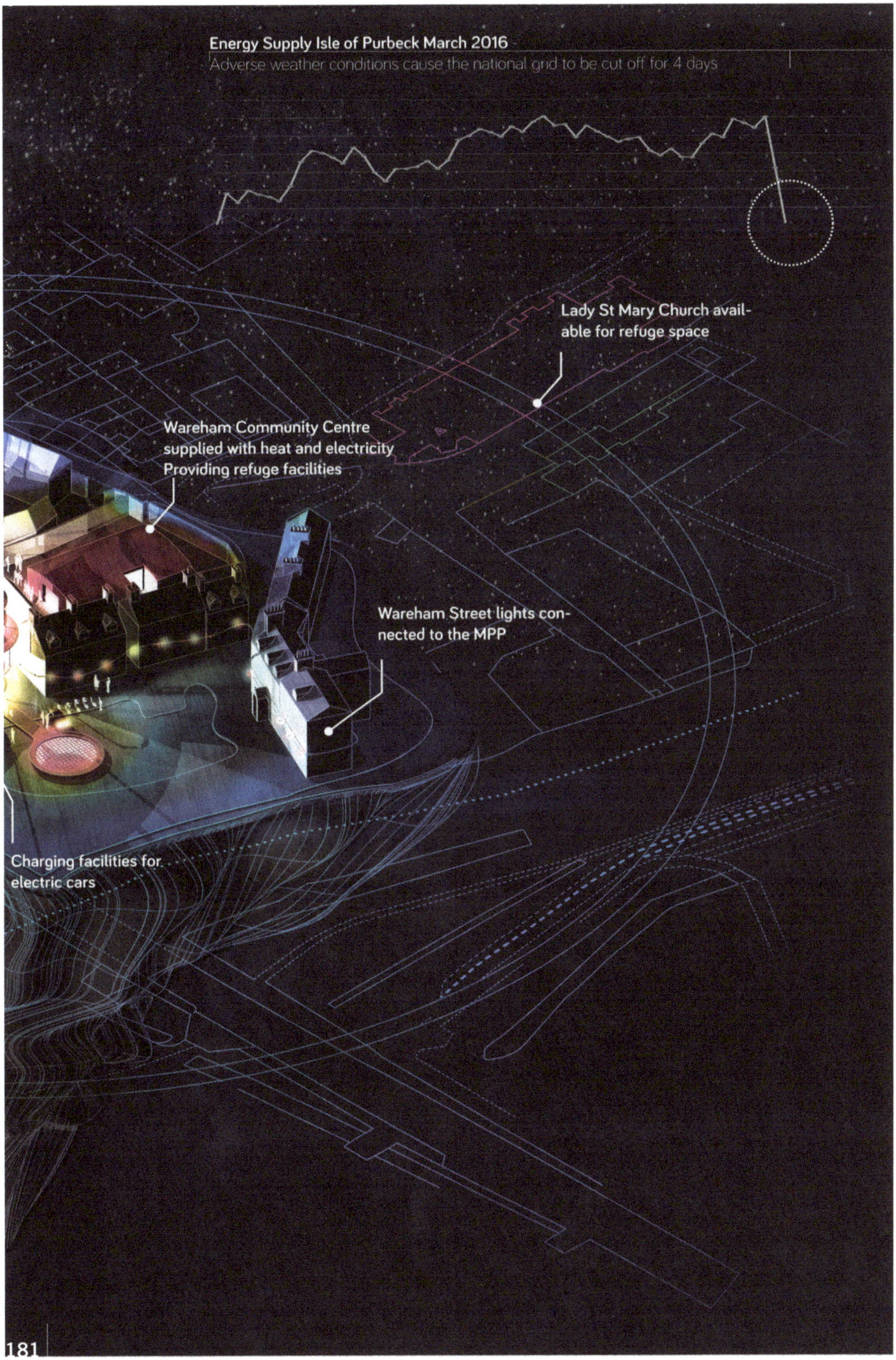
Energy Supply Isle of Purbeck March 2016
Adverse weather conditions cause the national grid to be cut off for 4 days
Lady St Mary Church available for refuge space
Wareham Community Centre supplied with heat and electricity Providing refuge facilities
Wareham Street lights connected to the MPP
Charging facilities for electric cars

MICROSEISMIC PAVILION

MICHAEL O'HANLON

In the process of fracking, the earth's properties are irreversibly altered and seismic waves produced have been known to reactivate nearby historic fault lines, causing earthquakes to occur more frequently and in locations that would otherwise be stable.

The Microseismic Pavilion explores the use and application of microseismic data as a tool for recording and experiencing the earth. It is a curated landscape experience where invisible data meets a physical landscape, experienced and comprehended as a soundscape. Located in the Isle of Purbeck, a number of seismic monitoring stations correspond to different geological strata and enhance bodily experience of the surface topography / landscape and geological instability. The data they gather is converted into a sonic experience of subterranean movement, such as the rumbling of earthquakes half a world away, the waves of ocean storms far out to sea and the twice-daily pull of moon and sun. In addition, data archives are exhibited through artistic visualizations and cartography. It is a place for contemplation to consider big questions that people share, and to rethink our connectivity with the earth in the era of the anthropocene.

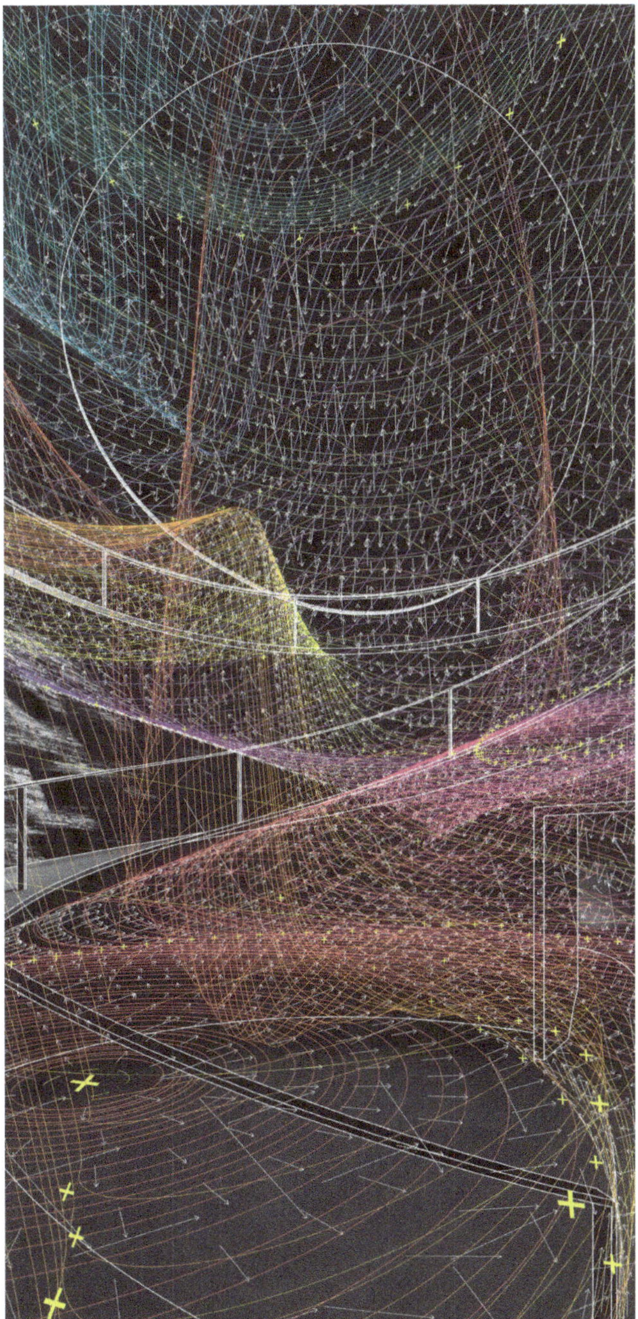

IMAGE CAPTIONS

P183 Visualisation of sonic reverberations in the Microseismic Pavilion

P184 [top] Microseismic Pavilion Diagrammatic Program

[bottom left] Topography and Landscape Features of the Isle of Purbeck

[bottom right] Geology of the Isle of Purbeck

P185 Nine Barrow Downs Ridge, Isle of Purbeck

The ridge of Nine Barrow Downs is comprised of various chalk and portland stone substrata; an historic faultline follows the topography of the ridge from east to west.

Source: Ian West (http://www.southampton.ac.uk)

P186-187 Imaginary Simulation of a Subterranean Section of Nine Barrow Downs

P188-189 Imaginary Simulation of the Subterranean Density of Nine Barrow Downs

P190 Microseismic Pavillion, Section

Annotated juxtaposition of local geologic and solar time, experienced visually and audibly through realtime sonic reverberations of microseismic earth movements.

P191 Microseismic Pavillion, visuaisation of sonic reverberations

The pavilion's interior reverberates according to emissions from microseismic recording devices.

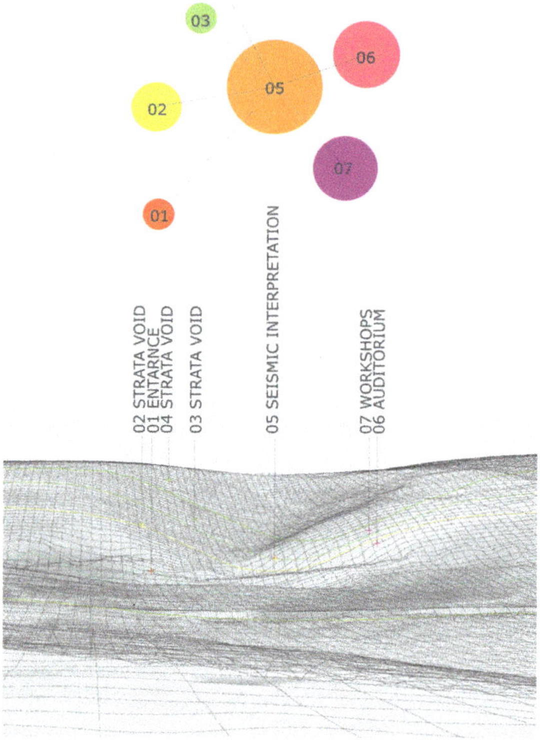

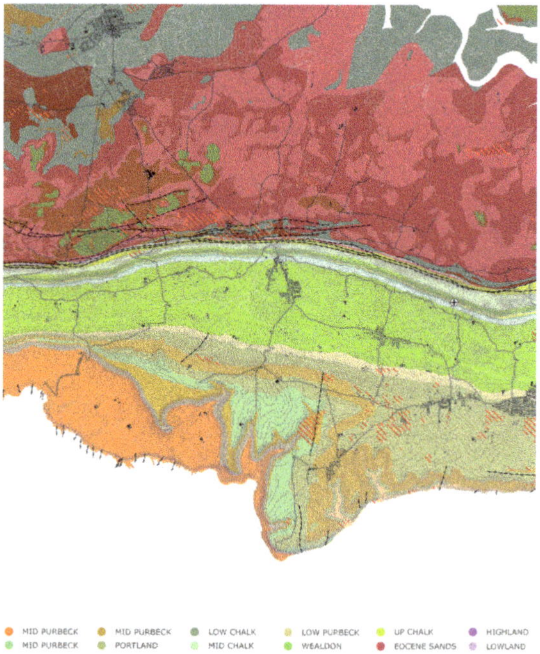

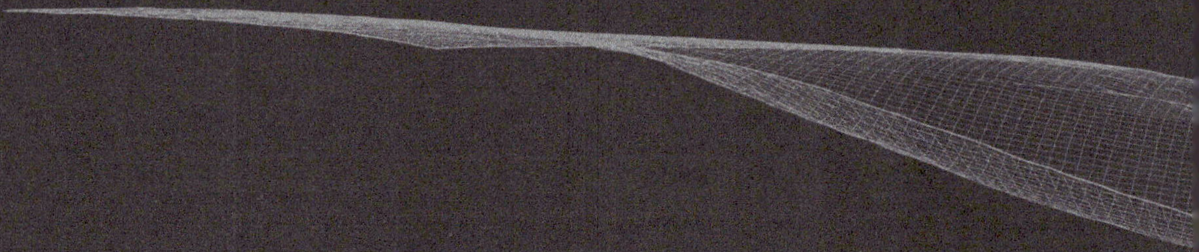

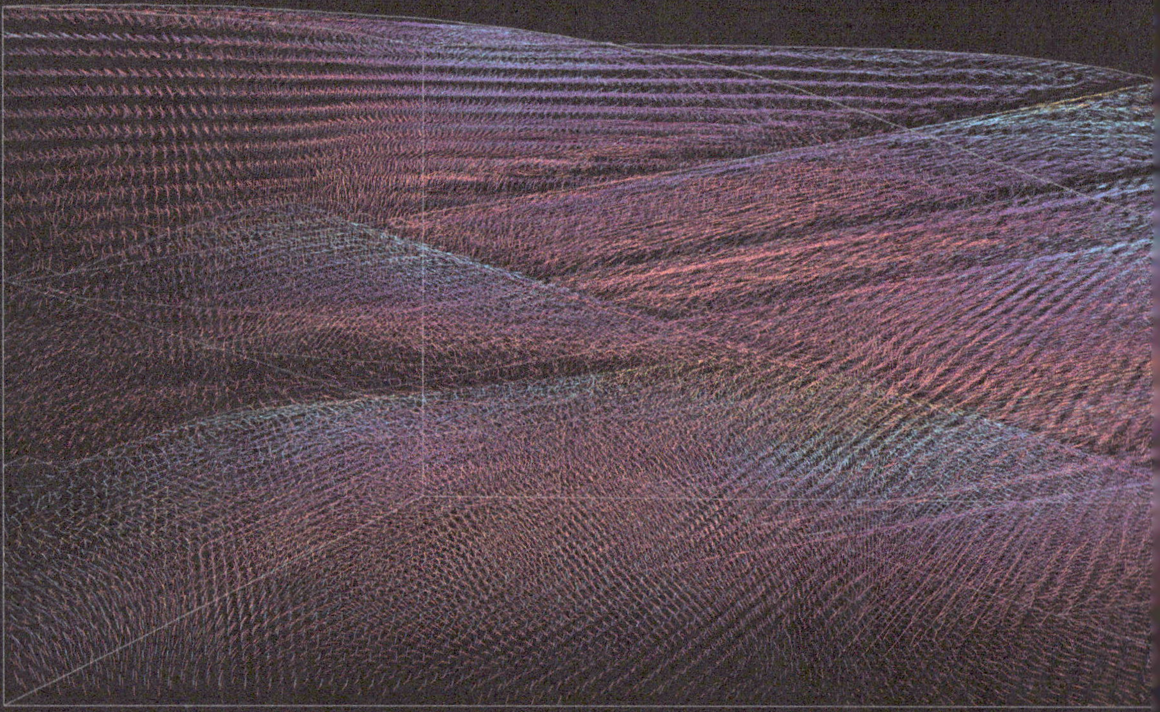

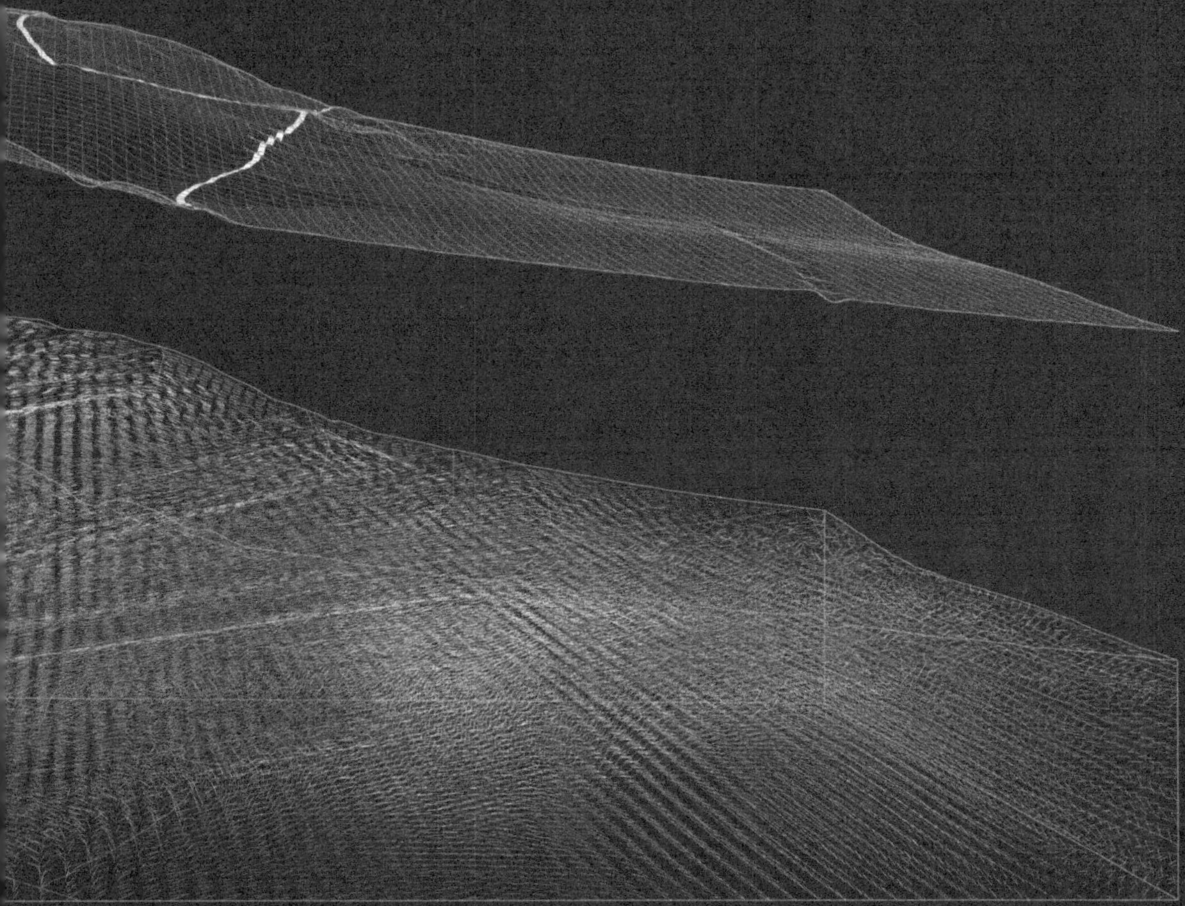

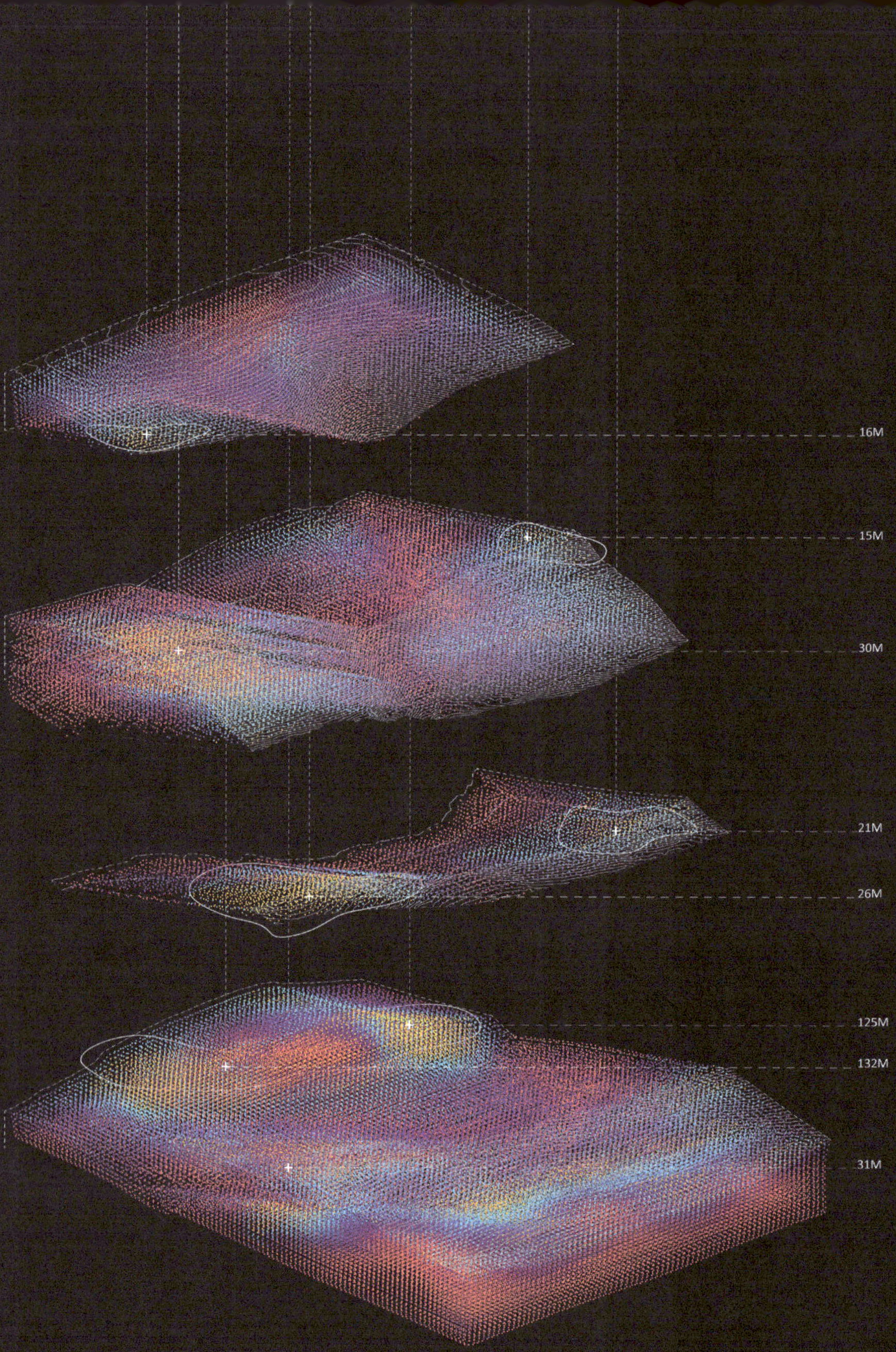
16M
15M
30M
21M
26M
125M
132M
31M

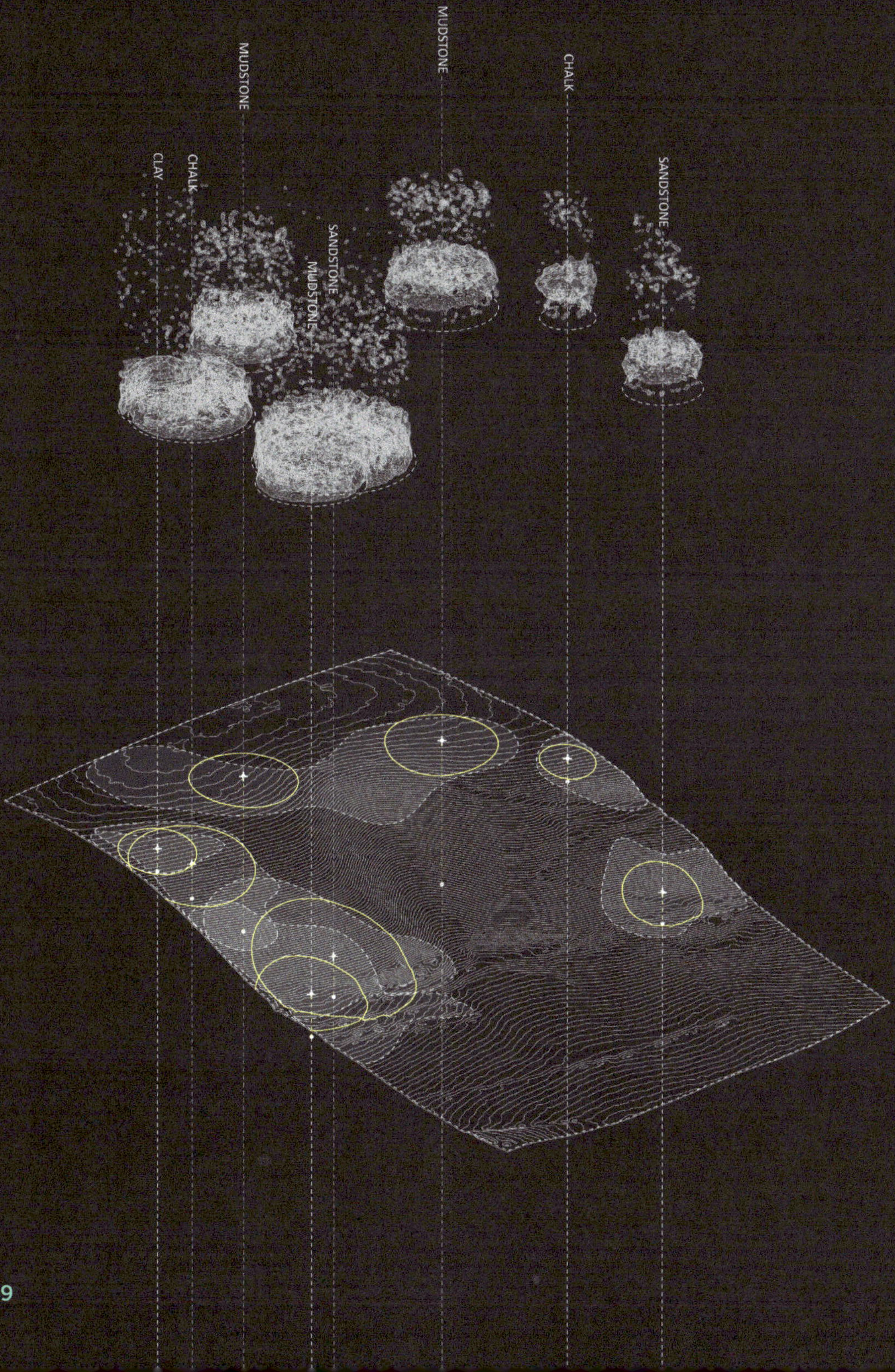
CLAY
CHALK
MUDSTONE
SANDSTONE
MUDSTONE
MUDSTONE
CHALK
SANDSTONE

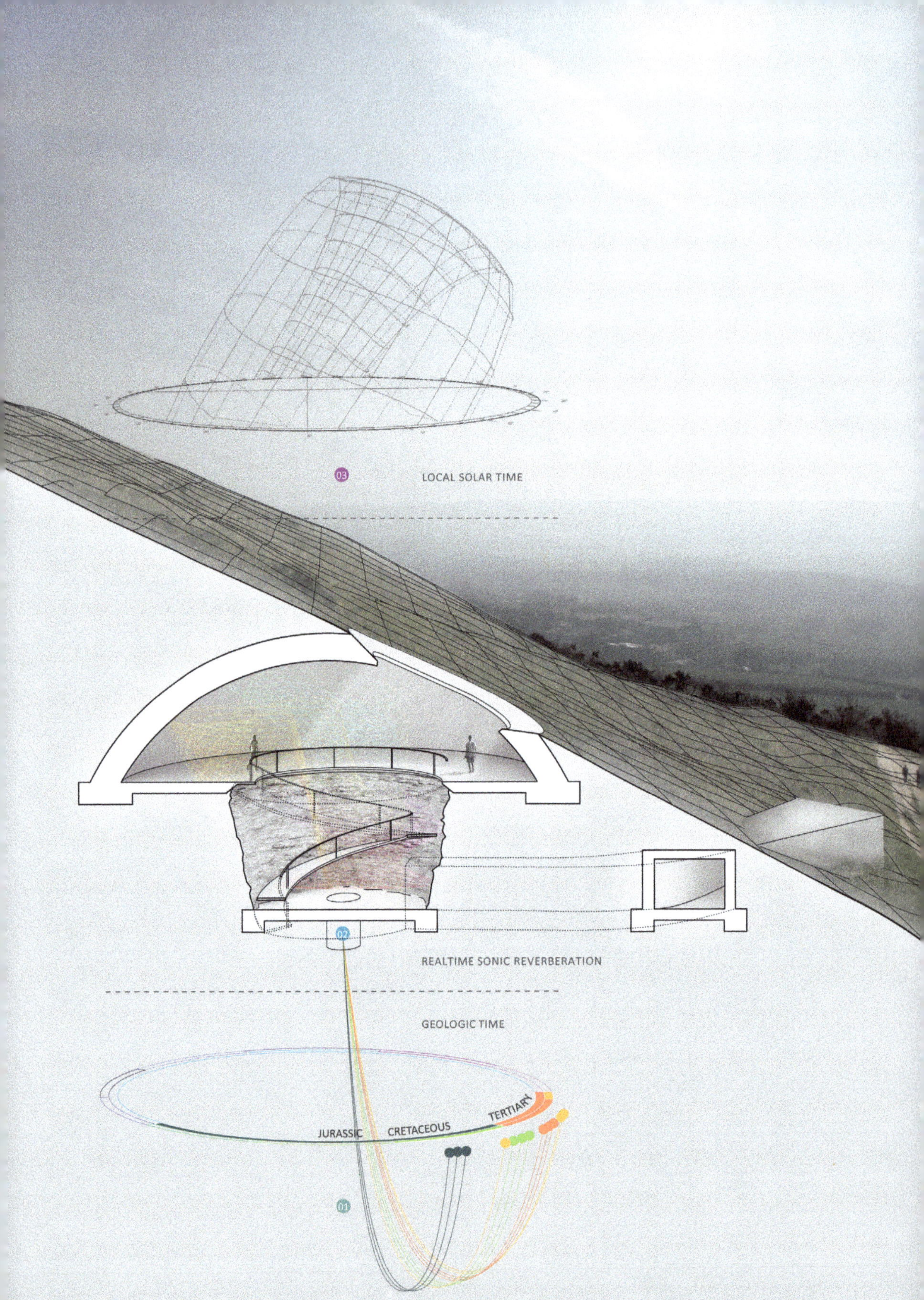

03
LOCAL SOLAR TIME
02
REALTIME SONIC REVERBERATION
GEOLOGIC TIME
JURASSIC
CRETACEOUS
TERTIARY
01

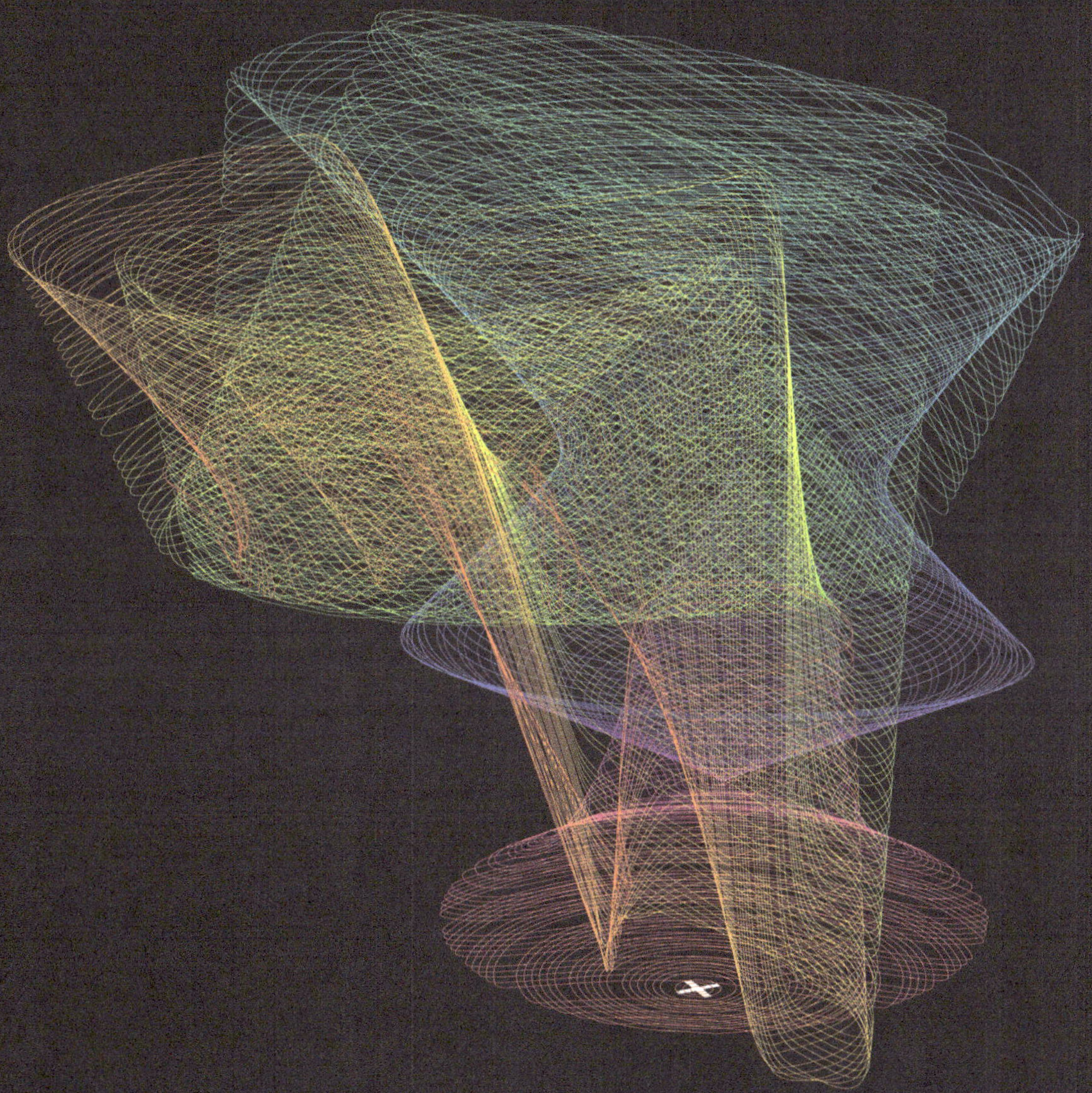

DESIGNING WITH ENERGY

2014/15

INTRODUCTION

LINDSAY BREMNER AND ROBERTO BOTTAZZI

In 2014/15, DS18 developed a way of thinking and a design methodology based on the non-linear behaviour and self-organising processes of matter. We explored how flows of energy and matter spontaneously bifurcate or change their state at certain critical moments under internal or external pressure. Computers provided windows into this world because non-linear behaviour can be visualised, simulated and manipulated in them. Bifurcation, attraction, repulsion, viscosity, stratification, phase space, mechanic phylum, genetic code - these were words that we grappled with as we explored ways of engaging, as designers and space formers, with the self-organising, more-than-human behaviour of matter and energy pervasive throughout nature.

Firstly, what is important is the spiritual principle itself for using new material for a dynamic architecture. Air, gas, fire, sound, odours, magnetic forces, electricity, electronics are materials. (01)

We began the studio by investigating these ideas computationally using the fluid simulation package RealFlow. (02) Students were asked to model the dynamic properties of material processes like erosion, deposition, sedimentation, compaction, condensation, eruption, shrinkage, swelling, contraction, evaporation, lamination, turbulence, freezing, thawing etc. We were looking to develop an understanding of material process in their complex and temporal quality, and the beginning of a design and aesthetic sensibility that would be further cultivated during the year. (03)

Social systems can also be understood as self-organising processes, as "ensembles of flows and the reservoirs driving those flows: water, metabolic energy, bonding pressures, action modes, population, trade, technology and so on." (04) Accounts of human development thought of in this way stress the role of flows and phase transitions in determining the social field. (05) In the studio we examined the landscape and history of the arid region of South Africa known as the Karoo and the reservoirs, flows and institutions of South Africa's energy economy in this light. We approached the government of

South Africa's proposal to introduce hydraulic shale-gas fracturing into the Karoo as a potential phase-changing moment, when solidified, stratified geological and social formations would be disrupted, bifurcated and transformed along any number of fault-lines.

South Africa's economy has grown rapidly since the end of the apartheid in 1994 and the country now has the largest economy in Africa with the highest energy consumption on the continent. It is also the leading CO2 emitter in Africa and the 14th largest in the world. It has limited proved reserves of oil and natural gas and uses its large coal deposits to meet 72% of its energy needs. It has a sophisticated synthetic fuels industry, producing petrol and diesel at its coal-to-liquids and gas-to-liquids plants. Most of its oil is imported from Middle East and West African producers and is locally refined. It is estimated that, at present rates, South Africa's coal reserves will run out in 50 years. A recent study by the U.S. Energy Information Administration found that the Karoo, a vast, arid interior part of the country holds notable shale gas resources, turning attention to the region as a potential alternative source of energy. In 2011 the South African government enacted a moratorium on the licensing and exploration of shale resources due to environmental concerns, but lifted it 18 months later. A number of international companies, Royal Dutch Shell included, submitted applications to explore the shale region. President Jacob Zuma said early in 2014 that shale gas could be a game changer and provide the country with a reliable alternative fuel to coal.

In order to familiarise ourselves with this environment, we undertook a research trip to South Africa in October 2014. (06) This began with a one day symposium at the University of Cape Town titled "Proposed Shale Gas Mining / Hydraulic Fracking: A Confluence of Science, Humanities, Law and Governance" organised by Professor Jan Glazewski of the Shale Gas Working Group and Institute of Marine and Environmental Law (IMEL) at UCT. (07) We then set off on a road trip through the Karoo. En route to the historic town of Graaff Reinet we visited a hillside outside Nelspoort, where a scattering of San rock engravings and gong rocks provided a window into Karoo archaeology. We stayed for three nights in Graaff Reinet, centre of the Camdeboo Municipality, where we went on a walking tour of Umasizakhe, an early morning game drive, enjoyed sundowners overlooking the Valley of Desolation and visited the village of Nieu Bethesda, the flash point of protests against fracking. We also met with Graaff Reinet anti-fracking lawyer, Derek Light, and Stefan Cramer, scientific advisor to the Southern African Faith Communities' Environment Institute on fracking.

This trip equipped us with experiences and information to take an informed position on the Karoo as an energy system. Students were asked to model how the Karoo might change as energy pressures were brought to bear on it. They were required to develop an energy strategy for the Camdeboo Municipality, taking into account the potential for energy infrastructure to contribute not only to the economic, but also to the political, social and

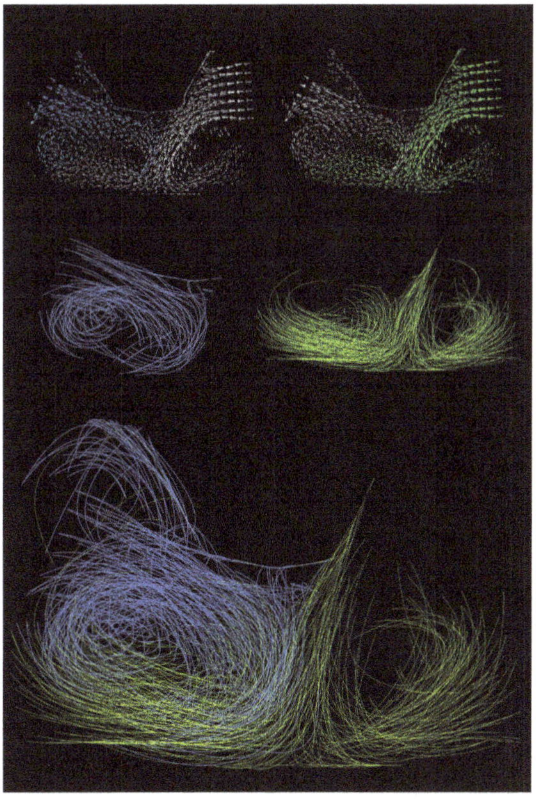

Simulation of ocean turbulence
Niall Green

cultural life of the region. Strategies were to be imaginative / inventive / critical and life-based, responding to the unique features of the Karoo landscape, its geological and human history and acknowledging the realities of global anthropocenic climate change. Key to this was a reading of the physical and socio-economic landscapes of the Camdeboo Municipality as energy fields and channelling, enhancing or capitalising on these energies to develop a scenario for its future. This was informed by economic data, demographic data, technical data, existing development objectives and other data necessary for developing a strategic energy masterplan for the Camdeboo Municipality. Within this, individual stances and projects were identified and further developed.

Strategic Energy Masterplans included in this publication are Camdeboo Solar Park (CSP), by Andrew Baker Falkner, John Cook, Michael O'Hanlon and Ben Pollock, Future Research for Experimental Energies (FREE) by Cheryl Choo, Matthew Hedges, Shiue Nee Pang and Iulia Stefan and the Waste Integration Initiative (WII) by Jared Baron, Sophie Fuller and Alice Thompson. Individual projects by Andrew-Baker Falkner, John Cook, Cheryl Choo, Shiue Nee Pang and Jared Baron follow their respective masterplans. The book concludes with Oscar McDonald's 'Wind Seed 01' is a system of off-grid commercial units powered by wind and Jack Thompson's 'Urban Generator' is the nucleus of an urban agriculture strategy for Umasizakhe, the apartheid era black residential township of Graaff Reinet.

Notes

(01) Y. Klein, 'The Evolution of Art Towards the Immaterial', in P. Noever, and F. Perrin (eds.), *Air Architecture: Yves Klein*, Ostfildern-Ruit: Hatje Cantz, 2004, pp. 44.

(02) RealFlow is the fluid simulation software developed by Next Limit Technologies, http://www.realflow.com/.

(03) The drawings resulting from these computer simulations are used throughout this book.

(04) M. DeLanda, *A Thousand Years of Nonlinear History*, New York: Zone Books, 1997.

(05) M. DeLanda, 'Nonorganic life', *Incorporations, Zone no. 6*, New York, Zone Books, 1992, p. 153.

(06) This research trip was partly sponsored by Johann Rupert's Reinet Foundation and the Faculty of Architecture and the Built Environment's International Committee.

(07) Speakers and papers presented were: Kevin Winter, Department of Environmental and Geographical Sciences, UCT, 'The geophysical context of the Karoo, with particular reference to water and geo-hydrology'; Jeremy Wakeford, Stellenbosch Sustainability Institute,'The South African Energy Context'; Saliem Fakir, WWF South Africa, 'The Economics of Fracking'; Timm Hoffman, UCT, 'Karoo Biodiversity'; David Morris, McGregor Museum, Kimberley, 'Karoo Archaeology & Palaeontology'; Tania Katzchner, Department of Planning, UCT, 'Situating shale gas mining: need and desirability, values and attitudes'; Lesley Green, Environmental Humanities, UCT, 'Fracking v Shale Gas: Perspectives from the Environmental Humanities'; Tracy Humby, University of Witwatersrand, 'Environmental Assessment in the minerals/gas extraction context: a critique'; Lisa Plit, Researcher, IMEL, UCT, 'The MPRDA: with specific reference to Petroleum Resources'; Loretta Feris, IMEL, UCT, 'The Karoo: 'A sense of place'; Anél du Plessis, North-West University, 'The governance of hydraulic fracturing in the Karoo: a local government perspective', and Dorota Rucinska, University of Warsaw, 'Benefits, risk and questions raised in the shale gas issue: perspectives from Poland'. Most speakers had agreed to contribute a chapter to a book provisionally titled: Fracking/shale gas mining in the Karoo: critical environmental and governance perspectives to be edited by J. Glazewski.

Energy Parliament

This building for Graaff Reinet stores energy produced by off-grid wind turbines for use during periods of low wind. Energy is stored in liquid metal batteries located underground. The surface above them provides terraces for community use; above this, a canopy of balloons inflated by heat loss from the storage batteries creates a visual indicator of how much energy is stored.

Rupert Calvert

CAMDEBOO SOLAR PARK

ANDREW BAKER FALKNER
JOHN COOK
MICHAEL O'HANLON
BEN POLLOCK

South Africa's historic reliance on coal as its main energy source as well an increasing energy demand and an ageing infrastructural network have led to unpredictable power cuts, which can have significant impacts, not just on businesses and the economy but also the quality of life, in both urban and rural areas. The Camdeboo Solar Park is a proposal to tackle these problems using Concentrated Solar Power (CSP). It is a component of a national energy strategy of five solar parks comprising 45 CSP plants spread out across the country. This will provide 55% of South Africa's energy through CSP by 2050, bringing it in line with global Long Term Mitigation Scenarios which, along with other predicted changes in energy production, aims to create a carbon neutral energy sector by 2050.

The proposal makes use of the fact that South Africa is within the top 3% of countries worldwide in terms of solar irradiance. Through a thorough analysis of these solar resources, as well as the existing economy, infrastructure and demographics, The Camdeboo Solar Park was developed to understand the factors which need to be accounted for when implementing these schemes, including local solar potential, land value and land use. The result of the data driven process is a masterplan situated within the Camdeboo Local Municipality consisting of a number of networked Concentrated Solar Power plants built in three phases. These are analysed and represented in terms of various factors such as efficiency, light, shadow and time. Camdeboo Solar Park, informed though a rigorous understanding of the existing national and local conditions, will allow for further speculation in terms of the cultural and social impacts of large scale roll out of solar power.

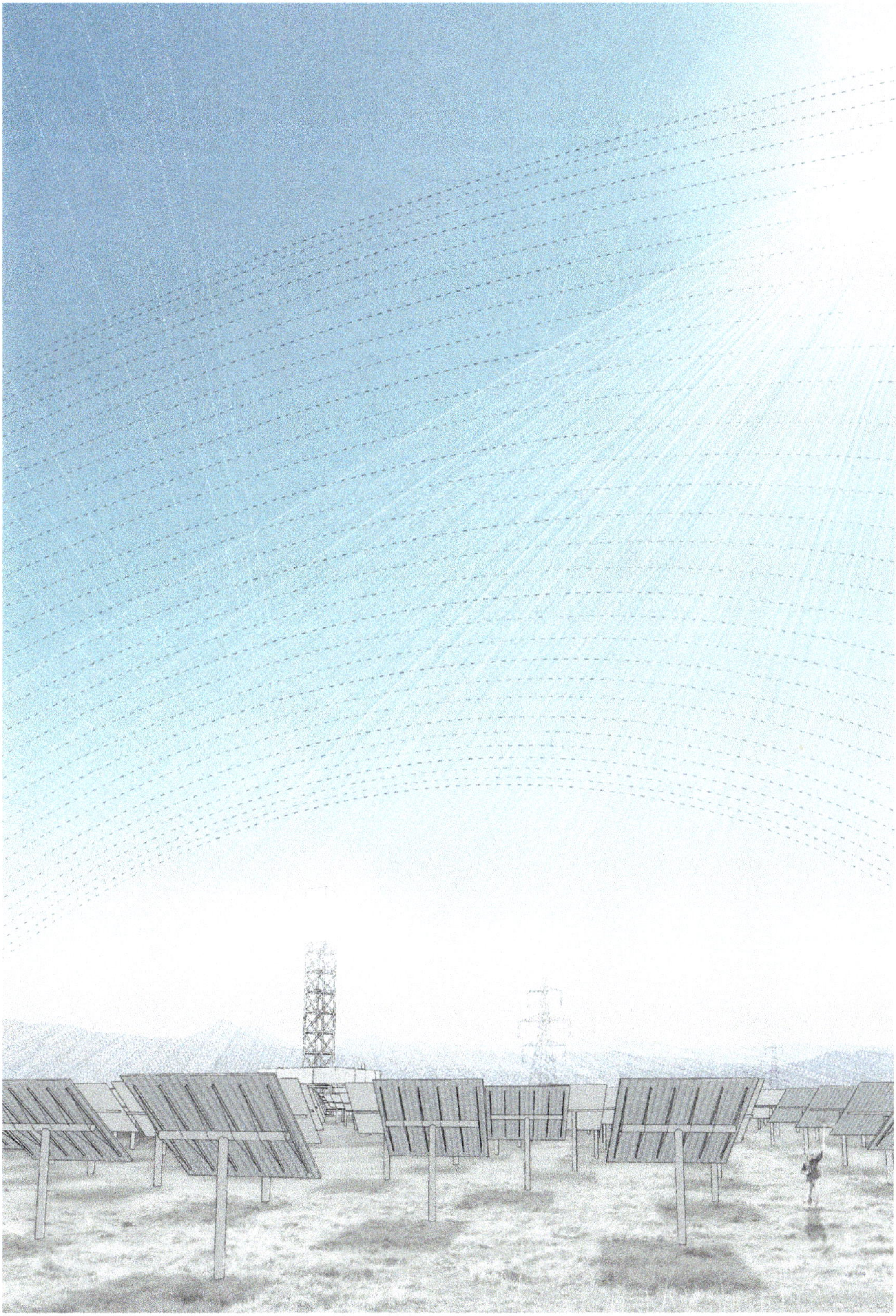

IMAGE CAPTIONS

P199 Camdeboo Solar Park

P200 South Africa's Electricity Crisis

South Africa sits at a complex crossroads between its current energy production capacity and energy requirements. Aging production facilities and distribution network, and an over reliance on diminishing coal supplies has led to an increase in power cut frequency. These cuts sweep across the municipalities, some experienced as scheduled 'load shedding,' others as blackouts. 2008 and 2014 saw the worst of these cuts, which look to continue as South Africa and its sole energy provider, Eskom, prepare to transfer its primary energy source and key export commodity, coal, to a more stable mix of sources ranging across renewables, nuclear, and natural gas.

Source: Cape Argus, 1 Nov 2014

MONDAY NOVEMBER 3, 2014 www.capeargus.co.za

Cape Argus

M EDITION

INDEPENDENT MY INDEPENDENT APP

Black in time

It's 2008 all over again as rolling cuts mean no power to the people

P201 South Africa

[top left] Global Horizontal Irradiance

[top right] Population Distribution

[botom left] Land Values

[bottom right] National Concentrated Solar Power Strategy

Sources: https://www.environment.gov.za/; https://africaopendata.org/group/south-africa; http://www.gap.csir.co.za/; http://www.gap.csir.co.za/images/images/total-gross-value-added-2009/view; http://www.southafrica.info/business/economy/sectors/automotive-overview.htm#.Vv2Lh_krLIU

P202 Camdeboo Local Municipality, Solar Terrain Overlay

Sources: http://itis.ngi.gov.za/itisportalinternet/; https://www.environment.gov.za/

P203 Camdeboo Local Municipality, Local Strategy

Sources: http://itis.ngi.gov.za/itisportalinternet/; https://www.environment.gov.za/

P204 Camdeboo Solar Park Masterplan

P205 [top left] Camdeboo Solar Park, Efficiency Analysis

[top right] Camdeboo Solar Park, Shadow Analysis

[bottom] Camdeboo Solar Park, Light Ray Analysis

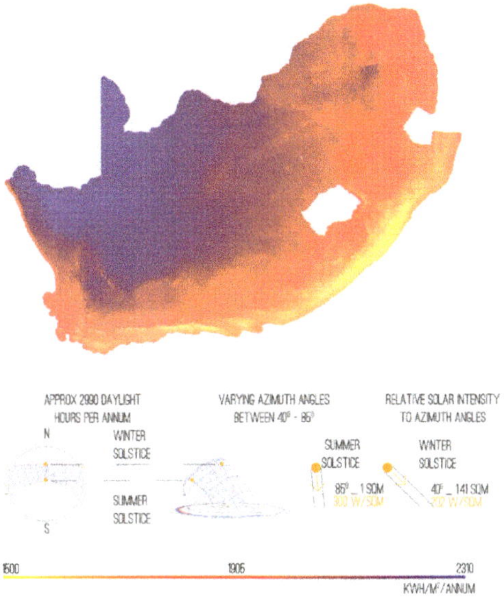
APPROX 2990 DAYLIGHT HOURS PER ANNUM
VARYING AZIMUTH ANGLES BETWEEN 40° - 86°
RELATIVE SOLAR INTENSITY TO AZIMUTH ANGLES
N
S
WINTER SOLSTICE
SUMMER SOLSTICE
SUMMER SOLSTICE
WINTER SOLSTICE
86° _ 1 SQM
40° _ 1.41 SQM
1500
1905
2310
KWH/M²/ANNUM

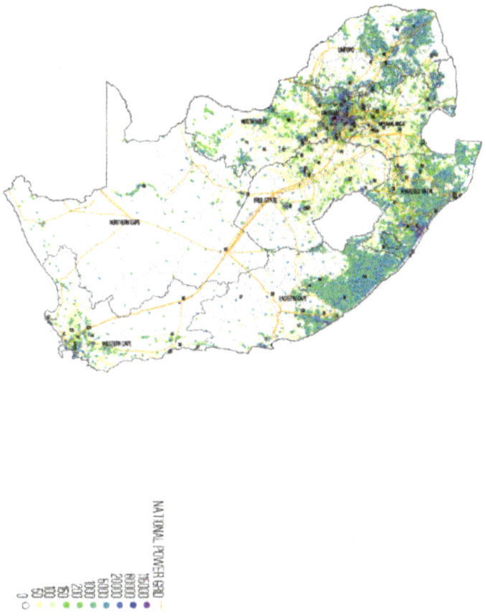
NATIONAL POWER GRID
0
10
50
100
150
250
1000
5000
20000
60000
150000

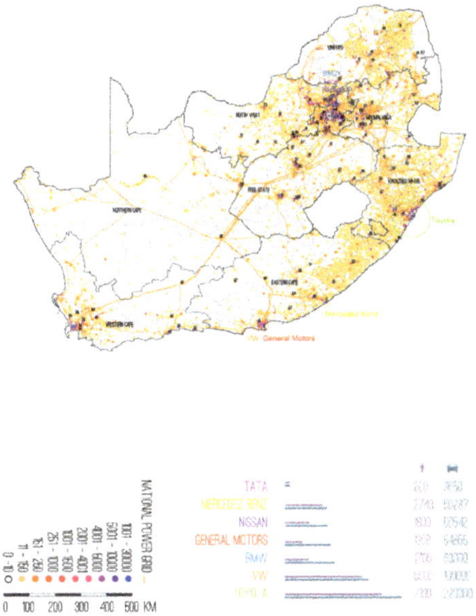
NATIONAL POWER GRID
0 - 10
11 - 150
151 - 250
251 - 1000
1001 - 1500
2001 - 4001
4001 - 5000
5001 - 10000
10001 - 30000
0 100 200 300 400 500 KM
TATA
MERCEDEZ BENZ
NISSAN
GENERAL MOTORS
BMW
VW
TOYOTA

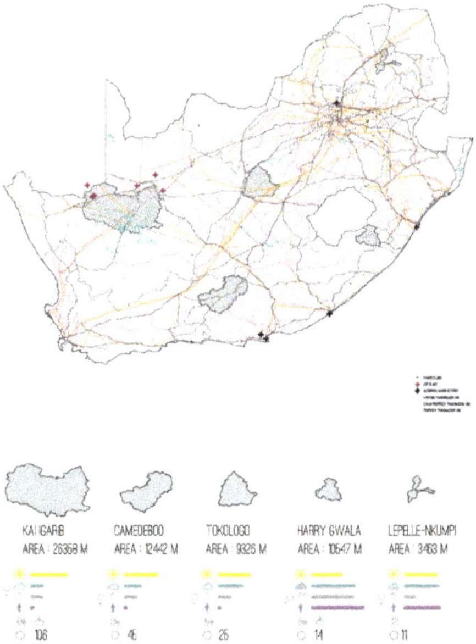
KAI !GARIB
AREA : 26358 M
106
CAMEDEBOO
AREA : 12442 M
45
TOKOLOGO
AREA : 9326 M
25
HARRY GWALA
AREA : 10547 M
14
LEPELLE-NKUMPI
AREA : 3463 M
11

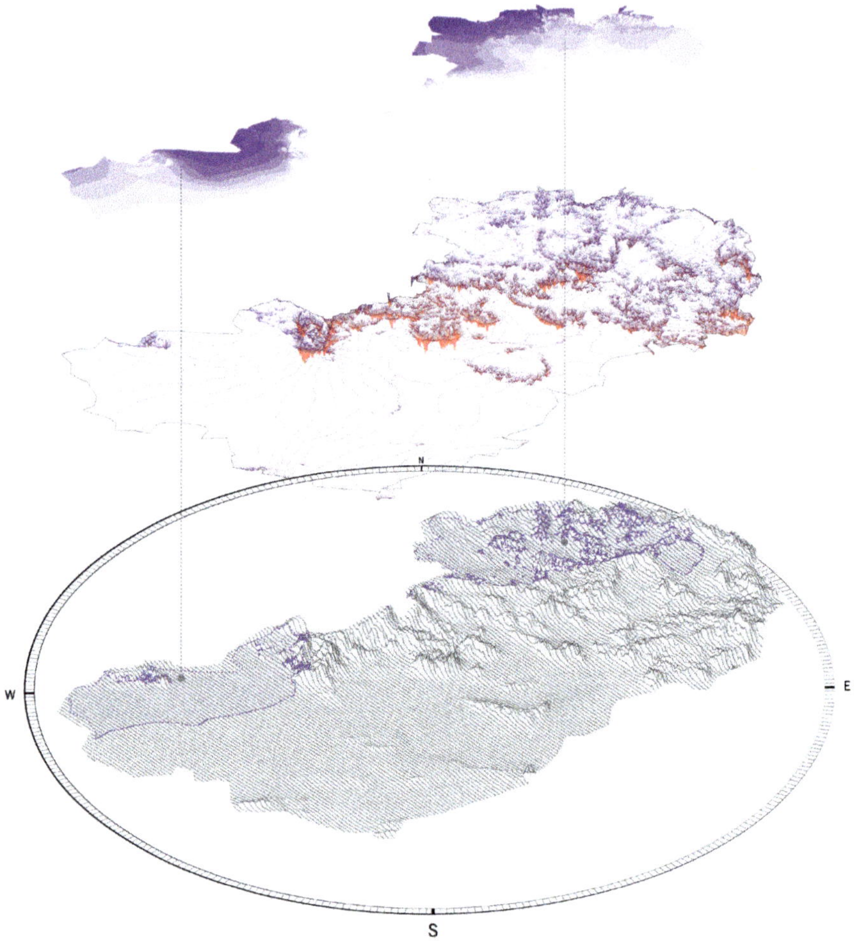
N
W
E
S

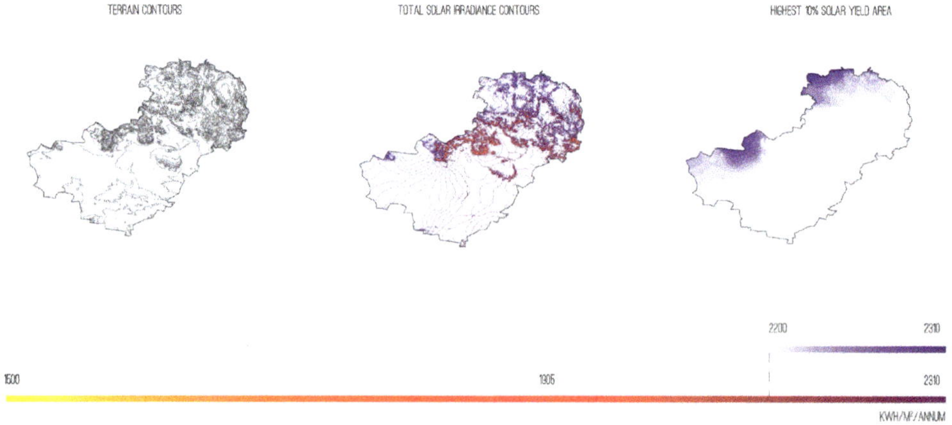
TERRAIN CONTOURS
TOTAL SOLAR IRRADIANCE CONTOURS
HIGHEST 10% SOLAR YIELD AREA
2200
2310
1500
1905
2310
KWH/M²/ANNUM

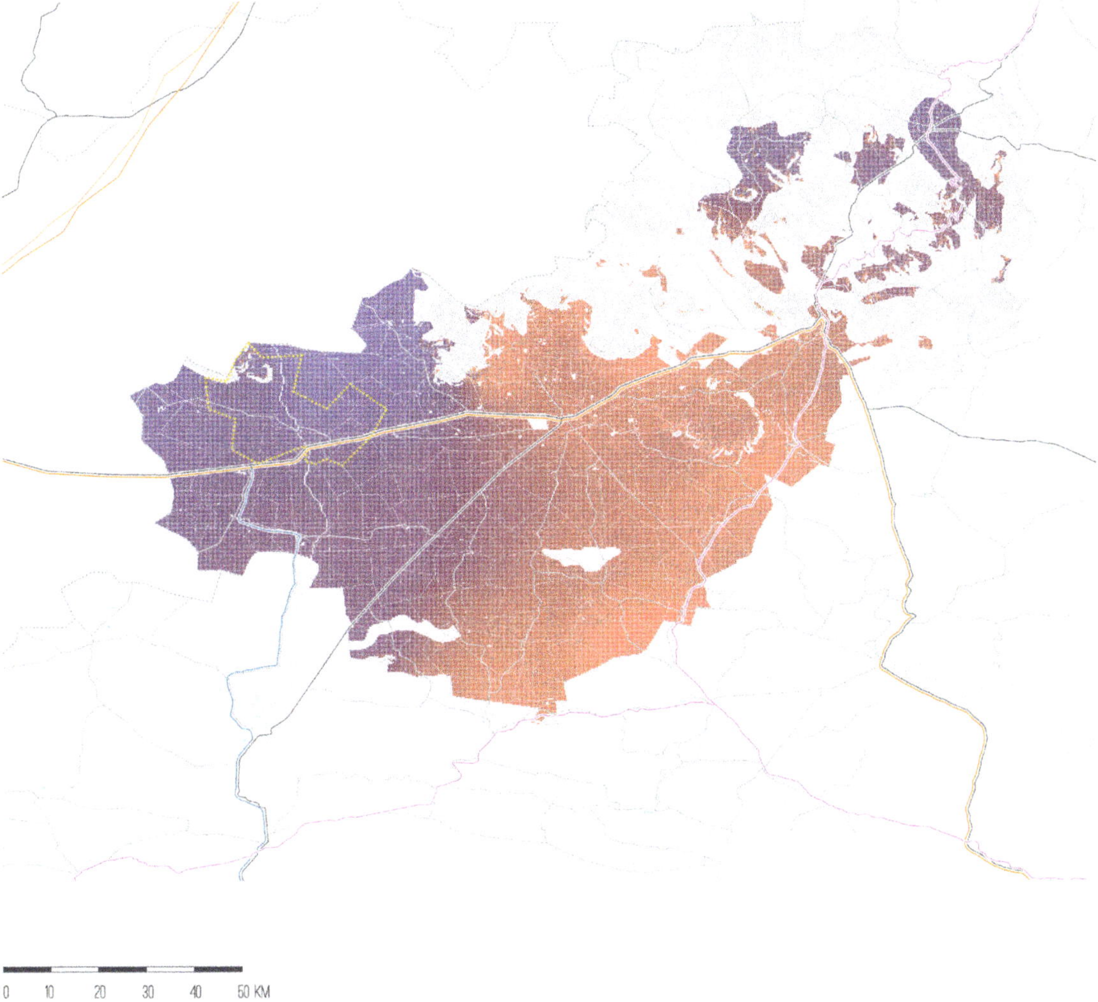

AREA OF HIGHEST
SOLAR RADIANCE

45 X 130MW
CSP PLANTS

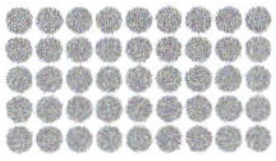

INTER-PLANT
INFRASTRUCTURAL CONNECTIONS

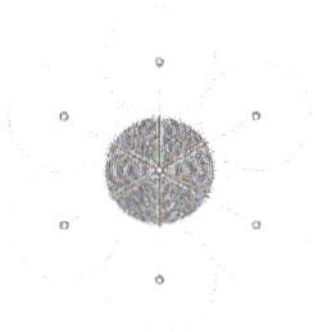

RIGHTS AND REGULATIONS

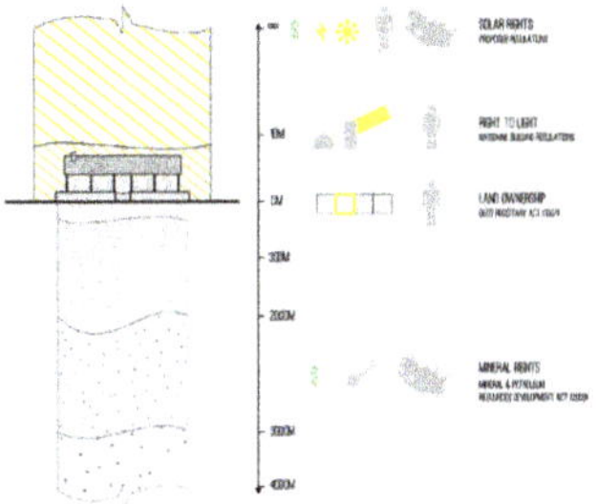

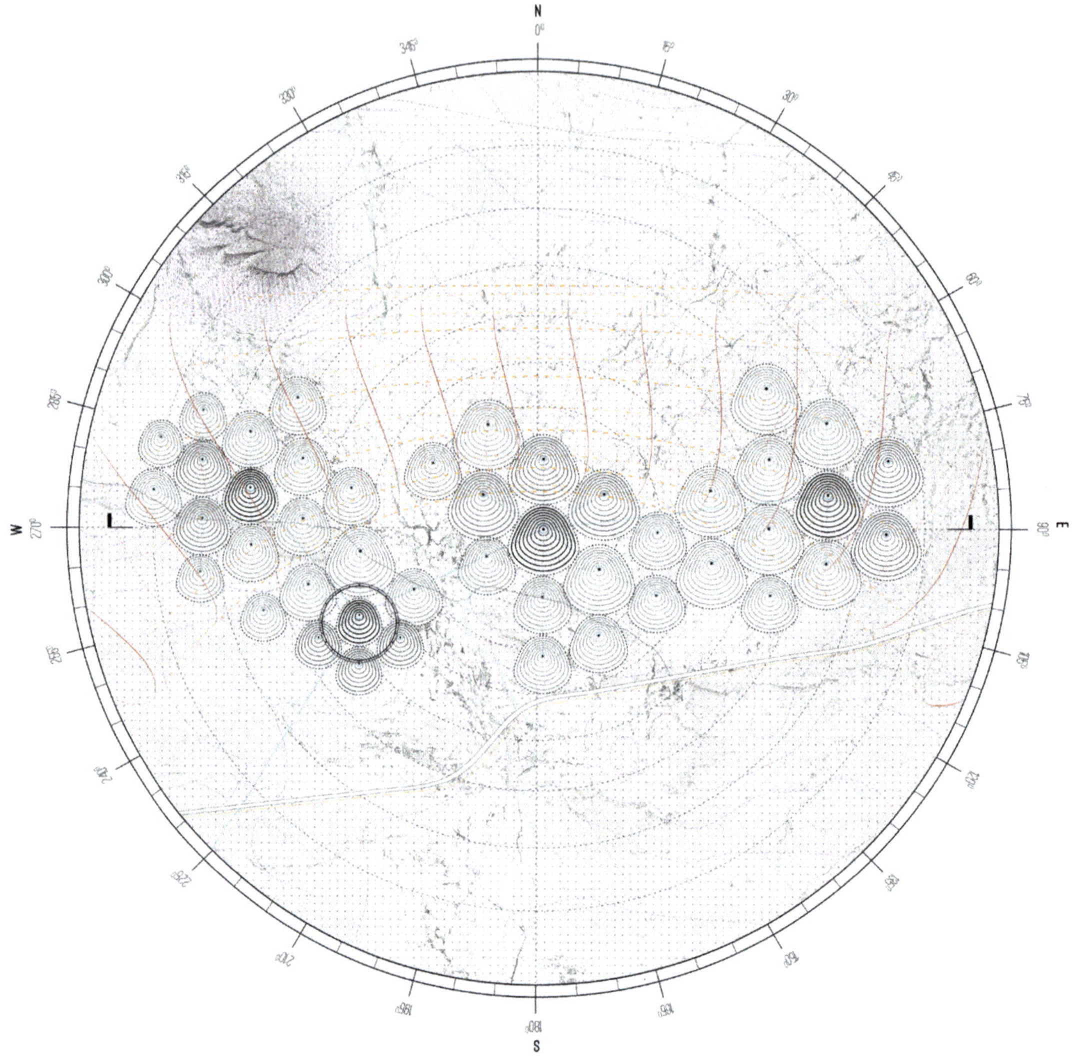
N
0°
15°
30°
45°
60°
75°
E
90°
S
180°
W
270°
285°
300°
315°
330°
345°

0
5KM

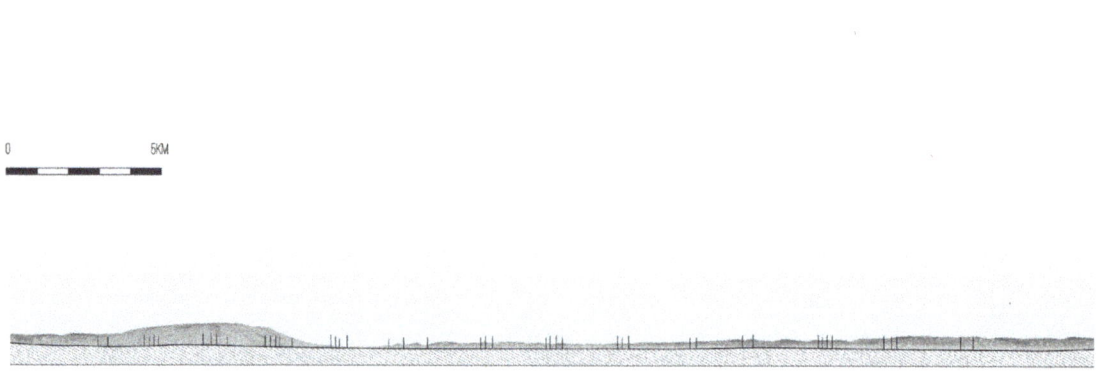

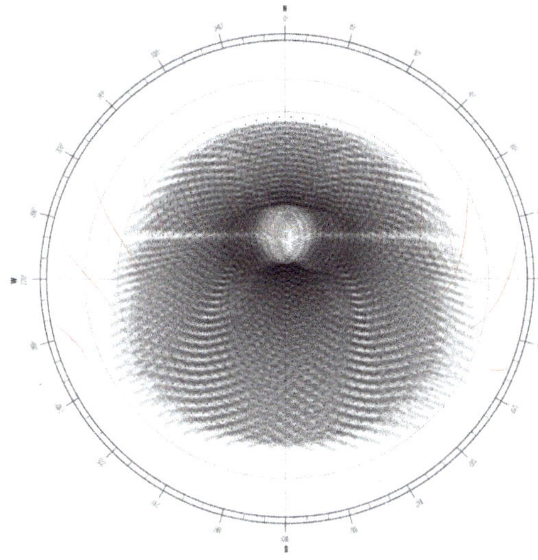

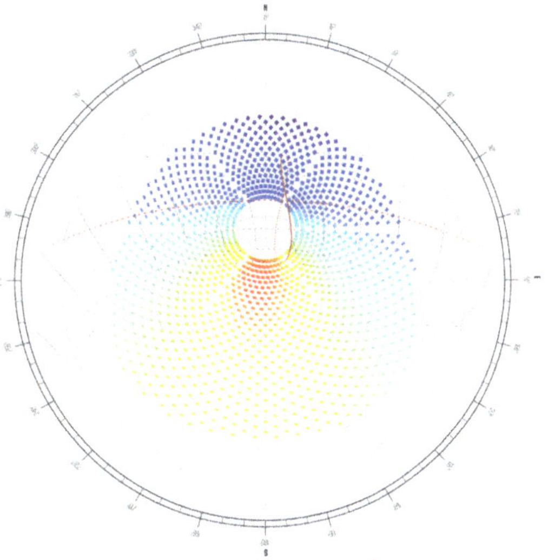

12:00

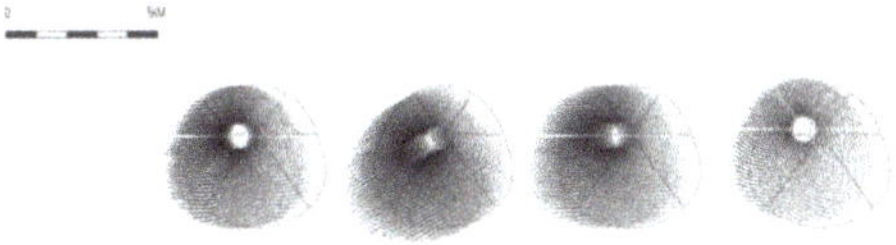

MARCH EQUINOX
JUNE EQUINOX
SEPTEMBER EQUINOX
DECEMBER EQUINOX

09:00
13:00
14:00
16:00

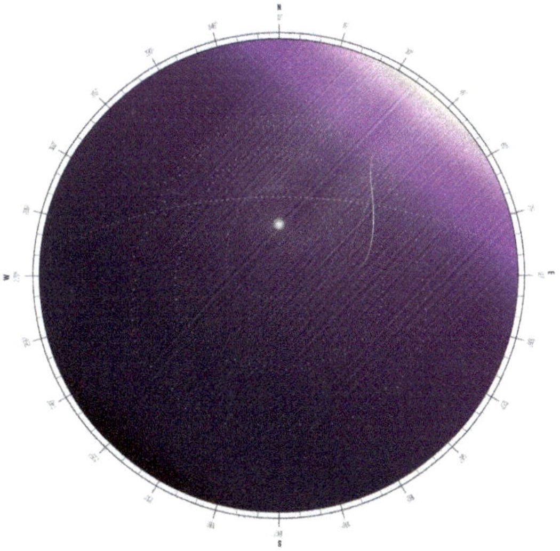

1200

NOMADIC CITY

ANDREW BAKER-FALKNER

A 2011 report published in Scientific American estimated that if you harnessed all of the sun's energy hitting the Earth's surface in just 88 minutes it would be equal to all of human society's energy consumption in an entire year.[01] While this would not be practically achievable it serves to highlight the massive potential energy source of the sun,, a source which, over the last 30 years, has become intertwined within all scales of human society from multinational research and development corporations to human scale individual installations. South Africa is within the top 3% of solar irradiance globally. A careful analysis and understanding of the environment and social structure led us to propose an energy masterplan of 45 industrial scale Concentrated Solar Power (CSP) plants to meet 55% the energy needs of the country. An important question became who and how these large infrastructural projects would be constructed. Due to the semi-arid and often remote areas required by CSP plants, the Nomadic Solar City houses the workers, rather than housing them in surrounding communities, which would place additional strain on already stretched local resources. The City is assembled prior to the realization of a CSP. The city is laid out over a regular grid, whose dimensions are dictated by heliostat components. These components become the building blocks of the City and help in creating a responsive and semi-temporary environment.

Note

(01) R. Naam, 'Smaller, cheaper, faster: Does Moore's law apply to solar cells?' *Scientific American Guest Blog*, March 16 2011, http://blogs.scientificamerican.com/guest-blog/smaller-cheaper-faster-does-moores-law-apply-to-solar-cells/

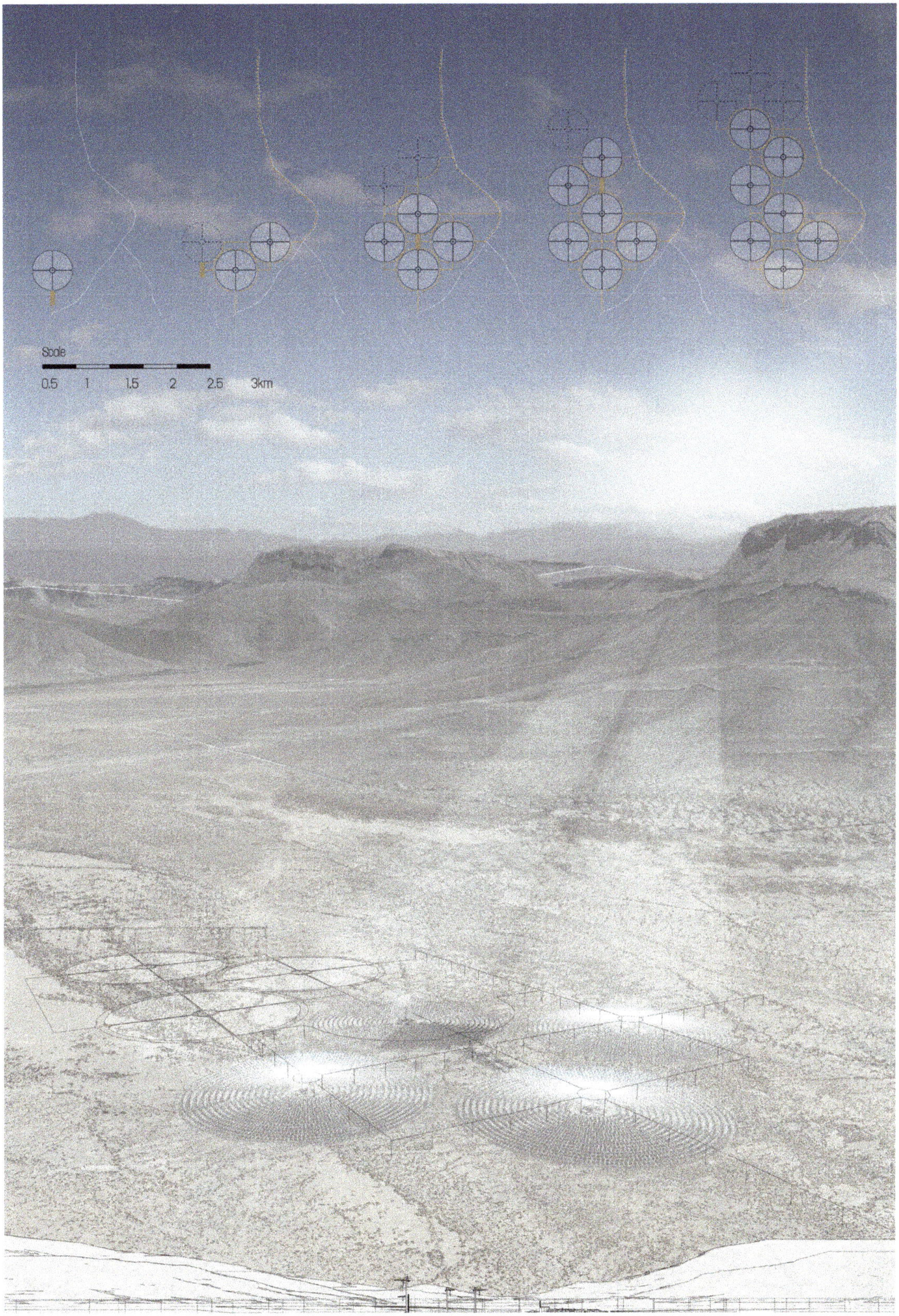
Scale
0.5
1
1.5
2
2.5
3km

IMAGE CAPTIONS

P207 Nomadic Solar City Masterplan

P208 Camdeboo Local Municipality, Environmental Analysis

P209 The Nomadic Solar City

The Nomadic Solar City moves with CSP plant construction sites. As each plant takes 4 years to construct, the city remains in place until the plant is complete and then moves to the next location. It is situated on the southern edge of the CSP plant to minimise solar gain from the north and also to face away from the central CSP receiver, which poses a potential hazard due the glare emitted.

P210 Nomadic Solar City, Concept Model

P211 Construction: Year 1, View From CSP plant

P212-213 Nomadic Solar City Water Strategy

P214-215 Nomadic Solar City, Sections

P216 Nomadic Solar City, Workers' Bar

P217 Nomadic Solar City, Living Module

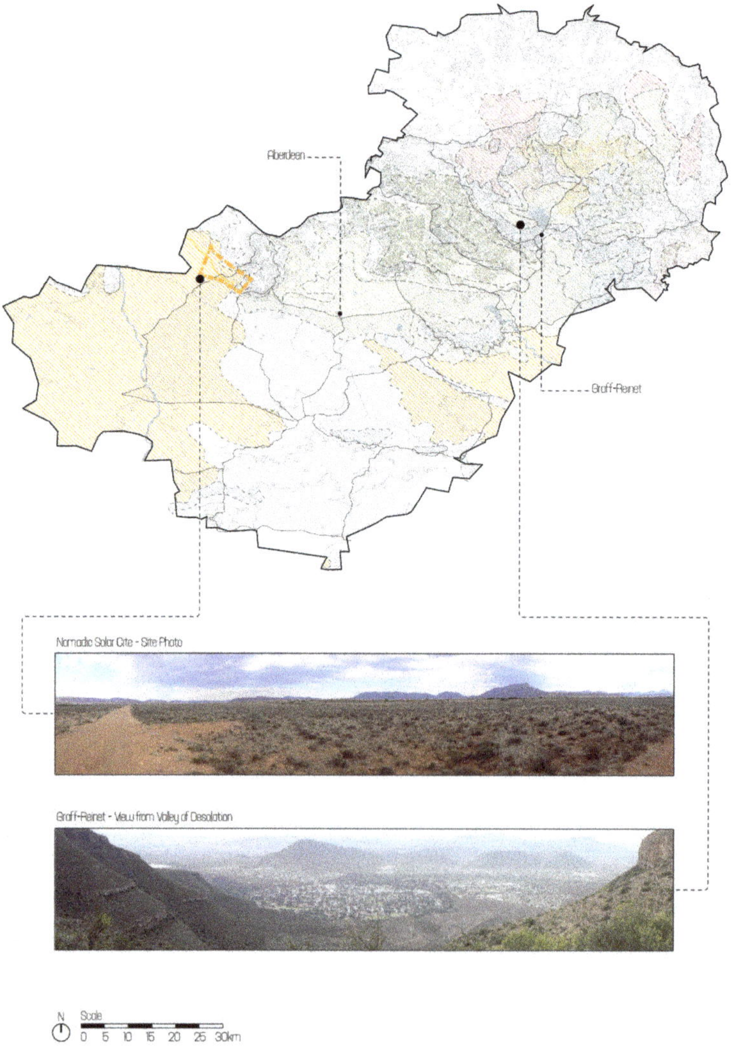

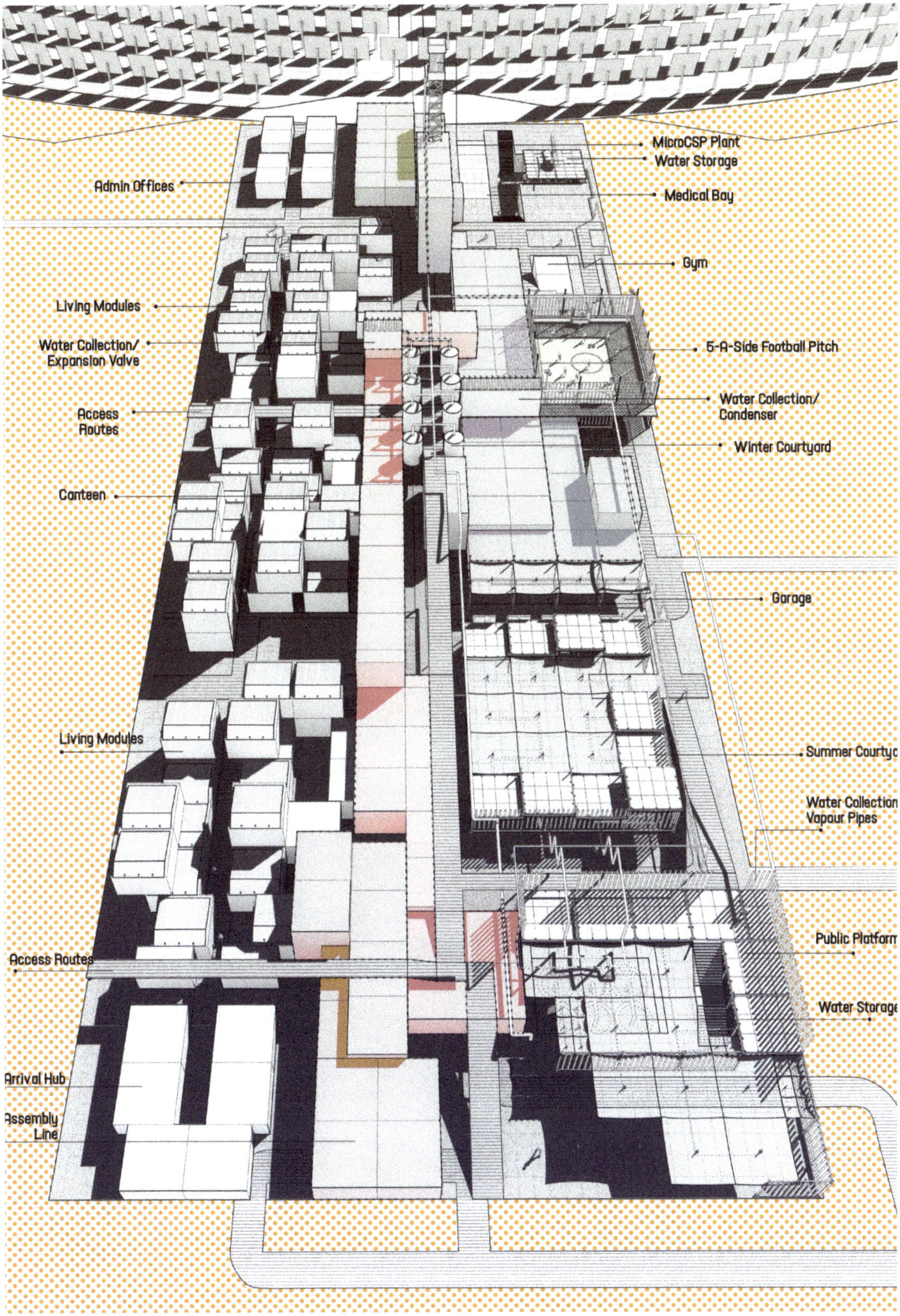

MicroCSP Plant
Water Storage
Medical Bay
Admin Offices
Gym
Living Modules
Water Collection/
Expansion Valve
5-A-Side Football Pitch
Water Collection/
Condenser
Access
Routes
Winter Courtyard
Canteen
Garage
Living Modules
Summer Courty
Water Collection
Vapour Pipes
Public Platform
Access Routes
Water Storage
Arrival Hub
Assembly
Line

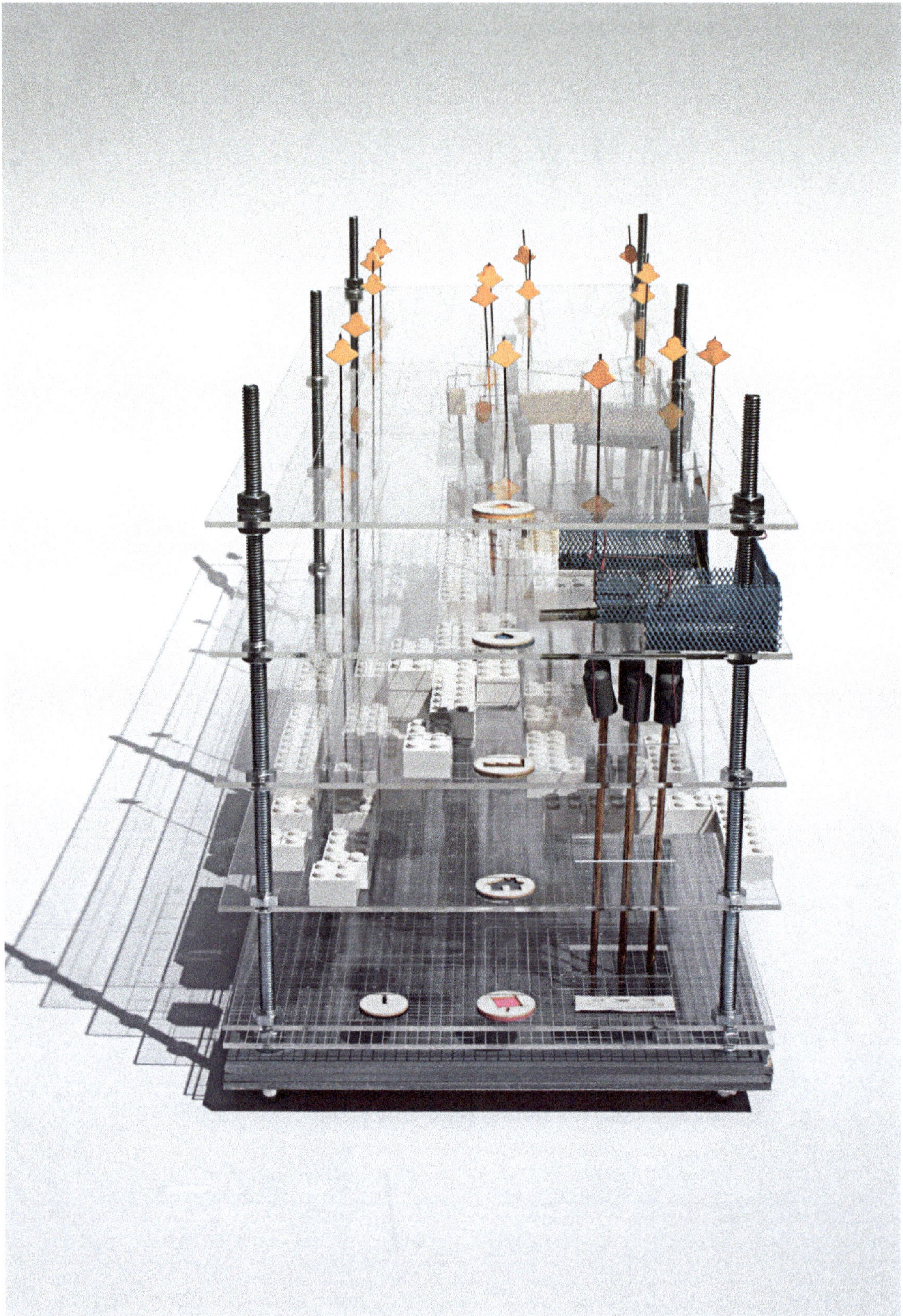

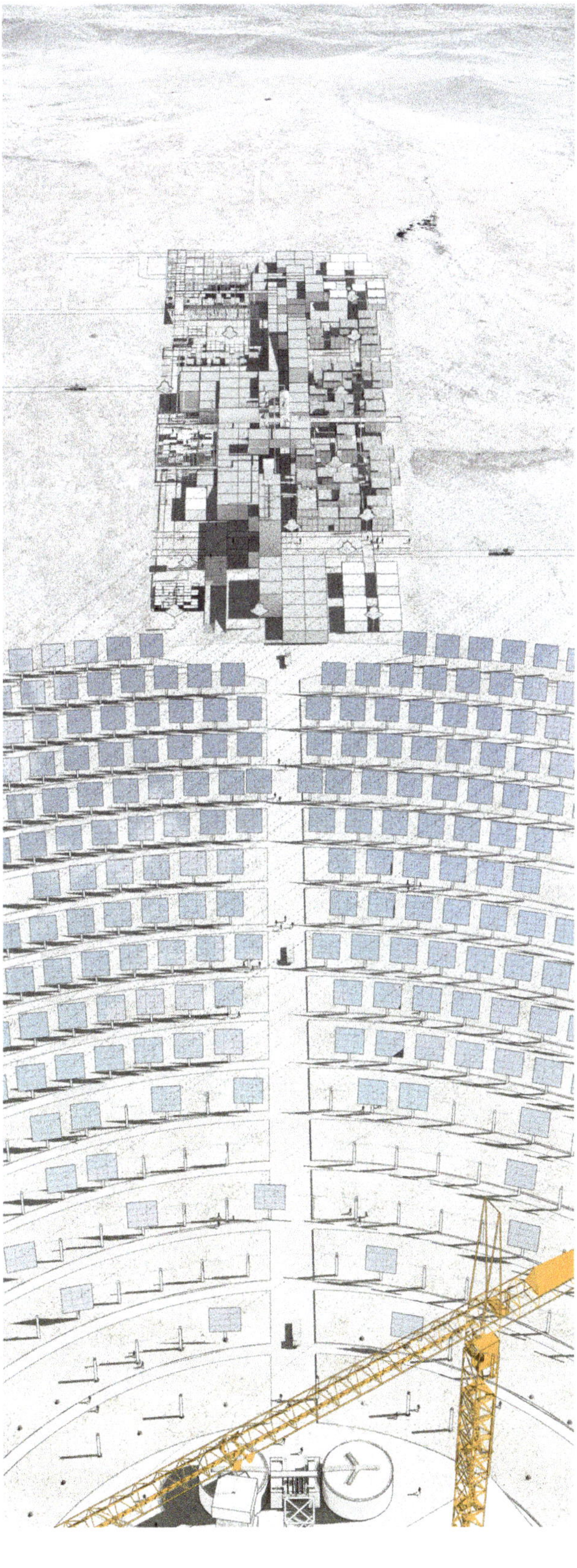

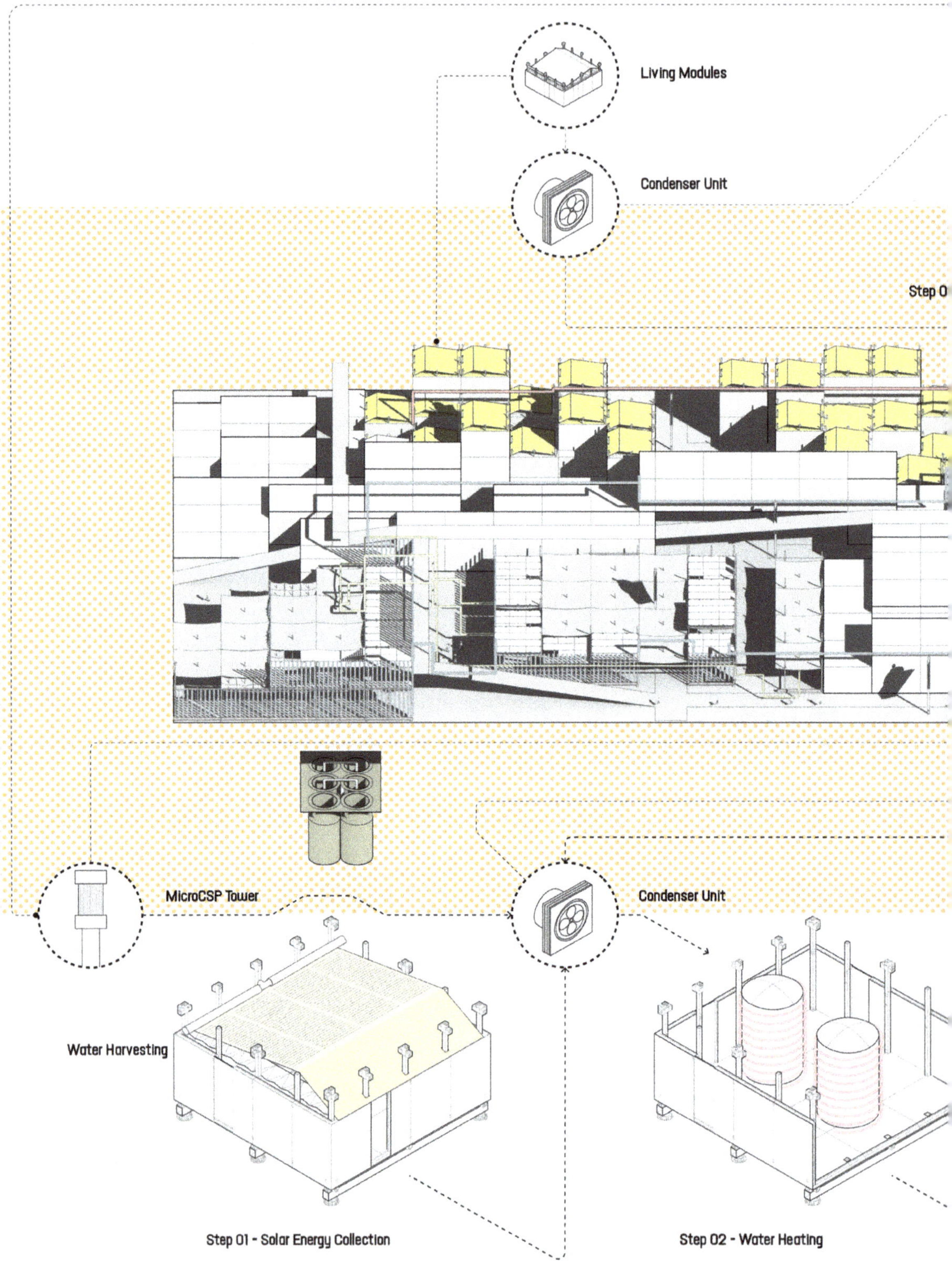
Living Modules
Condenser Unit
Step 0
MicroCSP Tower
Condenser Unit
Water Harvesting
Step 01 - Solar Energy Collection
Step 02 - Water Heating

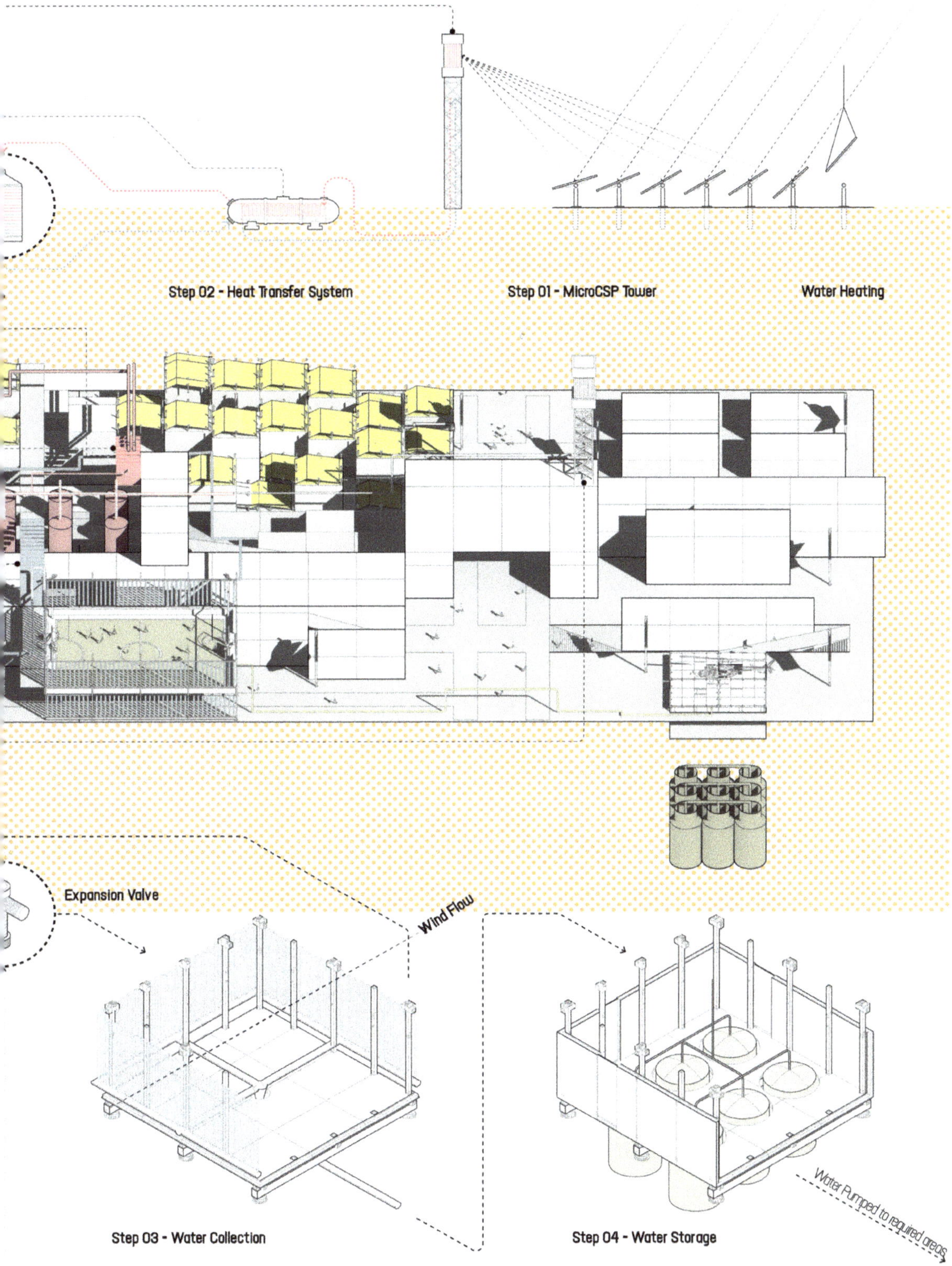

Step 02 - Heat Transfer System
Step 01 - MicroCSP Tower
Water Heating
Expansion Valve
Wind Flow
Step 03 - Water Collection
Step 04 - Water Storage
Water Pumped to required areas

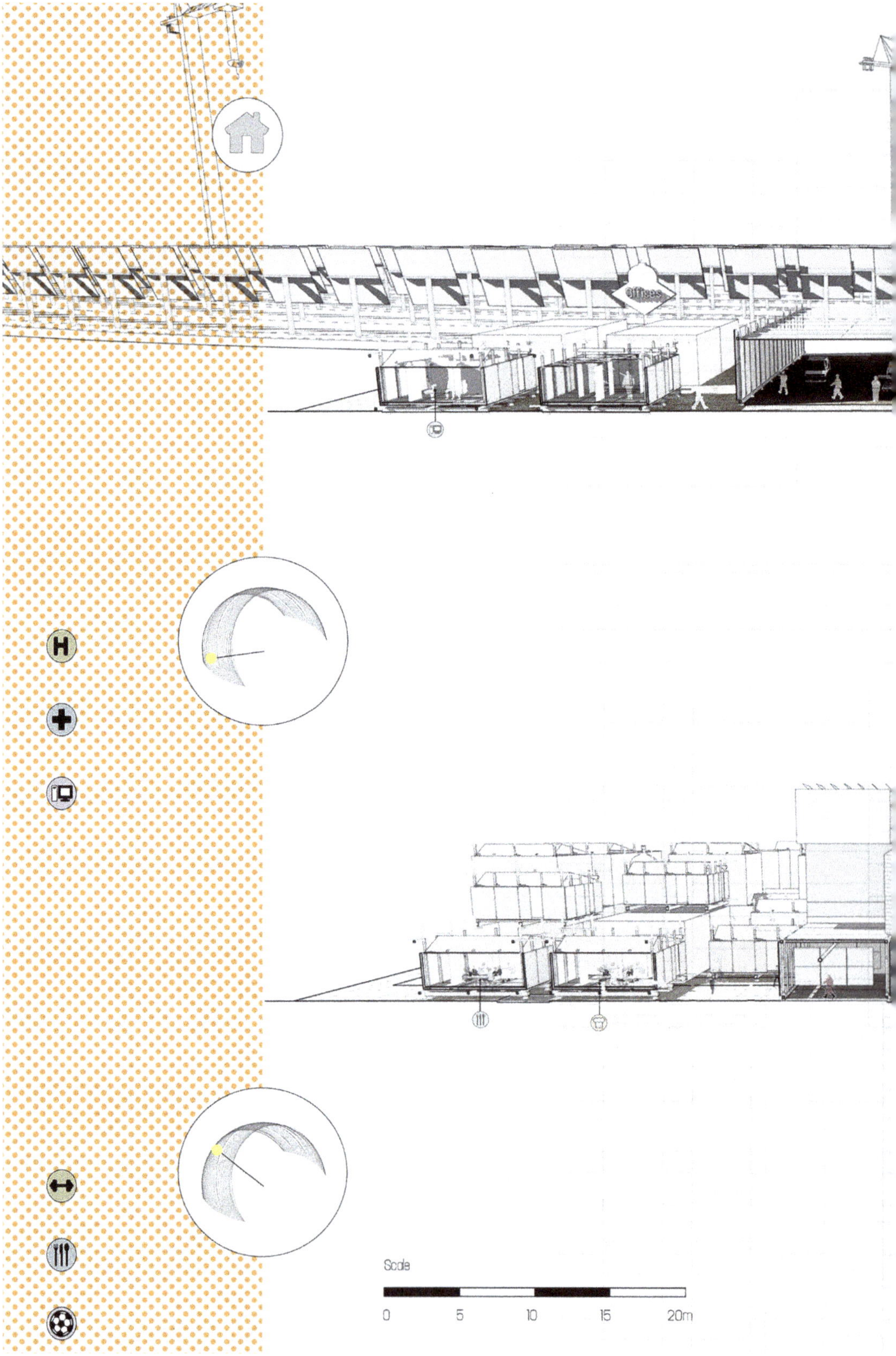
Offices
Scale
0
5
10
15
20m

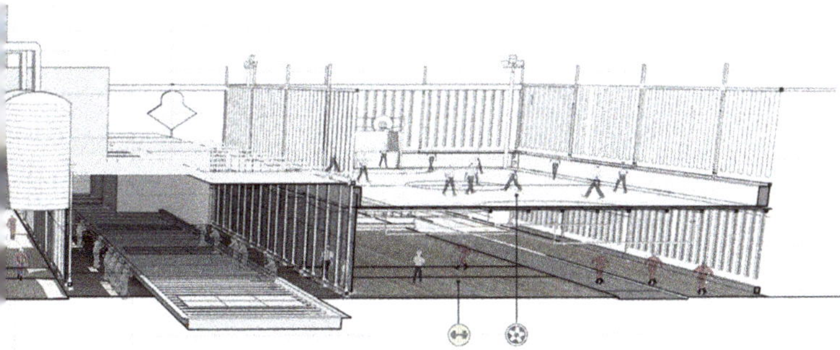

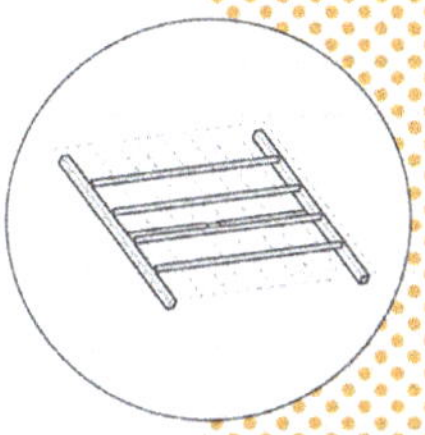

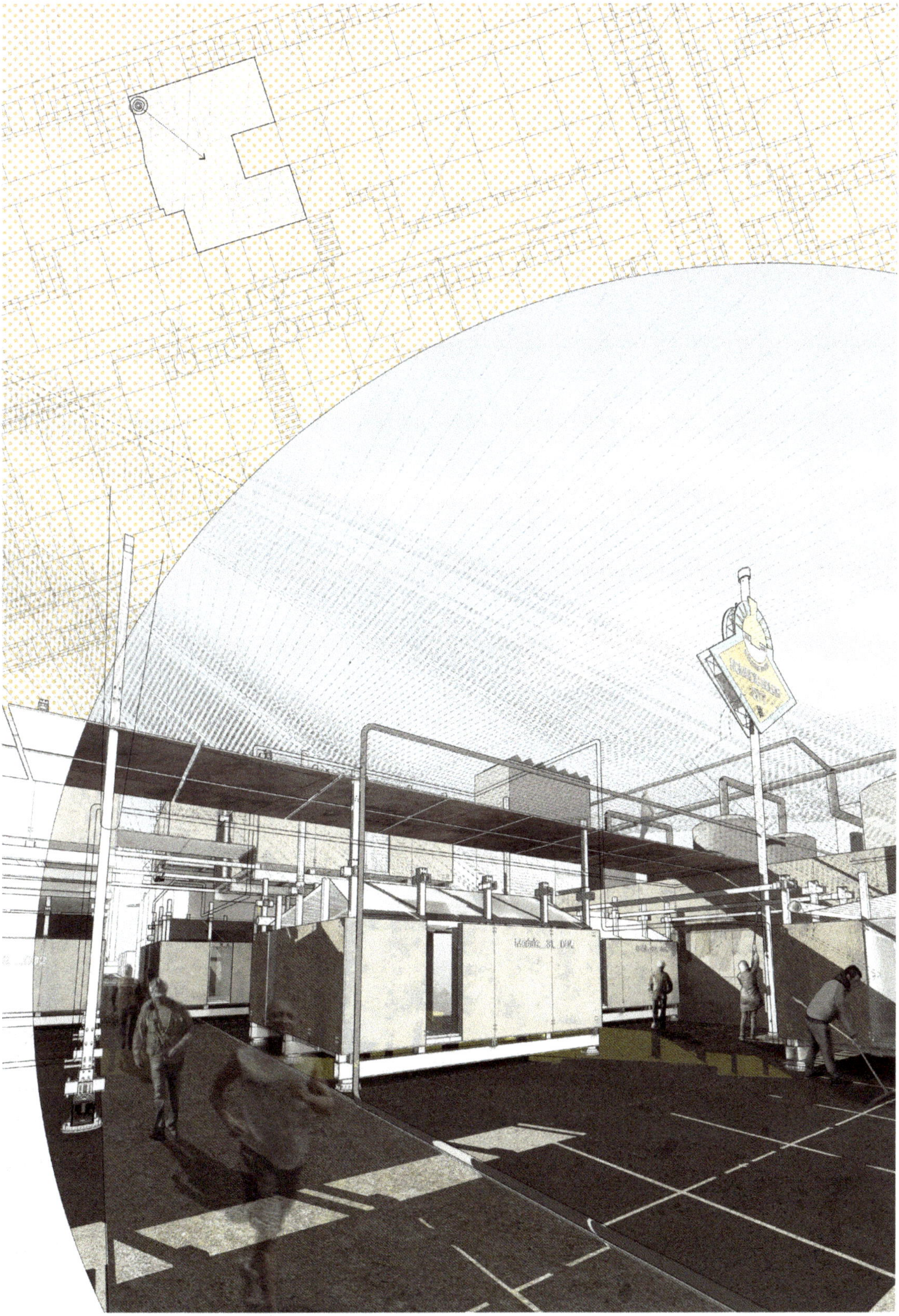

CAMDEBOO SOLAR ESTATE

JOHN COOK

In an age of dramatically rising population, diminishing fossil fuel resources and the alarming and all too visible consequences of climate change, the Camdeboo Solar Estate looks to provide an alternative energy strategy and agricultural resource for South Africa's burgeoning cities and rising energy demands. Located in a remote municipality in the semi-arid environment of the Karoo, the Camdeboo Local Municipality, its array of Concentrated Solar Power (CSP) plants are hybridised with the long practiced technique of terraced farming to enable a bountiful and economically prosperous wine industry. The masterplan arrangement, its axial pathways and internal orientations are calibrated to the positions of the celestial objects within our solar system at the time of opening in 2050. The proposal overlays agriculture, energy production and celestial movements and embeds a new economy of solar tourism within the region. The proposal will be experienced as a new form of urban / territorial restructuring at both landscape and building scale.

The Karoo welcomes you to ...
The Camdeboo Solar Estate

P219 The Camdeboo Solar Estate Poster

P221 Camdeboo Local Municipality Site Overview

The location of the solar estate was decided by analysing the Camdeboo's landscape through a number of factors: solar irradiance, physical terrain, environmental stability and infrastructural connectivity.

P222 Site, Scale + South Africa's Cosmic Context

The Camdeboo's precise location is explored a number of scales - from the comprehensible site scale of meters, to the position of our solar system within the cosmos, measured in parsecs. The relationship between Earth's rotational axis and its orbital route around the sun (illustrated here) dictates all life on earth - our sunlight hours, time-zones, tides, all organic growth and now, also, a potentially bountiful energy source.

P223 Solar Calendar [2016 - 2099]

This calendar illustrates the cyclic timings of significant solar events at the site's precise location from 2016 until 2099. The ever differing timings are due to the earth's slightly elliptical orbit around the sun, and the subsequent imprecision of our typical calendar year and measurement of time.

P224 Celestial Alignment of the Masterplan

The position of the planets/bodies within our solar system was simulated for particular solar events - such as the time of estate's opening on the solstice, 2050. The scale, orbits and orientation of the planets at this time would determine and embed themselves throughout the proposal - from the major axis and arrangement of the CSP masterplan, the scale of each plant, to the micro alignments of the walkways, buildings, and their astronomical instruments.

P225 CSP Plant Jupiter Core Building Plan

Ground Floor: Energy Processing

Opening in 2050, the Camdeboo Solar Estate begins operation as an array of CSP plants, harvesting the high levels of received solar irradiance to alleviate South Africa's rising energy demands and decrease the country's dependence on coal. At the same time, the installed energy and water infrastructure initiates a viable and fruitful wine estate and revives the once near extinct wine making industry of the Karoo. Energy processing and wine processing occurs simultaneously within the core building at the base of the CSP tower.

P226-227 Time-lapse Simulations

Operation [Summer Solstice, December 21st 2050]

Decomposition [Winter Equinox, June 21st 2200+]

After the 60 year life expectancy of CSP apparatus expires, and the predicted advances into more improved renewable energy sources are achieved, the energy infrastructure is de-constructed, and the estate remains operating as a series of vineyards. By 2200, the plant's has become obsolete, its building materials have deteriorated and the process of decomposition commences. By 2500 the progressive sinking of the inner core walls begin to reveal the astronomical instruments concealed within.

P228 [left] CSP Plant Jupiter: from Vineyard and Solar Terrace, Year 2050

View of the solar array and wine estate in operation during the year of opening, 2050.

[right] CSP Plant Jupiter: Plant Overview Mid Operation, Year 2050

Aerial view of the CSP Plant 'Jupiter', aligned with the corresponding celestial body's position at the summer solstice on December 21, 2050.

P229 CSP Plant Jupiter: Plant Ruins: Year 3000

As the building's ruins reconfigure and settle through decomosition, the hidden alignments of the ancient South African landscape of the Adam's Calendar are unveiled. In the long distant future, the Camdeboo Solar Estate's now ancient ruinous landscape is rediscovered. The mysterious constellations and orientations of its arrangement are again reinterpreted, to reveal the intricate relationships between its lost functions of energy, agriculture, and our place within the cosmos.

P230-231 The Camdeboo Solar Estate Tourist Guide

The proposal embeds a new economy of solar tourism within the region, allowing agriculture, energy production and celestial movements to be viewed and experienced whilst enhancing Camdeboo's existing tourist economy.

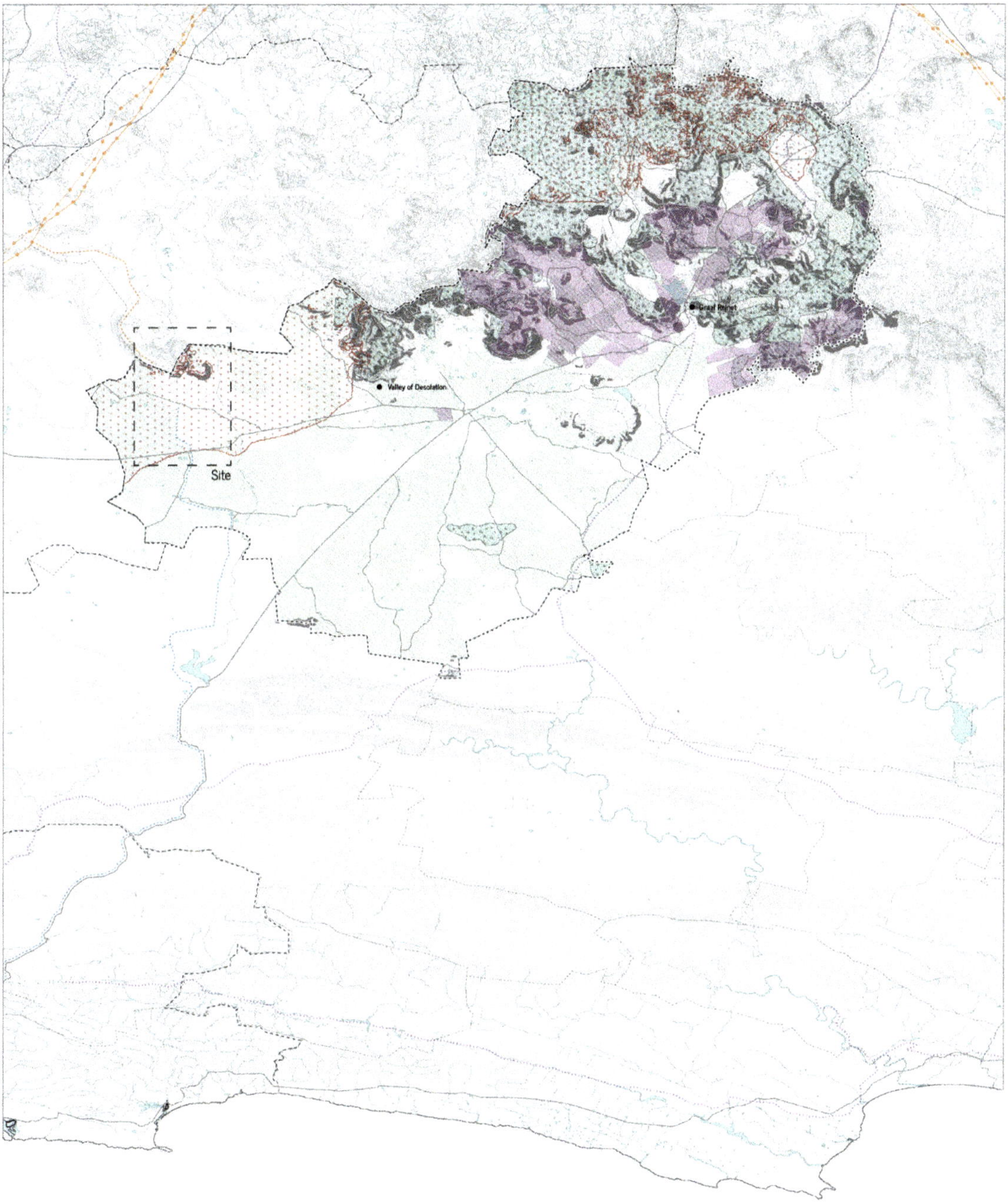

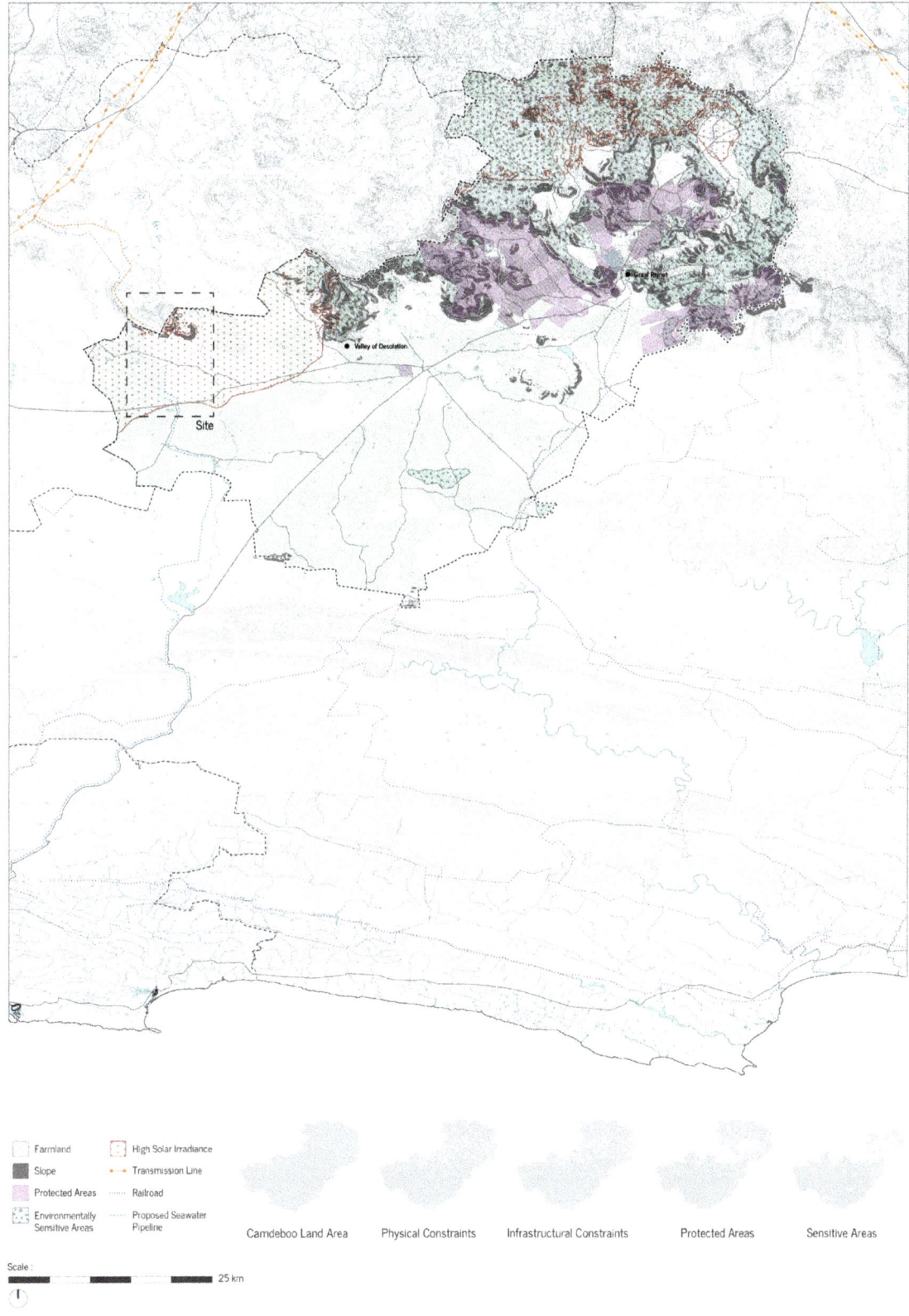

Camdeboo Land Area | Physical Constraints | Infrastructural Constraints | Protected Areas | Sensitive Areas

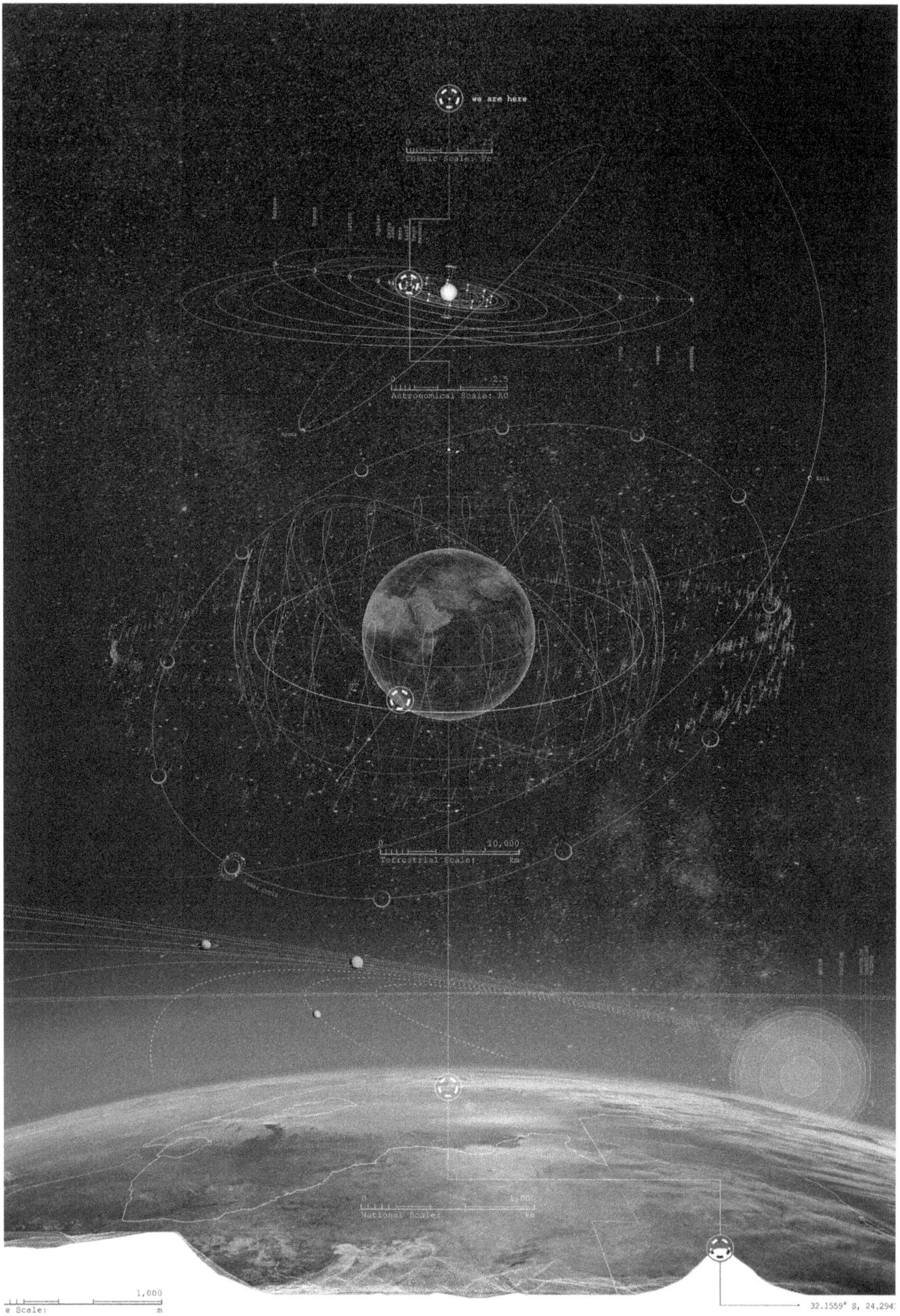
we are here
0
2.5
Astronomical Scale: AU
0
10,000
Terrestrial Scale: km
0
1,000
National Scale: km
1,000
e Scale: m
32.1559° S, 24.294

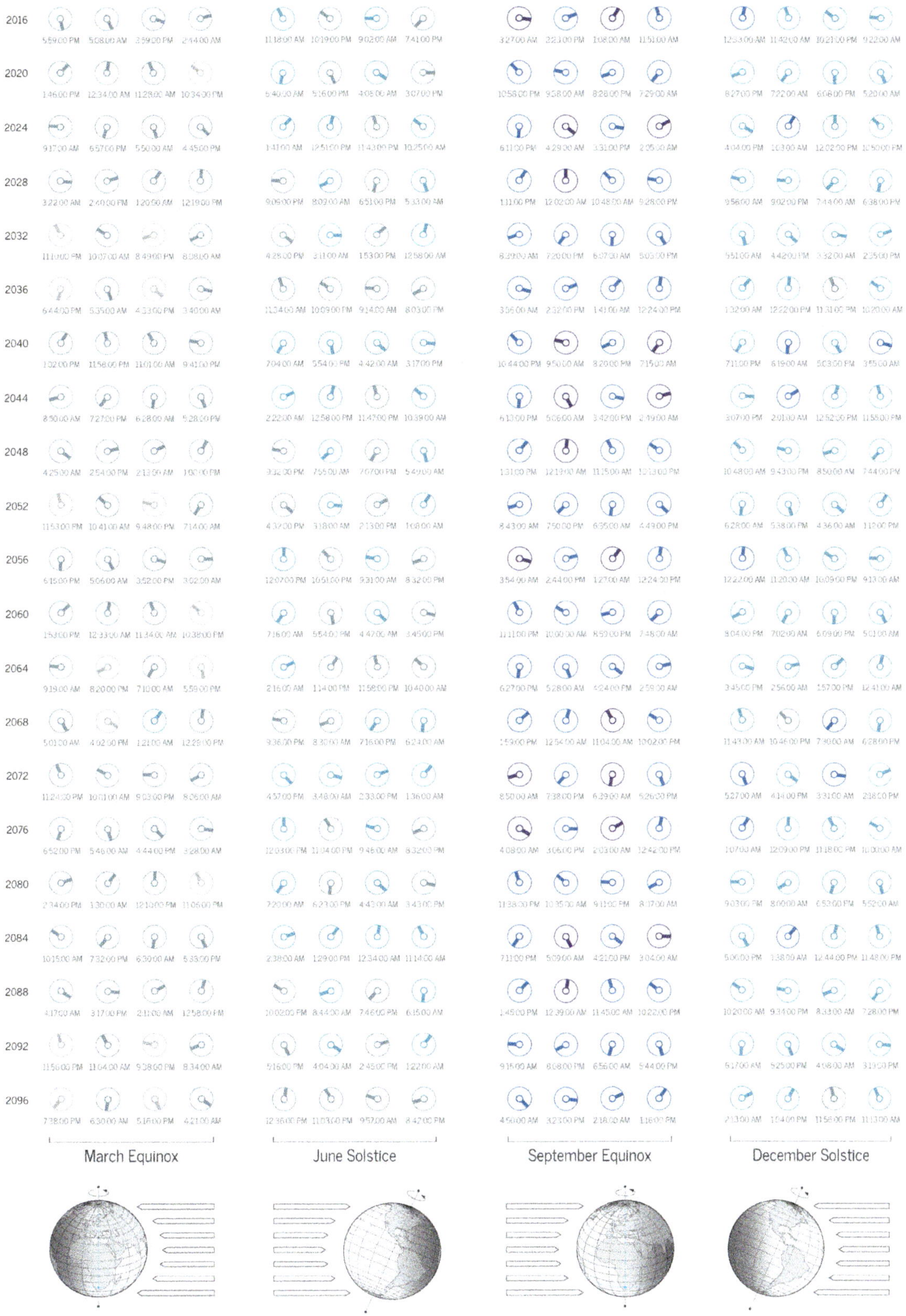
2016
2020
2024
2028
2032
2036
2040
2044
2048
2052
2056
2060
2064
2068
2072
2076
2080
2084
2088
2092
2096
March Equinox
June Solstice
September Equinox
December Solstice

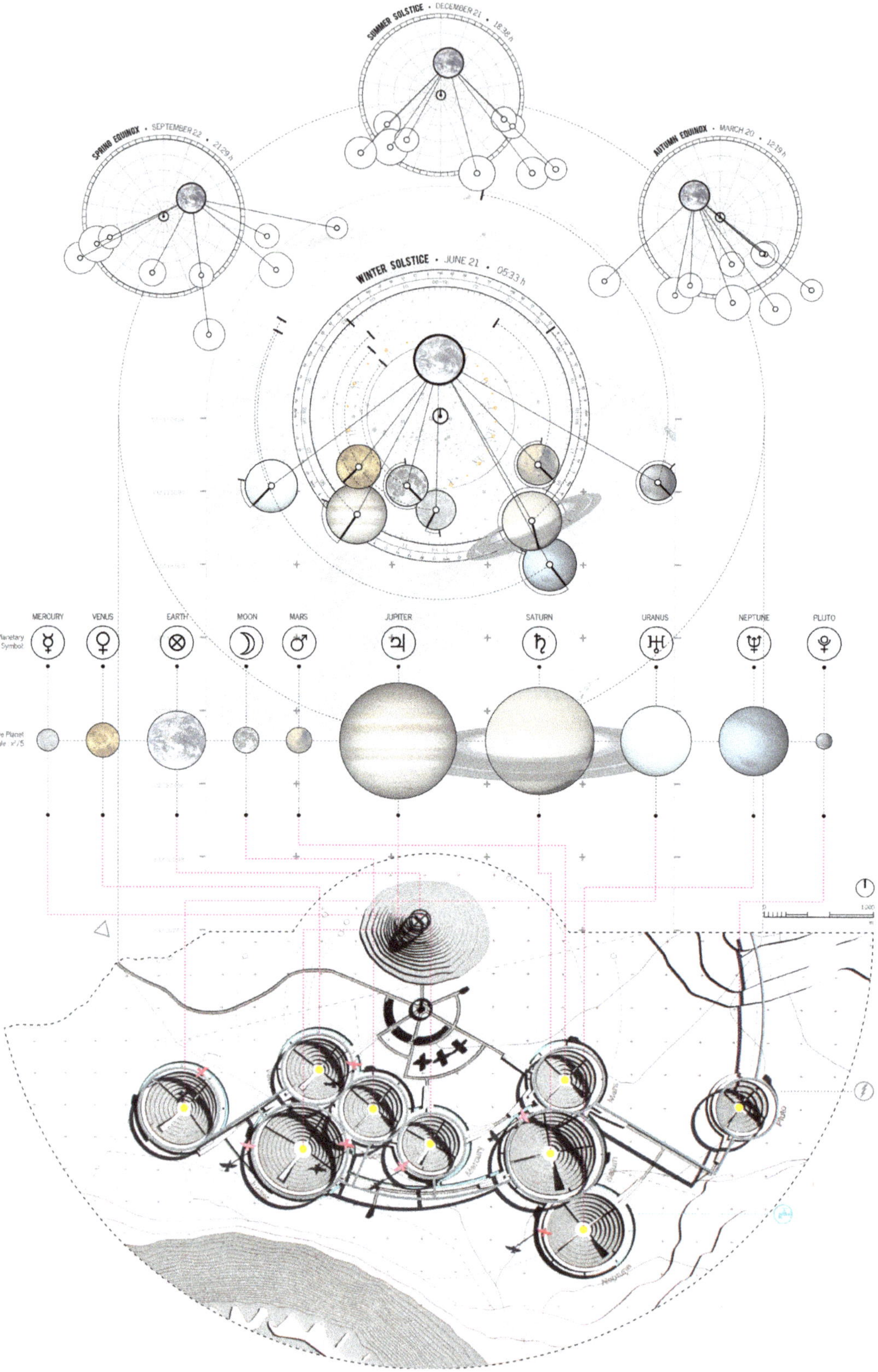

SUMMER SOLSTICE • DECEMBER 21 • 18:38 h
SPRING EQUINOX • SEPTEMBER 22 • 21:29 h
AUTUMN EQUINOX • MARCH 20 • 12:19 h
WINTER SOLSTICE • JUNE 21 • 05:33 h
MERCURY
VENUS
EARTH
MOON
MARS
JUPITER
SATURN
URANUS
NEPTUNE
PLUTO
Planetary Symbol
Relative Planet Scale

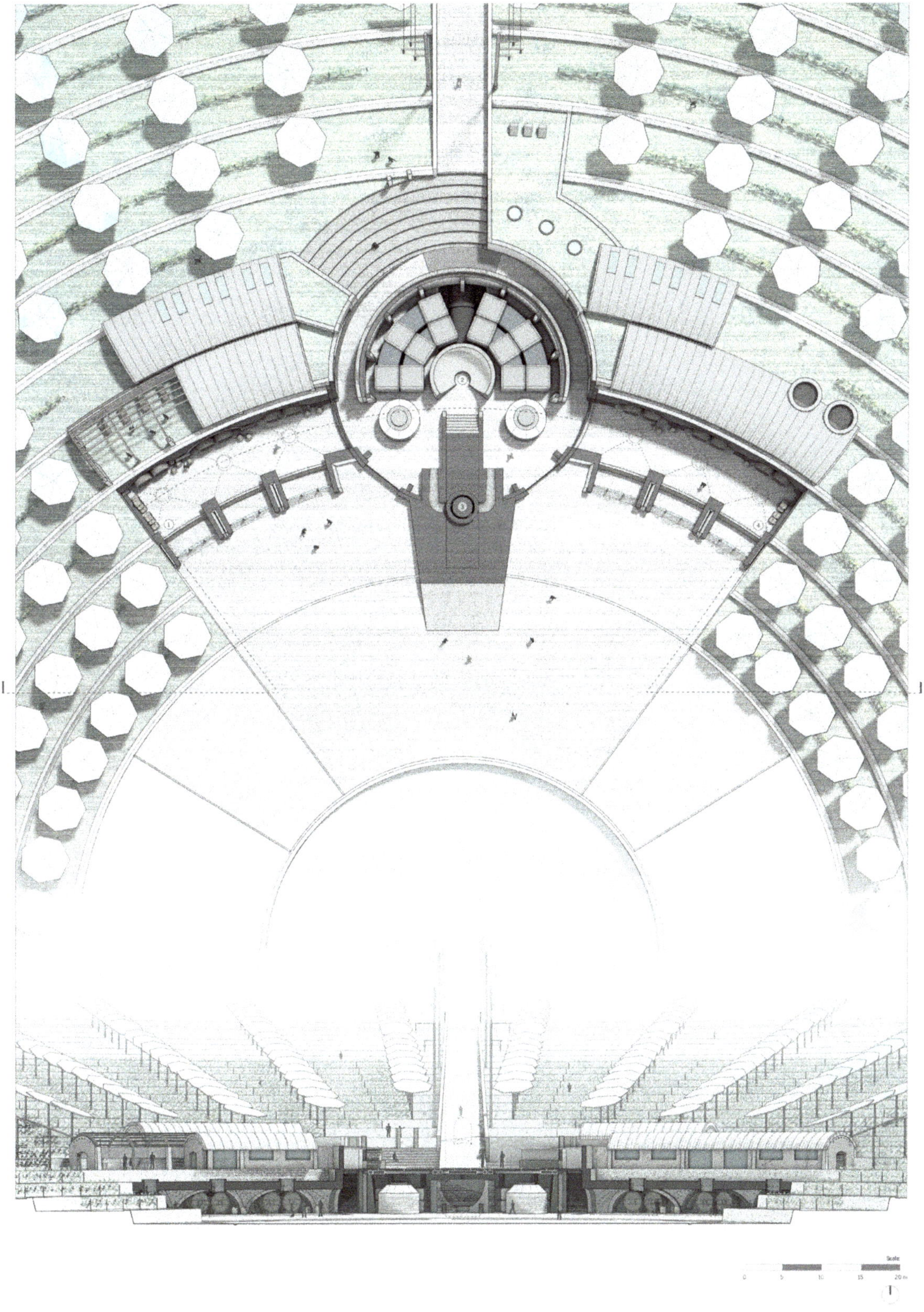
Scale
0
5
10
15
20 m

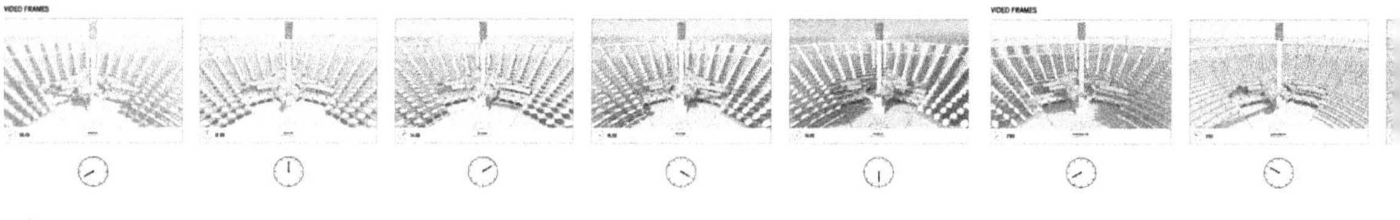
VIDEO FRAMES
VIDEO FRAMES

VIDEO FRAMES

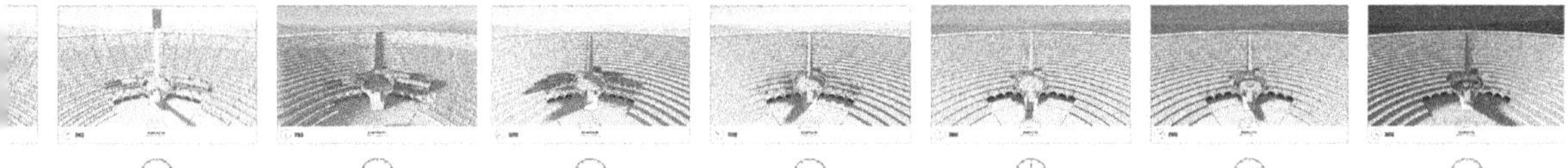

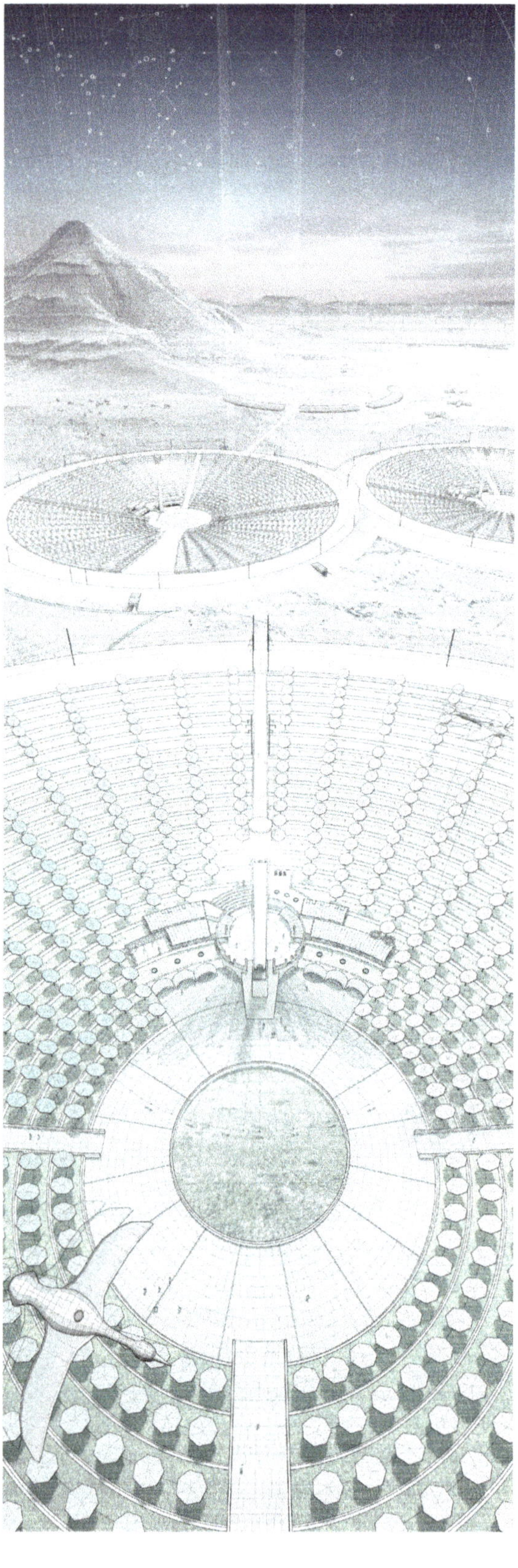

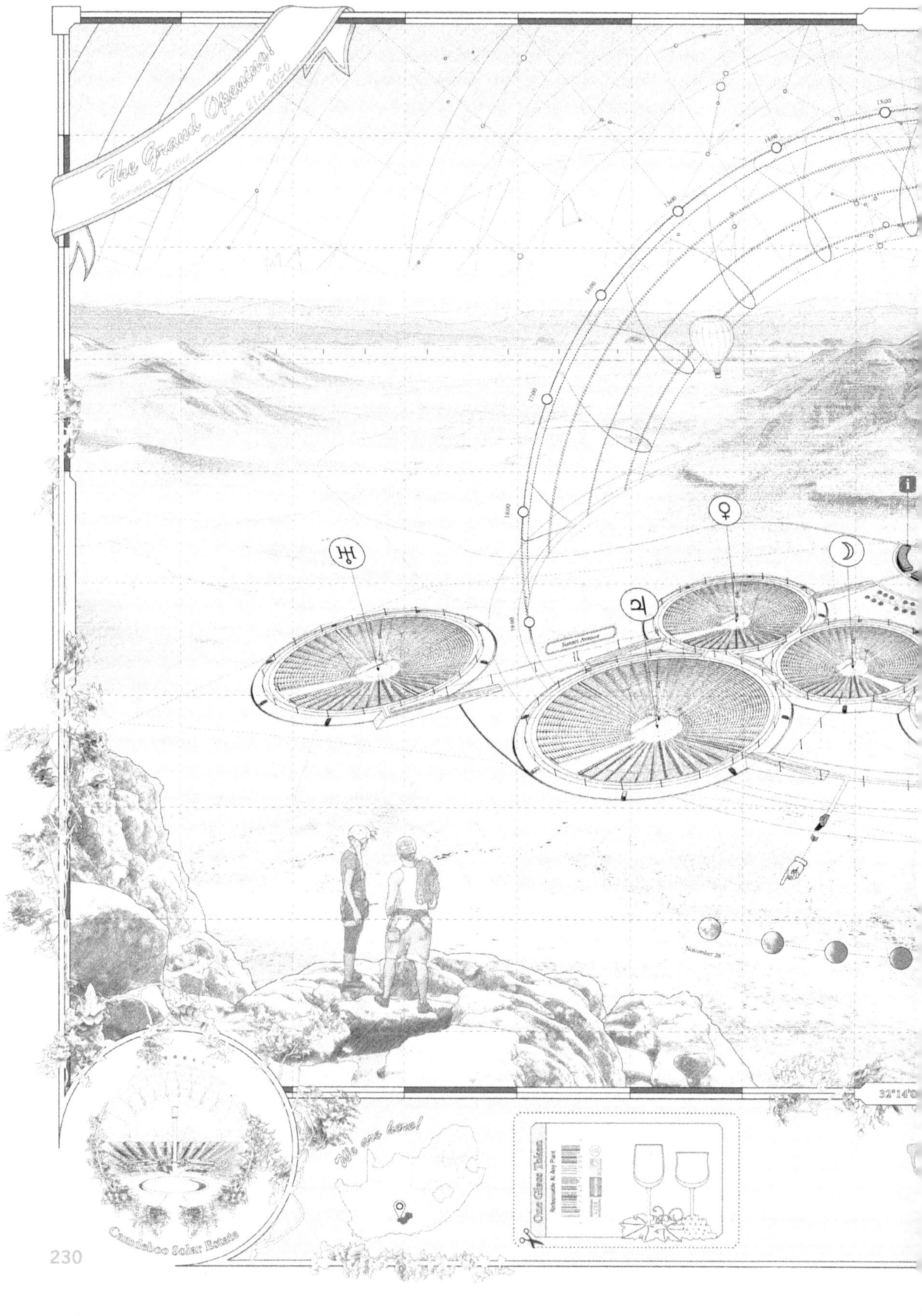

The Grand Opening!
Summer Solstice, December 21st 2050
1300
1400
1500
1600
1700
1800
1900
Sunset Avenue
November 28
We are here!
One Glass Token
Redeemable At Any Point
Camdeboo Solar Estate

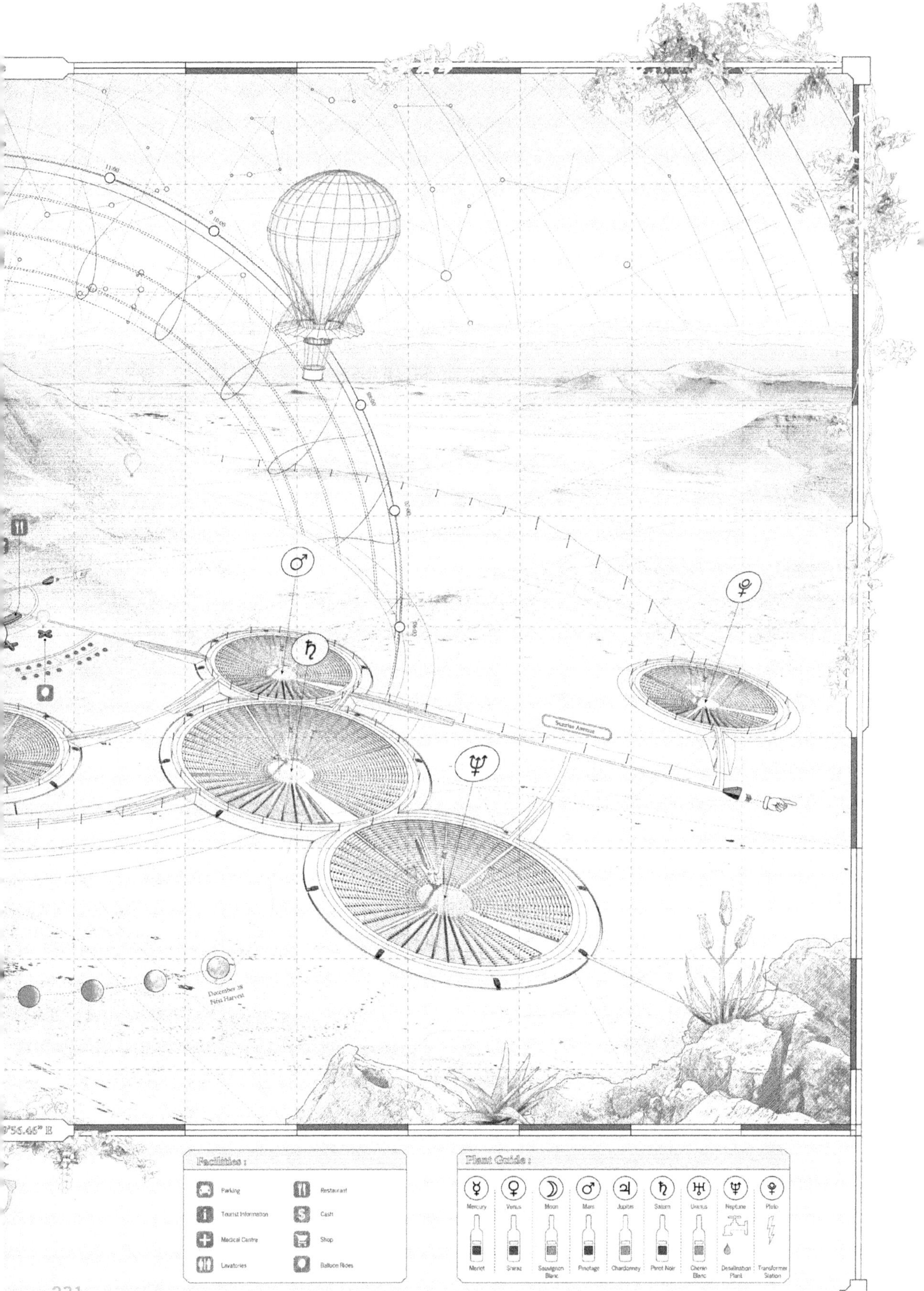
Sunrise Avenue
December 28
First Harvest
°56.46" E
Facilities :
Parking
Tourist Information
Medical Centre
Lavatories
Restaurant
Cash
Shop
Balloon Rides
Plant Guide :
Mercury
Venus
Moon
Mars
Jupiter
Saturn
Uranus
Neptune
Pluto
Merlot
Shiraz
Sauvignon Blanc
Pinotage
Chardonnay
Pinot Noir
Chenin Blanc
Desalination Plant
Transformer Station

WASTE INTEGRATION INITIATIVE

JARED BARON
SOPHIE FULLER
ALICE THOMPSON

This project attempted to provide an answer to the question of whether the Camdeboo Local Municipality's Waste Management System could be re-designed as a framework for socioeconomic, political and environmental transformation.

Waste is intrinsic to our energy economies, and is a tangible product of the times that we live, an era that has been coined the Anthropocene. Our rate of production and consumption has become evidently unsustainable, landfill is no longer a viable waste management technique and there is the need to develop alternative and innovative solutions to our growing waste issues. The Camdeboo, as a small rural municipality in central South Africa, struggles to appropriately manage its solid waste.

The current waste management system of Graaff Reinet, its neighbouring townships and surrounding areas were modelled and analysed to provide a foundation for our proposal. As a comparative tool, exemplary precedents of bottom up approaches to waste management practices around the globe (Cairo, Curitiba and Japan) were analysed. The culmination of this work resulted in a proposal that implemented the simple idea of separation at source and encouraged inhabitants to take ownership of their waste by providing a destination for waste that simultaneously served as a community space. This space was designed to bridge the conceptual, ideological and physical divisions that are the legacy of apartheid and to instigate integration of the divided local community.

By projecting peoples' existing movements and mapping them against the town's waste movements, our proposal realigned these to encourage points of contact and engage the inhabitants with both the issue of waste and the social and cultural rifts. To create an effective energy economy, considering the sparse population density of the region, the project required the cooperation of neighbouring municipalities to create a waste share. This empowered each municipality to champion a certain waste type to develop into a commodity.

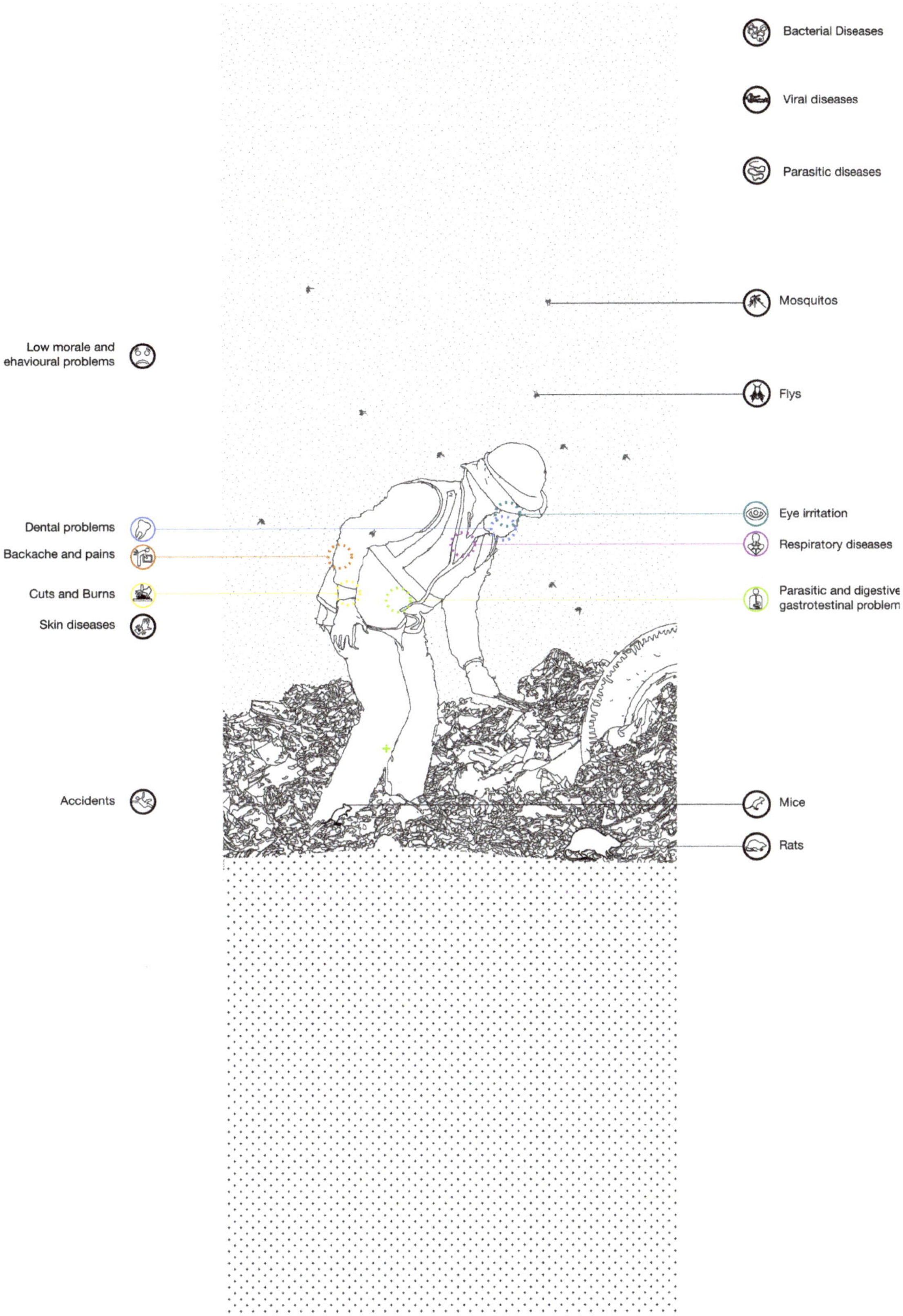
Bacterial Diseases
Viral diseases
Parasitic diseases
Mosquitos
Low morale and
ehavioural problems
Flys
Dental problems
Backache and pains
Cuts and Burns
Skin diseases
Eye irritation
Respiratory diseases
Parasitic and digestive
gastrotestinal problem
Accidents
Mice
Rats

P233 Landfill Health Issues

P235 The Camdeboo Local Municipality

The Camdeboo, highlighting the levels of unemployment (surplus / waste employment) of the main towns and townships. Many of these are similar to the South African average of 30% unemployment. Also marked on this map as red triangles are the three existing landfill sites as well as our proposed site in the north

P236 Graaff Reinet spatial segregation

This plan of Graaf Reinet shows the separation between the three racial groups' residential areas laid out during the apartheid era clearly defined

P237 [left] Waste volumes and types

The volumes of waste produced by households of different incomes is shown in the bar chart. Percentage of population of each income group is shown along the x-axis and waste volumes produced along the y. The Pie chart further illustrates the waste contribution of each income group as a percentage of total waste produced in the Graaff Reinet area. This information was extrapolated from waste volume and types available on the online census and using data provided by the Health Secretary of the Camdeboo Local Municipality

[top right] Existing waste flows in Camdeboo

Currently there is a household waste collection system which services 99% of Graaff Reinet. However this system is insufficient and poorly services, with much of the domestic rubbish lying strewn across the towns, particularly in the townships, and many of the public skips overflowing, with waste uncollected.

[bottom right] Proposed waste flows in Camdeboo

This proposed waste management scheme for Graaff Reinet combines aspects of three global precedents () with the existing scheme to create a more efficient waste management scheme. This idea combines a collection system but also encourages people to separate at source (with an incentive) and deliver their waste to a waste collection point where all the waste from the towns of Graaff Reinet, Aberdeen and Nieu Bethesda are sorted into organic and recyclable components. Reusable products are then be sold on and some items of waste are recycled at this point, utilising the craft and construction skills of the local people

P238 [top] Graaff Reinet residents

[bottom left] Gerhard's daily interactions with waste

[bottom right] Abel's daily interactions with waste

P239 Proposed interactions through waste

The tracking of proposed waste movement was overlaid on the movement patterns of selected local people, to highlight where these two systems intersected. This provided an understanding of how social relations could be intensified around the management of waste

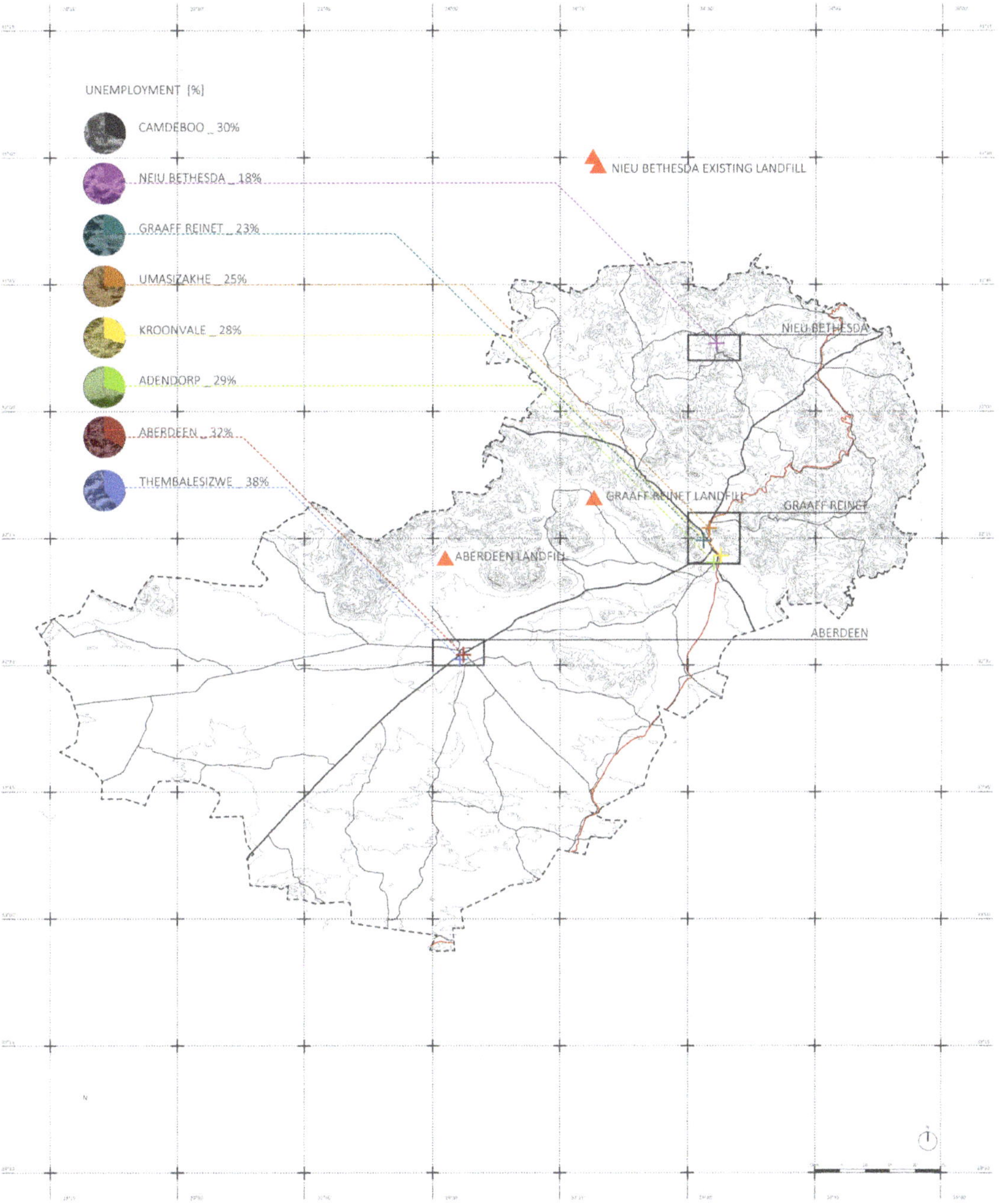
UNEMPLOYMENT [%]
CAMDEBOO _ 30%
NEIU BETHESDA _ 18%
GRAAFF REINET _ 23%
UMASIZAKHE _ 25%
KROONVALE _ 28%
ADENDORP _ 29%
ABERDEEN _ 32%
THEMBALESIZWE _ 38%
NIEU BETHESDA EXISTING LANDFILL
NIEU BETHESDA
GRAAFF REINET LANDFILL
GRAAFF REINET
ABERDEEN LANDFILL
ABERDEEN

EASTERN CAPE
SOUTH AFRICA
CAMDEBOO
EASTERN CAPE
GRAAFF REINET
CAMDEBOO

GRAAFF REINET
KROONVALE
GRAAFF REINET
KEY
ROADS
CONTOURS
RAILWAY
RIVERS
TOWN CENTRE
BLACK TOWNSHIP
COLOURED TOWNSHIP
CHURCH
INDUSTRIAL
SCHOOL
N

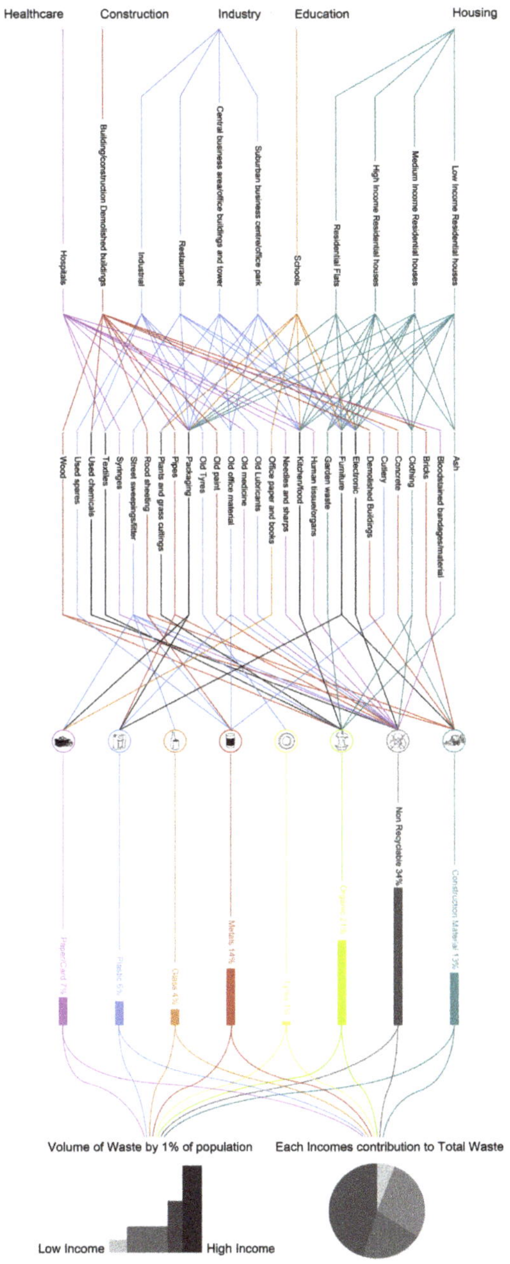

Healthcare
Construction
Industry
Education
Housing
Hospitals
Building/construction Demolished buildings
Industrial
Restaurants
Central business area/office buildings and tower
Suburban business centre/office park
Schools
Residential Flats
High Income Residential houses
Medium Income Residential houses
Low Income Residential houses
Wood
Used spares
Used chemicals
Textiles
Syringes
Street sweepings/litter
Roof sheeting
Plants and grass cuttings
Pipes
Packaging
Old Tyres
Old paint
Old office material
Old medicine
Old Lubricants
Office paper and books
Needles and sharps
Kitchen/food
Human tissue/organs
Garden waste
Furniture
Electronic
Demolished Buildings
Cutlery
Concrete
Clothing
Bricks
Bloodstained bandages/material
Ash
Paper/Card 7%
Plastic 5%
Glass 4%
Metals 14%
Non Recyclable 34%
Construction Material 13%
Volume of Waste by 1% of population
Each Incomes contribution to Total Waste
Low Income
High Income

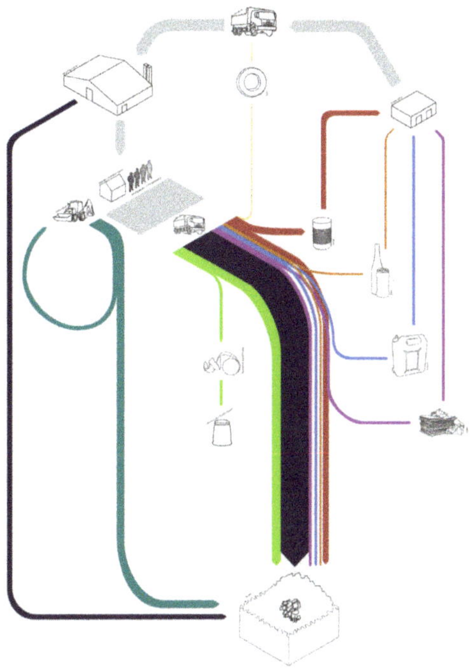

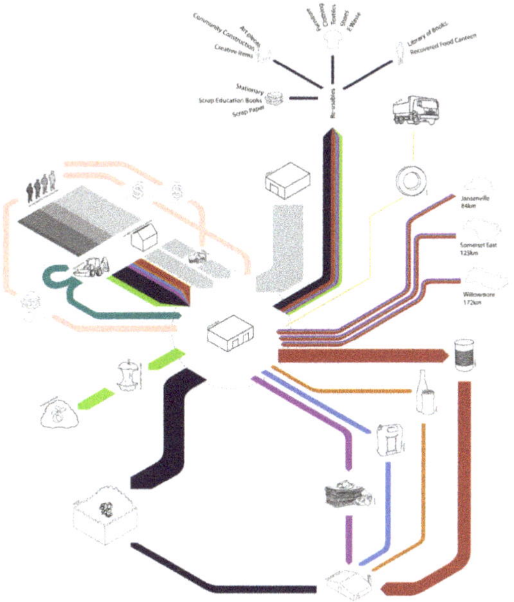

Community Construction
Art pieces
Creative items
Furniture
Clothing
Textiles
Shoes
E Waste
Library of Books
Recovered Food Canteen
Stationary
Scrap Education Books
Scrap Paper
Re-usables
Jansenville
84km
Somerset East
123km
Willowmore
172km

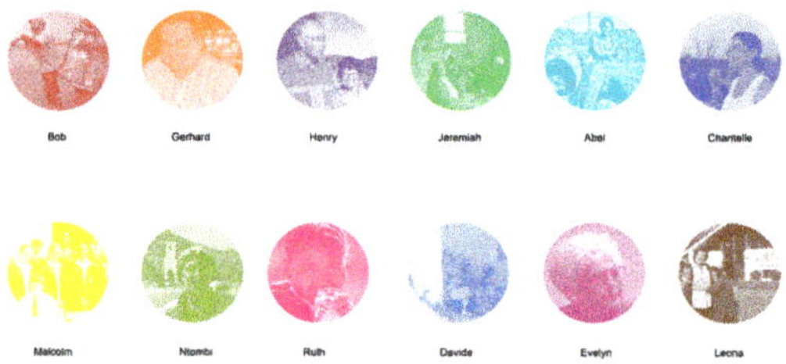
Bob
Gerhard
Henry
Jeremiah
Abel
Chantelle
Malcolm
Ntombi
Ruth
Davide
Evelyn
Leona

Gerhard's Day
7-8am
8am-2.30pm
3-4pm
4-5pm
Gerhard's House
Community Hub
Union High School
Supermarket

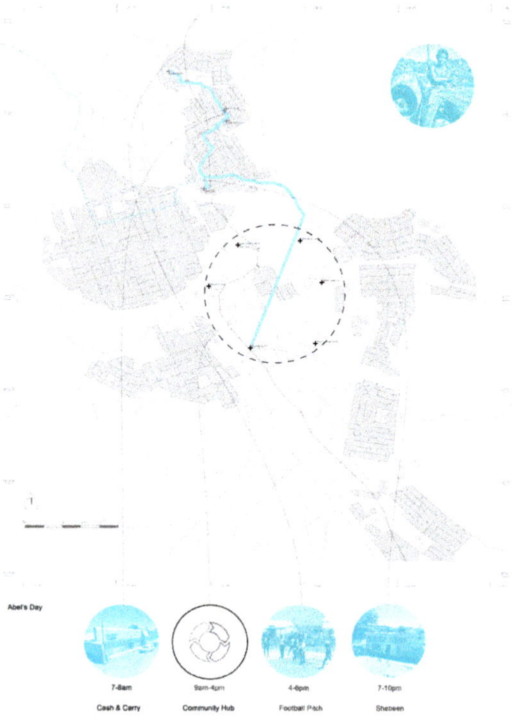
Abel's Day
7-8am
9am-4pm
4-6pm
7-10pm
Cash & Carry
Community Hub
Football Pitch
Shebeen

Community Centre and Canteen | Education Centre | Artist Residency | Reprocessing Centre | Sorting Centre | Resale Market

CULTIVATED WASTELANDS COOPERATIVE

JARED BARON

Wasted products, wasted space and a wasted labour force all converge on the small South African town of Graaff Reinet. Waste is often discussed as a dirty reminder of our excessive consumerism and unsustainable world. This project however, attempts to explore the opportunities that lie within our waste, focusing on the potential of composting organic waste. Composting is the human form of mulching; the natural process of returning organic materials to the earth to provide soil and crop fertility.

The Cultivated Wastelands Co-operative proposes a scenario where organic waste in Graaff Reinet is utilised to fertilise the wastelands around a disused railway line that used to serve as a buffer between racial groups during the apartheid era. This is transformed into a productive agrarian belt that aims to physically and socially link local communities who remain spatially segregated. A series of architectural components are developed to allow the co-operative to grow, process and trade the crops and livestock that were identified key economic opportunities for the project – rooibos tea, agave, ostriches and thatching grass. The components are designed as Open Source Architecture that anyone can access to adapt or improve to suit other economic, social or environmental conditions. The program is then distributed along the disused railway tracks that is re-purposed as a linear distribution and harvesting network.

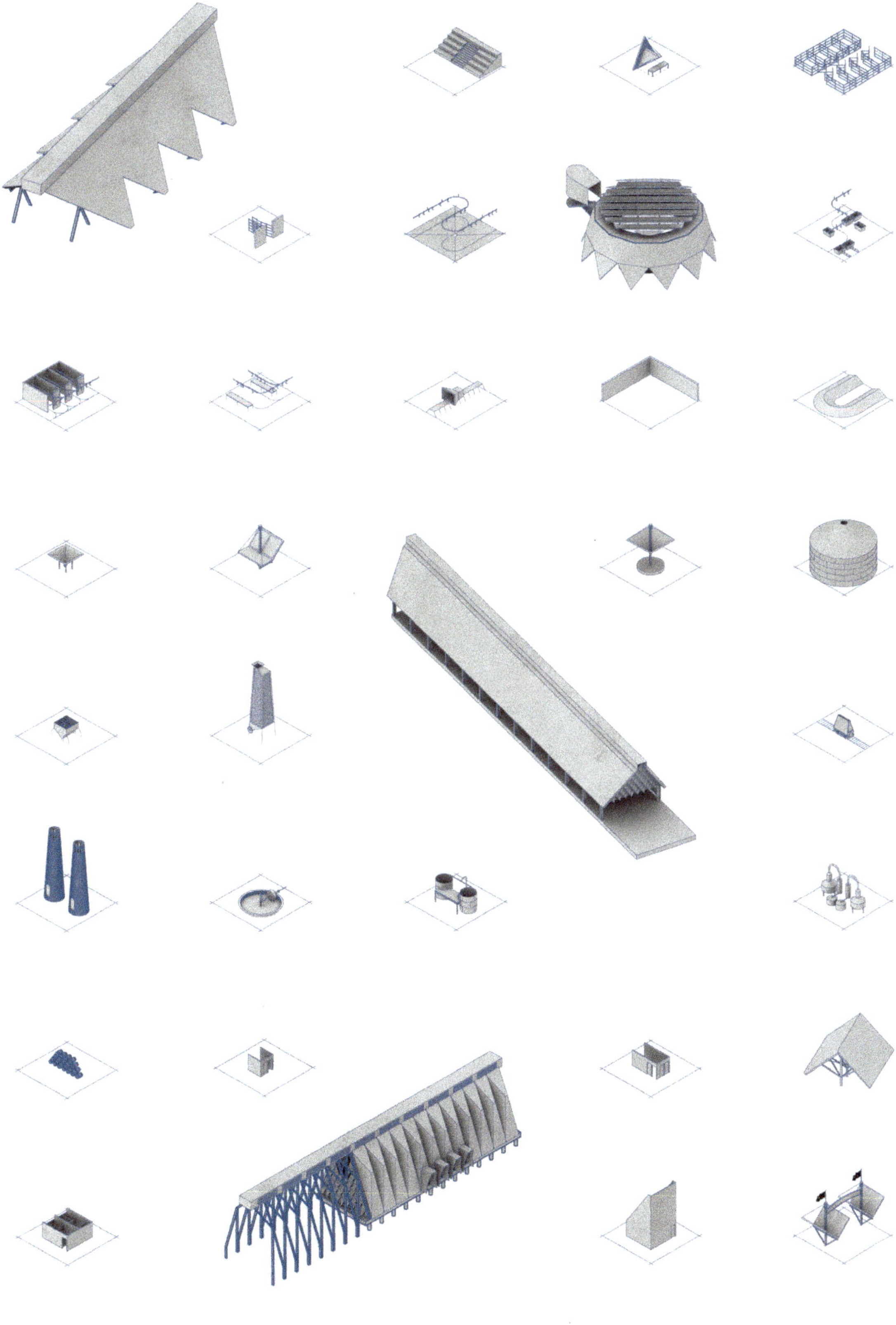

IMAGE CAPTIONS

P241 Open Source Architecture Component Catalogue

P242 Model of Wasteland 0

P243 Pioneer Crops and Livestock

Each of the crop and livestock chosen to instigate the scheme each have a specific relationship with the area. It is intended that these crops form a pioneer agrarian co-operative that will later be able to introduce further experimental crops.

P244 Apartheid era Wastelands in Graaff Reinet

The project contains 7 chapters, but focuses on a developed scheme for chapter 1 as an initiation of the new agrarian economy. 7 Wastelands are highlighted along the railway line's trajectory through the town, sprouting from the initial development at Wasteland 0.

P245 Aerial Perspective of the Cultivated Wastelands Co-operative

P246 The Compost Refinery: Exploded Axonometric

The compost refinery is the architectural symbol of the very premise of the Cultivated Wastelands Co-operative: of social and economic energy stemming from the utilisation of the areas' organic waste. This facilitates the fertilisation of the barren wastelands into a carefully managed dryland farming economy

P247 The Compost Refinery

P348-249 The Process Hub

P250 The Exchange: Exploded Axonometric

The Exchange functions as the pavilion of idea and knowledge sharing, where co-operative members are able to influence the project's economic and social trajectory through a direct democratic system. The members discuss and vote on topics such as field division strategy, price fixing and distribution of fertiliser, labour and attention

P251 The Exchange

P252-253 The Public Hub

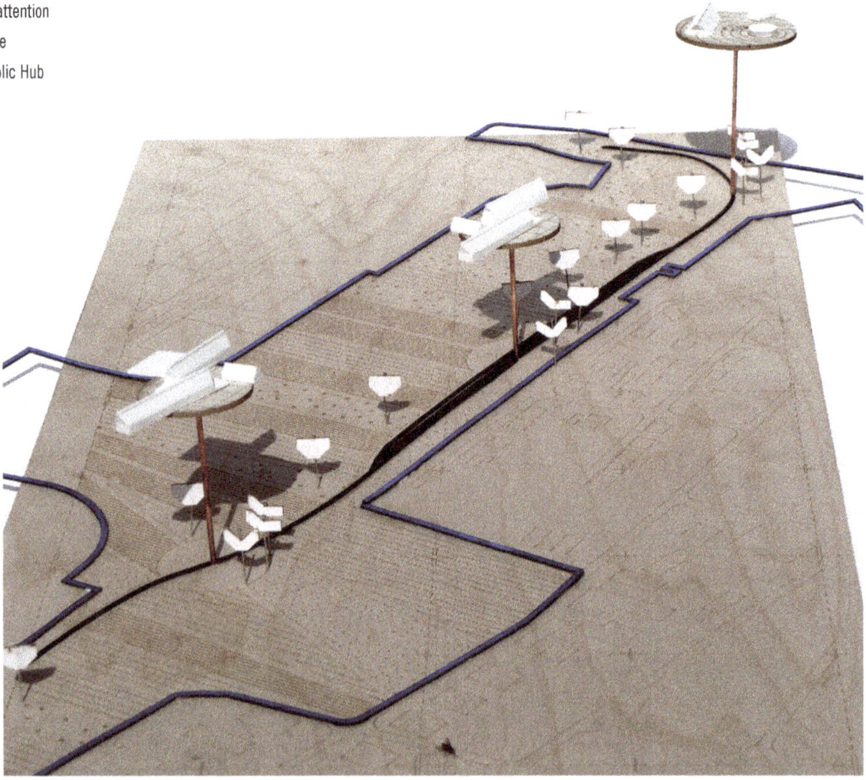

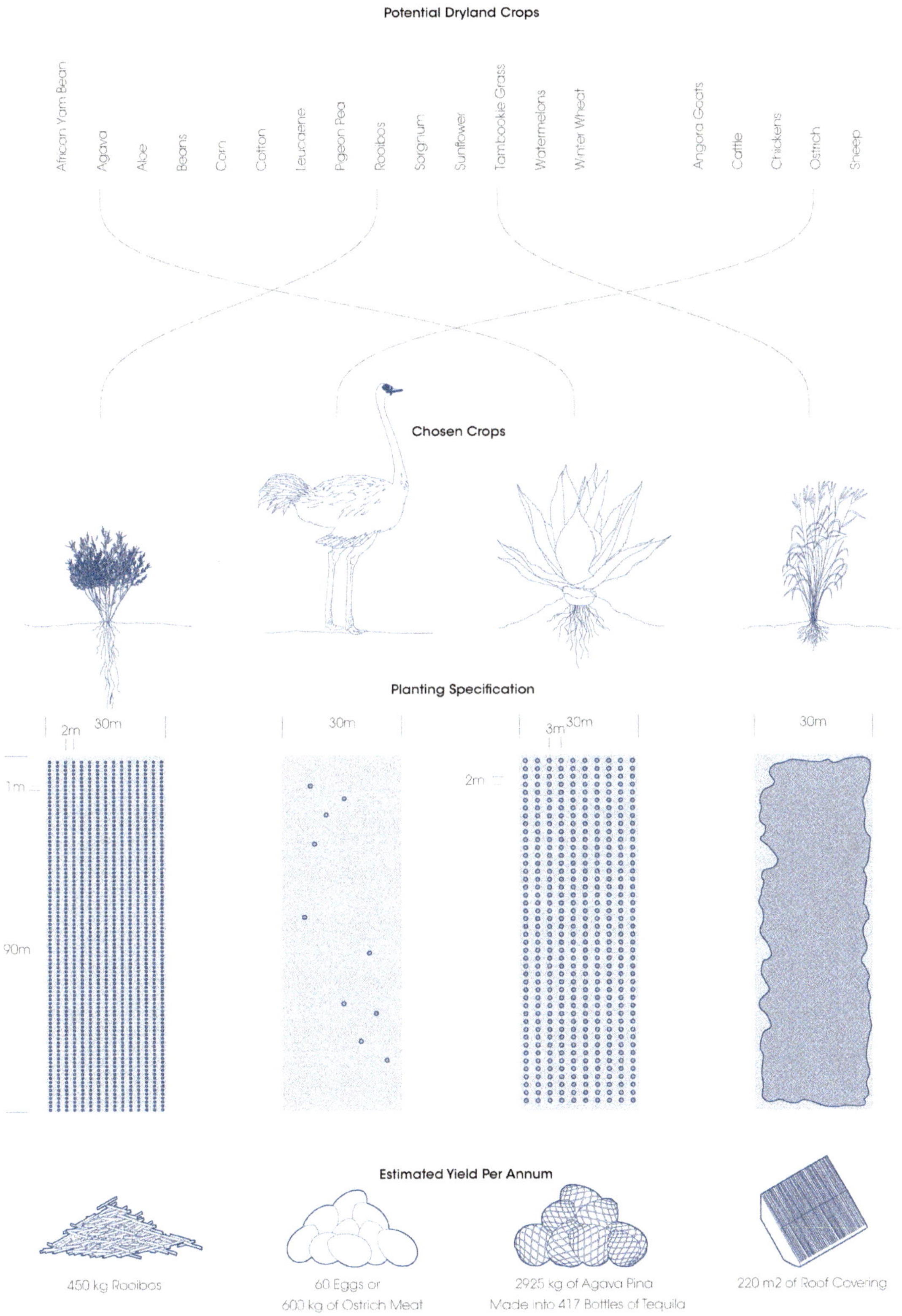
Potential Dryland Crops
African Yam Bean
Agava
Aloe
Beans
Corn
Cotton
Leucaene
Pigeon Pea
Rooibos
Sorgnum
Sunflower
Tambookie Grass
Watermelons
Winter Wheat
Angora Goats
Cattle
Chickens
Ostrich
Sheep
Chosen Crops
Planting Specification
30m
2m
1m
90m
30m
3m
2m
30m
30m
Estimated Yield Per Annum
450 kg Rooibos
60 Eggs or
600 kg of Ostrich Meat
2925 kg of Agava Pina
Made into 417 Bottles of Tequila
220 m2 of Roof Covering

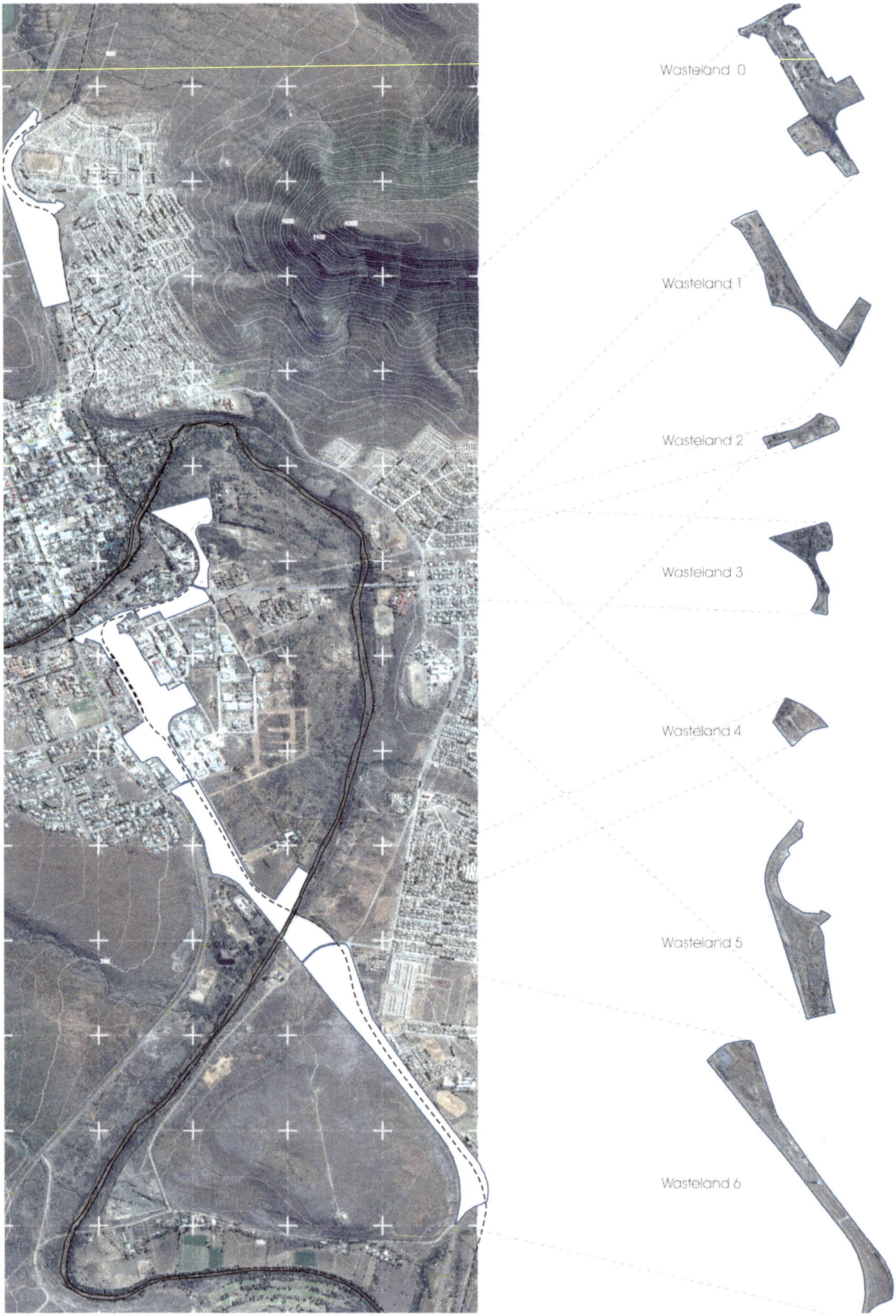
Wasteland 0
Wasteland 1
Wasteland 2
Wasteland 3
Wasteland 4
Wasteland 5
Wasteland 6

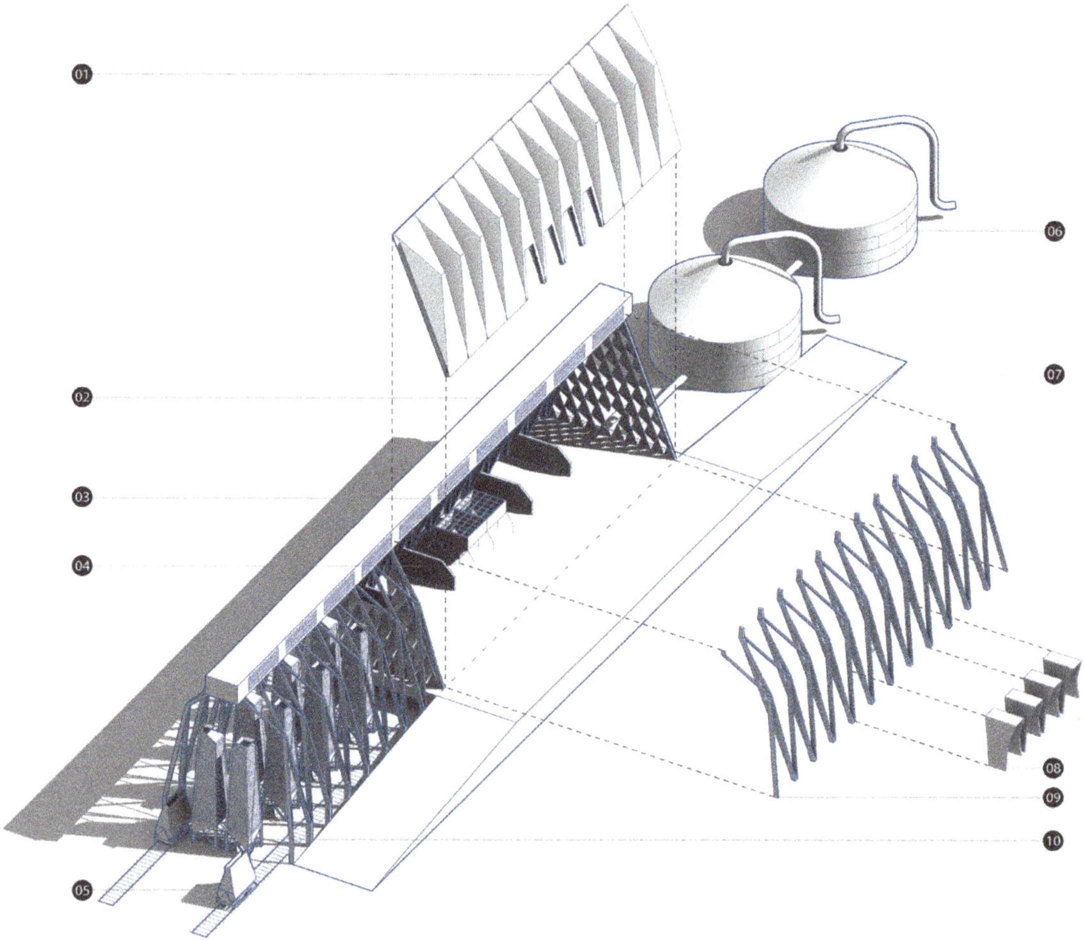

1 Solar Harvesting
2 Recycled Steel Sheet Roofing
3 First Floor Slab
4 Thatched Facade
5 Stair Core to Public Gallery
6 Toilet Block
7 Steel Structural Cassette
8 Steel Floor Beams
9 Feature Steel Column
10 Congregational Seating

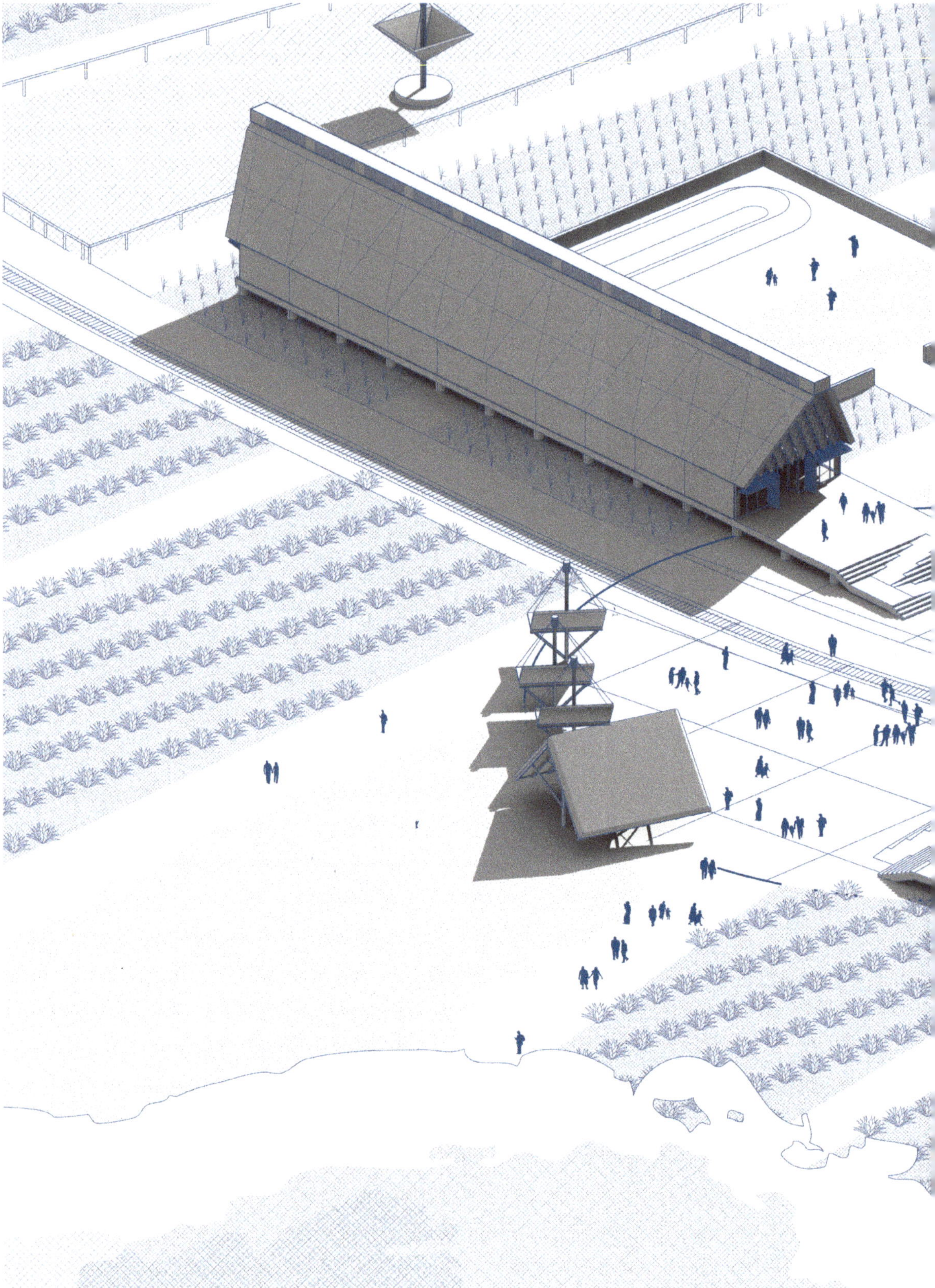

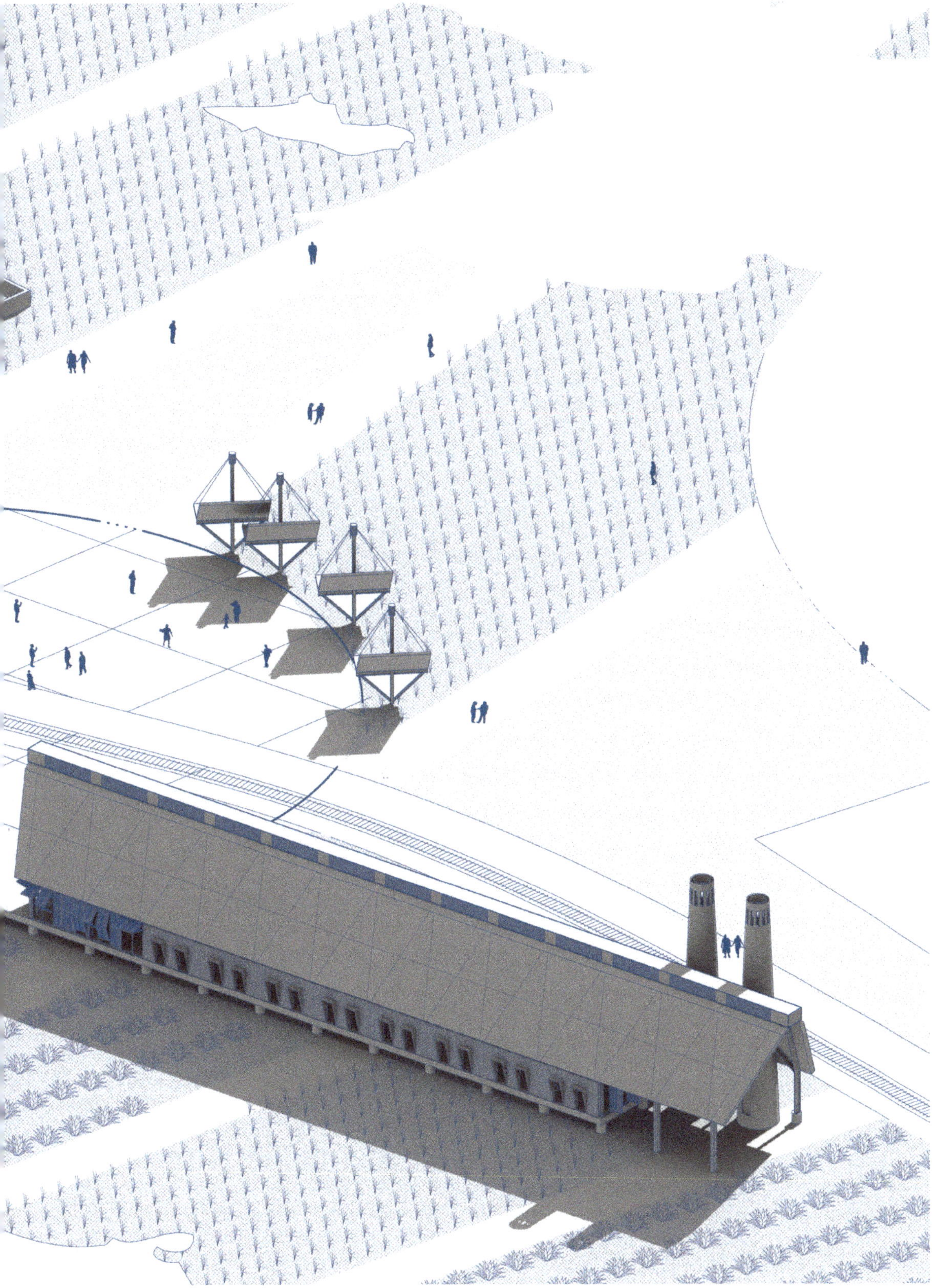

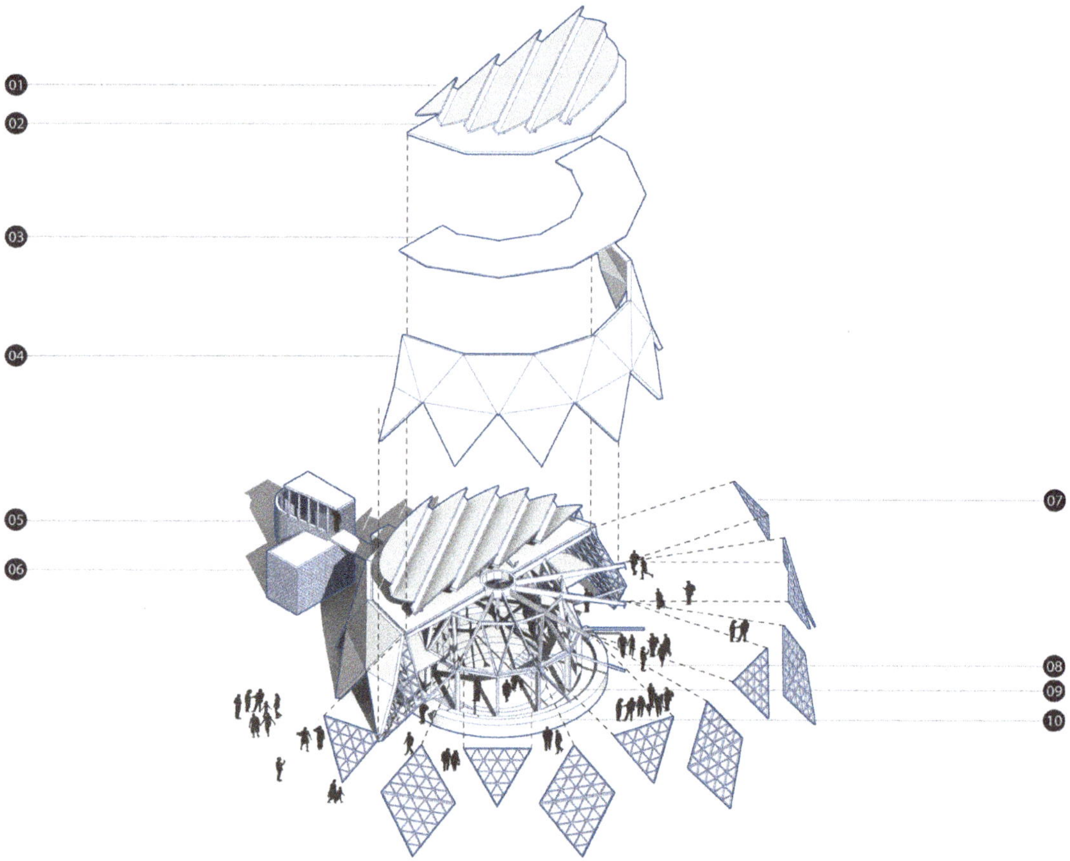

1 Thatched Roof
2 Recycled Steel Sheet Ridge Cap
3 Cutting
4 Plant Unit
5 Sieving and Packaging
6 Rest-Room Unit
7 Bioplastic Facade Panels
8 Public Tea House
9 Tea House Terrace
10 Sweating Yard
11 Drying Yard
12 Steel Feature Columns
13 Rainwater Harvester
14 Thatch Shutters
15 Mycelium Bricks

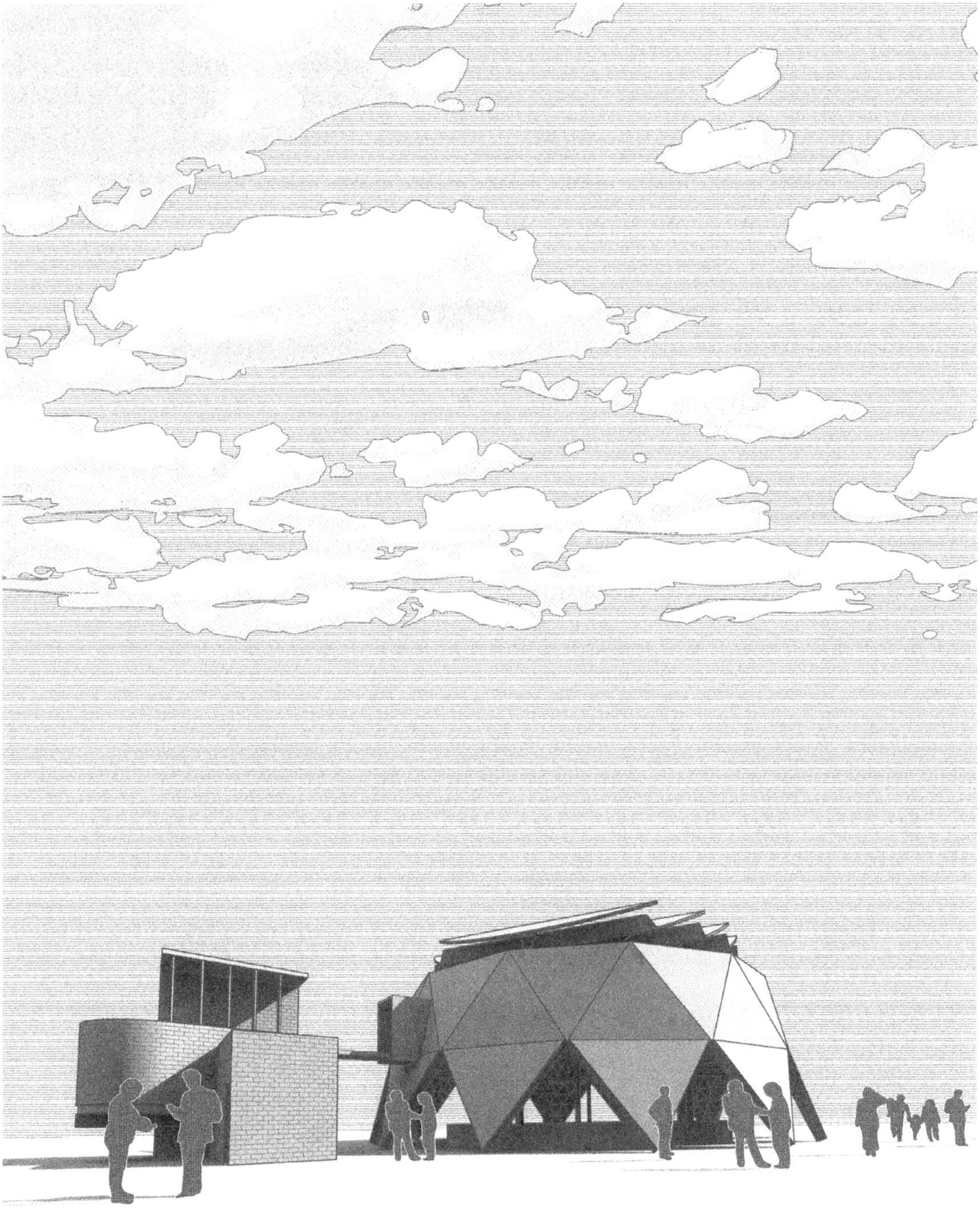

FUTURE RESEARCH FOR EXPERIMENTAL ENERGIES (FREE)

CHERYL CHOO, MATTHEW HEDGES
SHIUE NEE PANG, IULIA STEFAN

FREE is a conceptual energy strategy that integrates architecture and energy systems in the Karoo, South Africa. Stemming from investigating relationships between the indigenous San peoples and energy, the strategy reads the semi-arid landscape of the Karoo as potential energy sources, fully tapping into its natural features, such as atmosphere, sun, wind, water and earth and introducing new experimental energies such as cosmic energy and lightning. The main aim of FREE is to transform the Camdeboo Local Municipality in the Karoo into a research hub for future renewable energies, while, at the same time, to achieving an energy efficient local community by 2030. FREE also acts as an agent that integrates science and technology into the everyday lives of the people of Camdeboo by interfacing artists and scientists and local and international organisations. Additionally, FREE plays an important role in constructing a link between the spirituality of the San people's past and the science of the future.

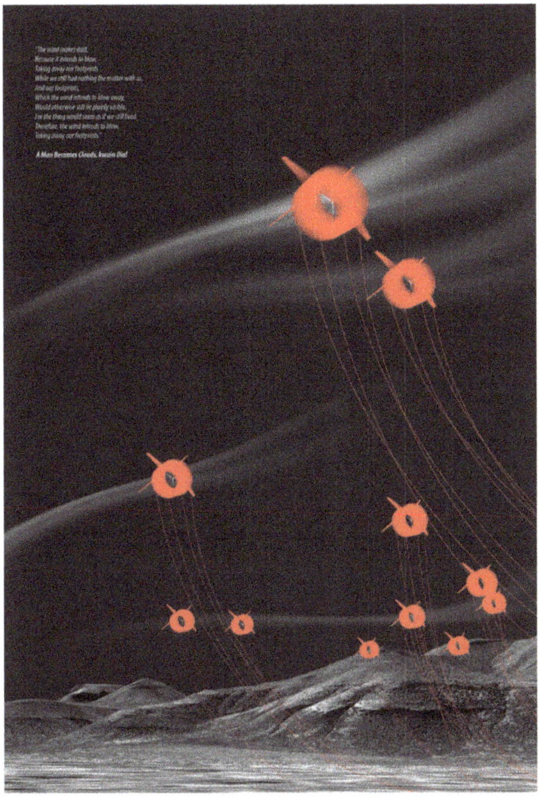
Taking away our footprints.
Which the wind intends to blow away,
Would otherwise still be plainly visible.
A Man Becomes Clouds

ENERGY LANDSCAPE
SPACE-BASED SOLAR POWER
E1
Craddock Mountains, Guy Butler

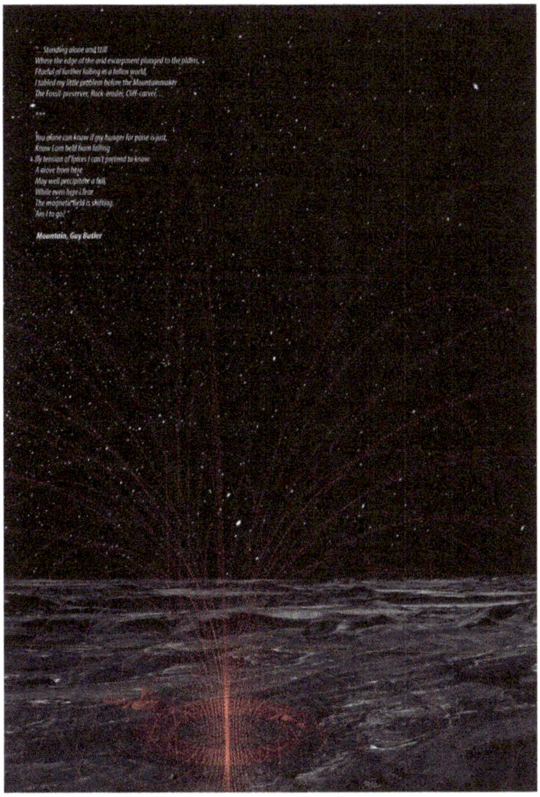
"… Standing alone and still
Where the edge of the arid escarpment plunged to the plains,
Fearful of further falling in a fallen world,
I tabled my little problem before the Mountainmaker
The Fossil-preserver, Rock-eroder, Cliff-carver…
You alone can know if my hunger for poise is just,
Know I am held from falling
By tension of forces I can't pretend to know.
A move from here
May well precipitate a fall,
While even here I fear
The magnetic field is shifting.
Am I to go?"
Mountain, Guy Butler

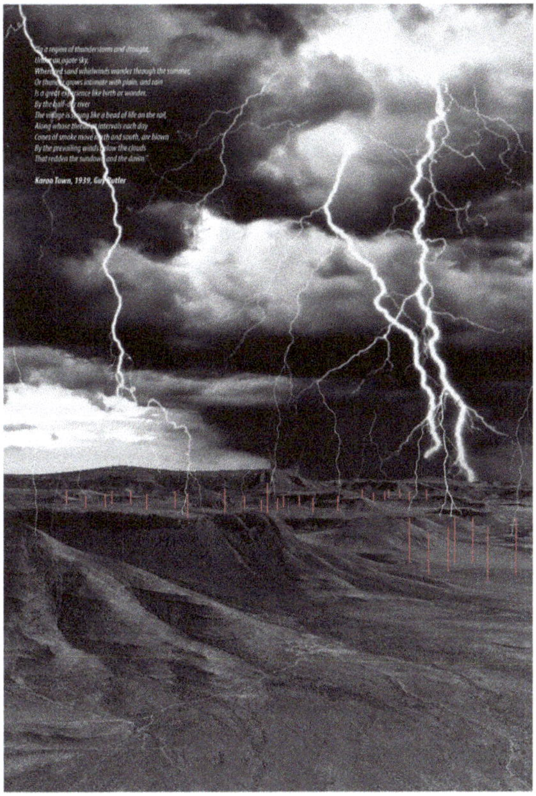
Karoo Town, 1939,

IMAGE CAPTIONS

P255 Experimental Energy Technologies
[top left] Buoyant Airborne Turbines, Shiue Nee Pang
[top right] Space Based Solar Power, Iulia Stefan
[bottom left] Electromagnetic Energy, Cheryl Choo
[bottom right] Lightening, Matthew Hedges

P257 HistoricalContext: The original inhabitants of the Karoo, the SAN
Knowledge of the landscape and its energies embedded in spiritual beliefs and practices.

P258-259 Tables of experimental energy technologies
A menue of possible energy technolgies to harvest renewable energy resources in the Karoo.

P260 [top left] Map of a potential energy development strategy for the Camdeboo Local Municipality
[top right] Map of the Camdeboo Local Municipality's energy landscape potential, highlighting protected areas
[bottom left] Sectional diagram of experimental energy iechnologies and their optimal elevational position on the earth or in its atmosphere
[bottom right] Potential bodies, institutions, associations and agents
The international, national and local agencies, both private and public sectors, that could be tapped to create an integrated global funding network aimed at collectively securing renewable energy for future generations.

P261 FREE Masterplan

HISTORICAL CONTEXT

THE SAN PEOPLE

The San are hunter gatherers that have lived within the landscape of Camdeboo . They provide insight into the use of the landscape and its resources.

Land Occupation

The San never settled in the same campsite, but used the same waterholes as they moved around. This ensured that the land did not become exhausted. They also believed that no individual owned the land; everyone has the right to use it. Land is regarded as sacred and a gift from God because.

Politics/ Social

The San were an egalitarian society. They made decisions among themselves by consensus,with women treated as relative equals.

Economy

San economy was a gift economy, based on giving each other gifts regularly rather than on trading or purchasing goods and services.

SAN OCCUPATION IN AFRICA

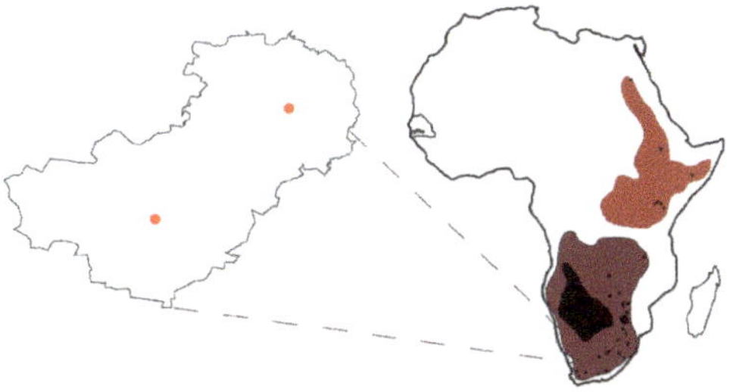

- Modern distribution of the San
- Probable distribution at the end of 17th century
- Probable additional distribution during the late stone age
- Sites with skeleton remains as credible sources for locating the San

SAN ENERGY MENU

SOURCE

WATER

Water was very important in the dry regions in SA where the San lived. Droughts may last many months and waterholes may dry up

FIRE

The most important part of the San's lives is fire. Their lives revolved around their fire because it provided warmth, light and a way to cook food.

GI!

The San sought this power and used it for the benefit of their community. It allowed for the healing of the sick and for the healing of divisions within society.

INSTRUMENTS

Rainmaking Dance

Cagn is said to have created the moon which holds special significance to the San people; the phase of the moon dictated when rainmaking rituals were to be performed.

Sip Wells

When waterholes dry up. San people would use sip wells. To get water this way, a San scrapes a deep hole where the sand is damp. Into this hole is inserted a long hollow grass stem.

Ostrich Eggs

To collet and store their precious liquid the San used the shell of ostrich eggs and than stored them under the ground where it would stay cool.

Stick Fire

Men are responsible for making fire by using two fire sticks that they carry with them at all times.

Fire Dances

Fire dances are held through the night. Sound travels best in dense cool air. Then, land was empty and quiet - very few things made loud noises. In the dry cool air the sound of people holding a trance dance can be heard almost twenty miles.

Trance

To enter the spirit world, trance has to be initiated by a shaman through the hunting of a tutelary spirit or power animal this was done through rhythmic dancing, music, sensory deprivation and hyperventilation.

Rock Engraving

The rock art of the San was believed to be rich in GI! special power. The San read the landscape as an energy field, often going to barely accessible places (ie mountain tops) to perform rock art making. The images likely derived from visions they had experienced at special ceremonies

ENERGY MENU
SOURCES OF ENERGY

SOLAR

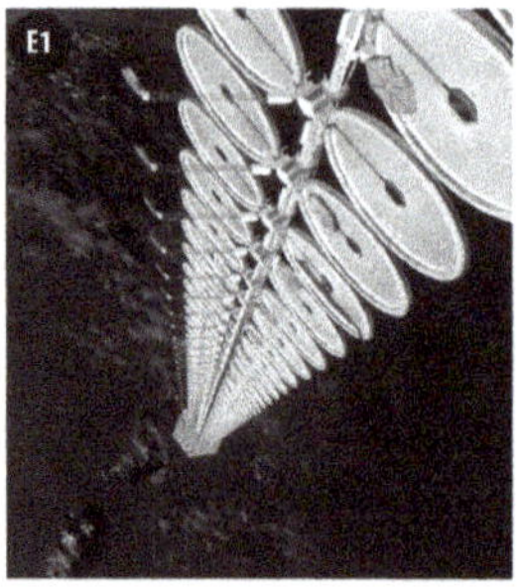

Sun-based Solar Power

Source : Solar
Geography : Space/ valleys
Scale of Application: Large scale /long distances
Influencing Factors : Solar intensity
Energy Output : Estimated 6-8 times more power than a comparable solar cell on the Earth's surface

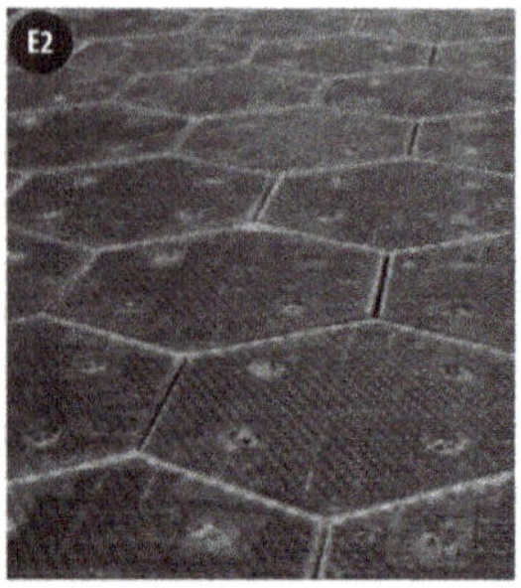

Solar Panels Road

Source : Solar
Geography : Urban / roads
Scale of Application: Large scale / long distances
Influencing Factors : Sunlight / daylight hours
Energy Output : 70kwh per square meter per year

Hydrogen Leaf

Source : Solar/ hydrogen
Geography : Urban
Scale of Application: Small scale
Influencing Factors : Sunlight / daylight hours
Energy Output : Unknown

SOLAR

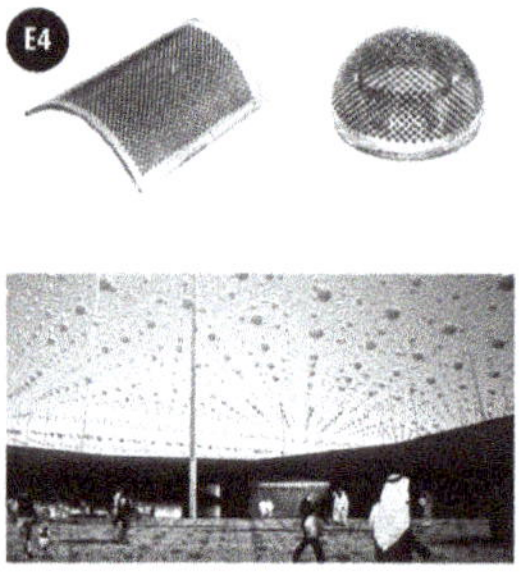

Sphellar Solar Cells

Source : Solar
Geography : Landscape/ urban/ expressways
Scale of Application: Micro to large
Influencing Factors : Solar intensity
Energy Output : 1 module = 0.6-0.66mW/ 0.481-0.484V

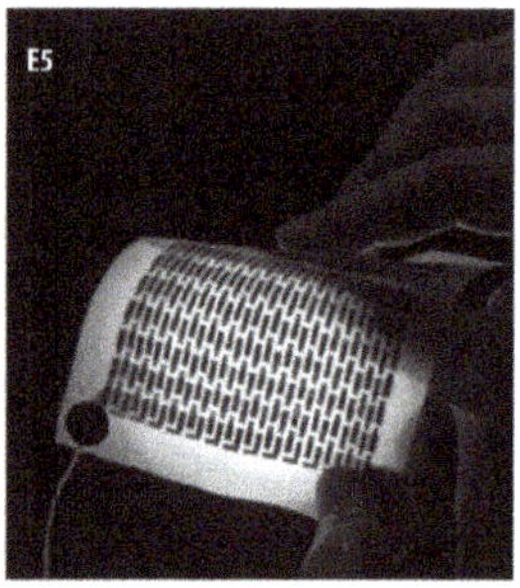

3D-Print Organic Solar Cells

Source : Solar
Geography : Urban
Scale of Application: Large scale / local
Influencing Factors : Sunlight / daylight hours
Energy Output : 10x lower than mainstream

Hybrid Solar Technology

Source : Solar/ thermal
Geography : Plain landscape
Scale of Application: Large Scale/ local
Influencing Factors : Sunlight / daylight hours
Energy Output : 100kW of power/ 170kW thermal power

BIO

LIGHTNING

HYDRO

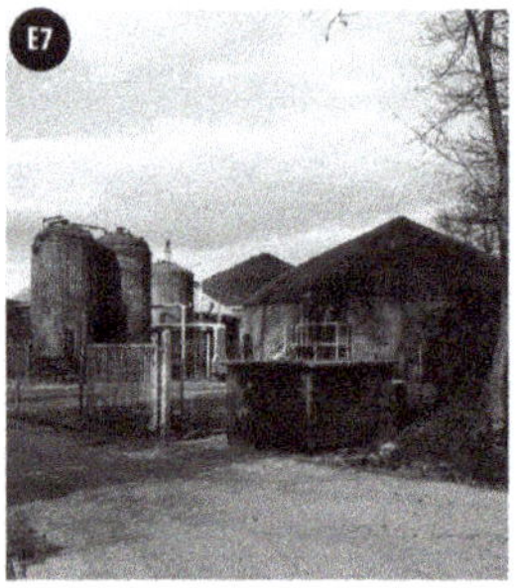

Bio-Gas/ Animal Waste

Source : Animal waste/ methane
Geography : Farmland/ Agricultural
Scale of Application: Medium scale
Influencing Factors : Temperature/ waste amount
Energy Output : 100 cow farm > 30kW enough to power 35+ houses

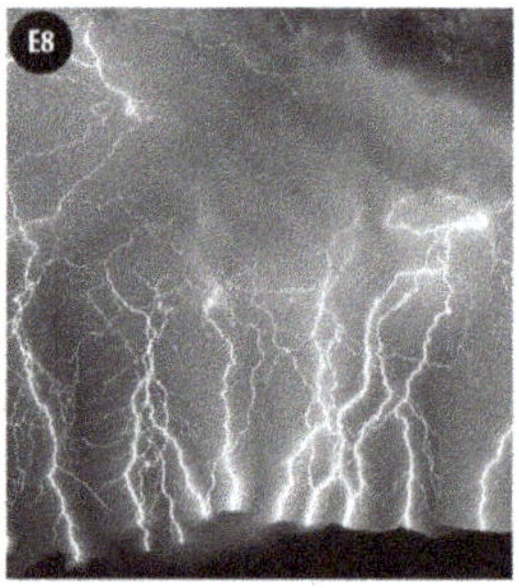

Lightning Energy

Source : Lightning/ Storm
Geography : Storm areas
Scale of Application: Very large scale/ mountainous terrains
Influencing Factors : Lightning C\charge
Energy Output : 1 bolt > 30000 homes per day

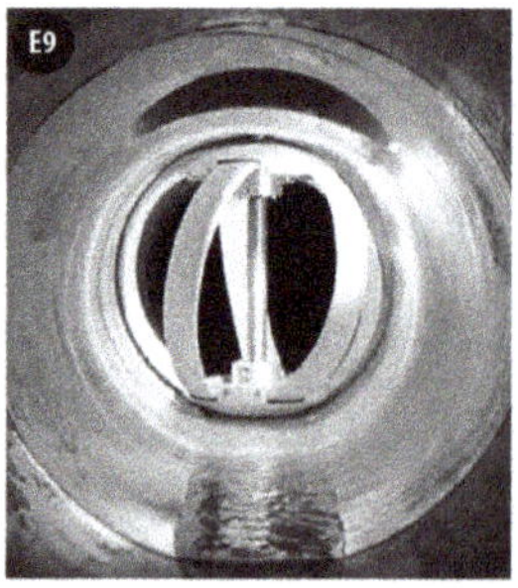

In-pipe hydropower/ Pyco-Hydro

Source : Water flow
Geography : Urban pipes/ rivers
Scale of Application: Very small scale
Influencing Factors : 1 turbine > 100kW

ENERGY MENU
SOURCES OF ENERGY

NUCLEAR

PIEZOELECTRIC

Piezoelectic Road

Source : Cars moving/ piezo crystal
Geography : Urban/ roads
Scale of Application: Large scale /long distances
Influencing Factors : Traffic
Energy Output : 1km>400kW power

Fusion Energy

Source : Lithium/ helium/ deuterium
Geography : Remote area
Scale of Application: Large scale
Influencing Factors : Temperature/ magnetics
Energy Output : 17.6MeV (megaelectron volts) of energy per reaction

Cold Fusion

Source : Hydrogen/ nuclear reaction
Geography : Remote
Scale of Application: Very large scale
Influencing Factors : Temperature
Energy Output : Unknown

COSMIC

Testatika Machine

Source : Electrostatic
Geography : n/a
Scale of Application: Medium scale
Influencing Factors : Electromagnetic
Energy Output : Up to 30000kW

Wardenclyffe Tower

Source : Ionosphere/ cosmic
Geography : High altitude/ remote
Scale of Application: Very large scale
Influencing Factors : Zero-point energy field
Energy Output : Unknown

Apparatus for the Utilisation of Radiant Energy

Source : Ionosphere/ cosmic
Geography : High altitude
Scale of Application: n/a
Influencing Factors : zero-point energy field
Energy Output : Unknown

WIND

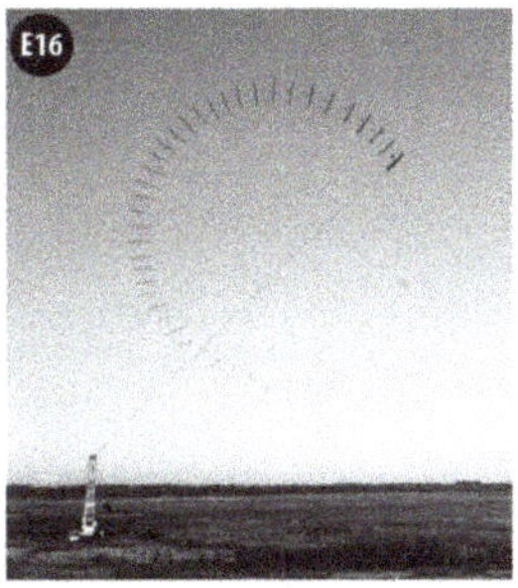

Airbourne Wind Turbine (kite)

Source : Wind
Geography : Open field
Scale of Application: Large scale/ tall
Influencing Factors : Wind speed/ altitude
Energy Output : up to 1800TW

Airbourne Wind Balloon

Source : Wind
Geography : Open field/ high altitude
Scale of Application: Large scale/ tall
Influencing Factors : Wind speed/ altitude
Energy Output : up to 800TW

Airbourne Wind Turbine (kite)

Source : Wind
Geography : Open field
Scale of Application: Large scale/ tall
Influencing Factors : Wind speed/ altitude
Energy Output : up to 3kW/ hour

LANDSCAPE INTENSITY
SUMMARY

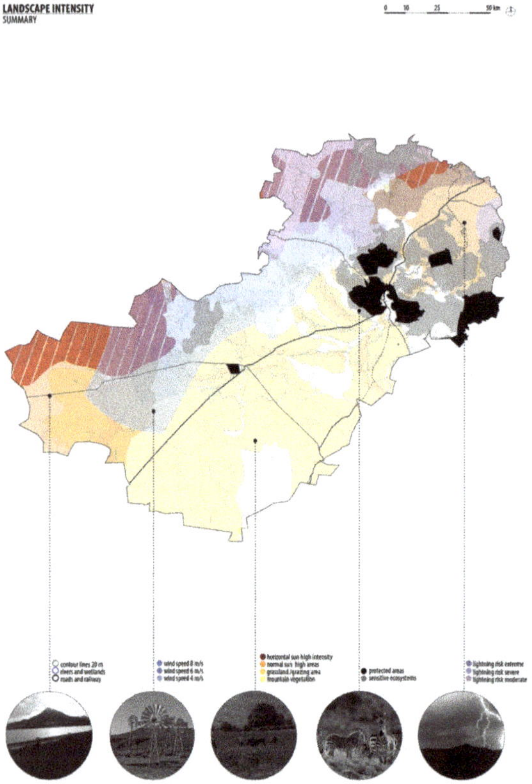

LANDSCAPE ENERGY POTENTIAL
HIGHEST CONCENTRATION AREAS

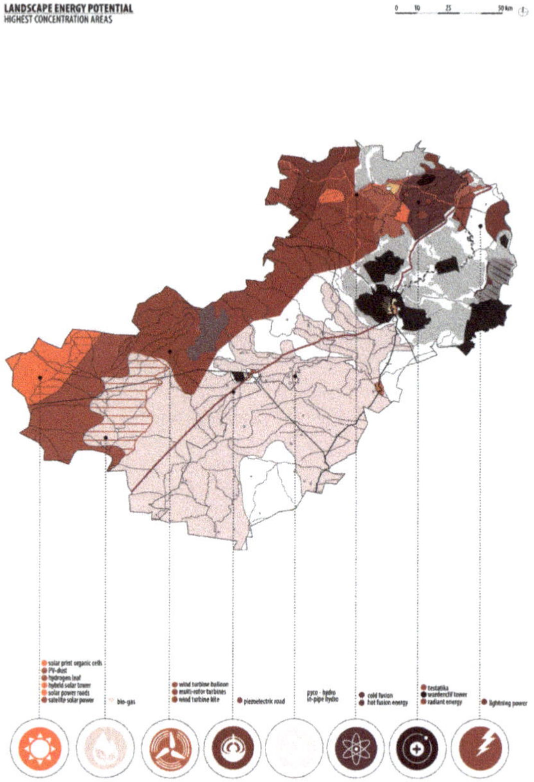

ENERGY MASSING DIAGRAM
SECTION

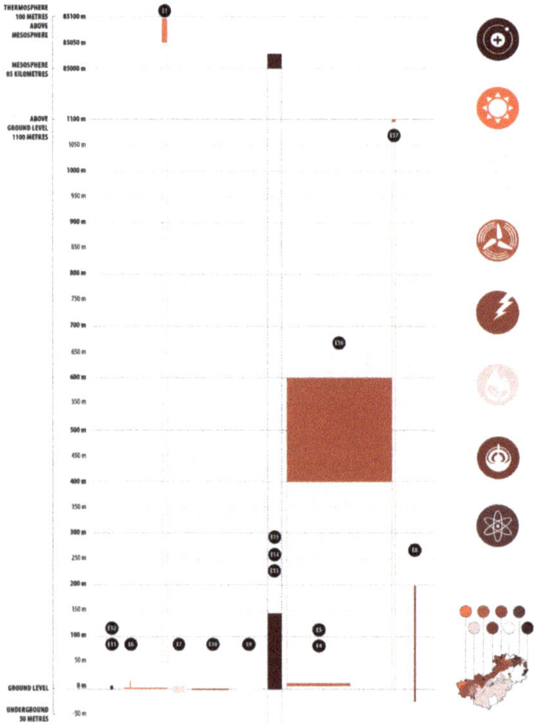

FUNDING BODIES & AGENTS
SUSTAINABLE ENERGY FUTURE

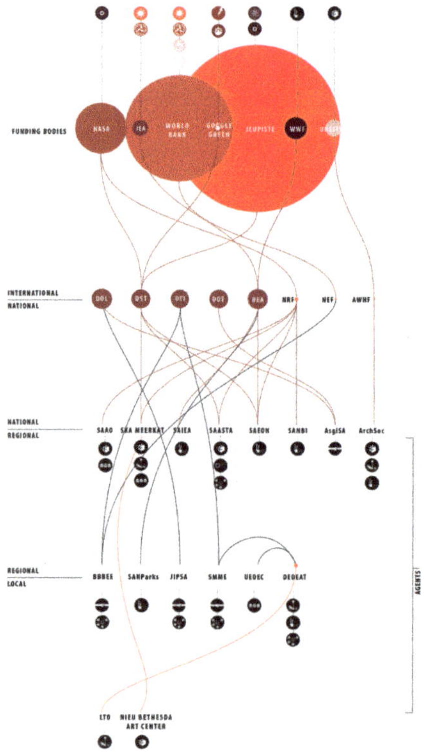

FREE STRATEGY MASTERPLAN

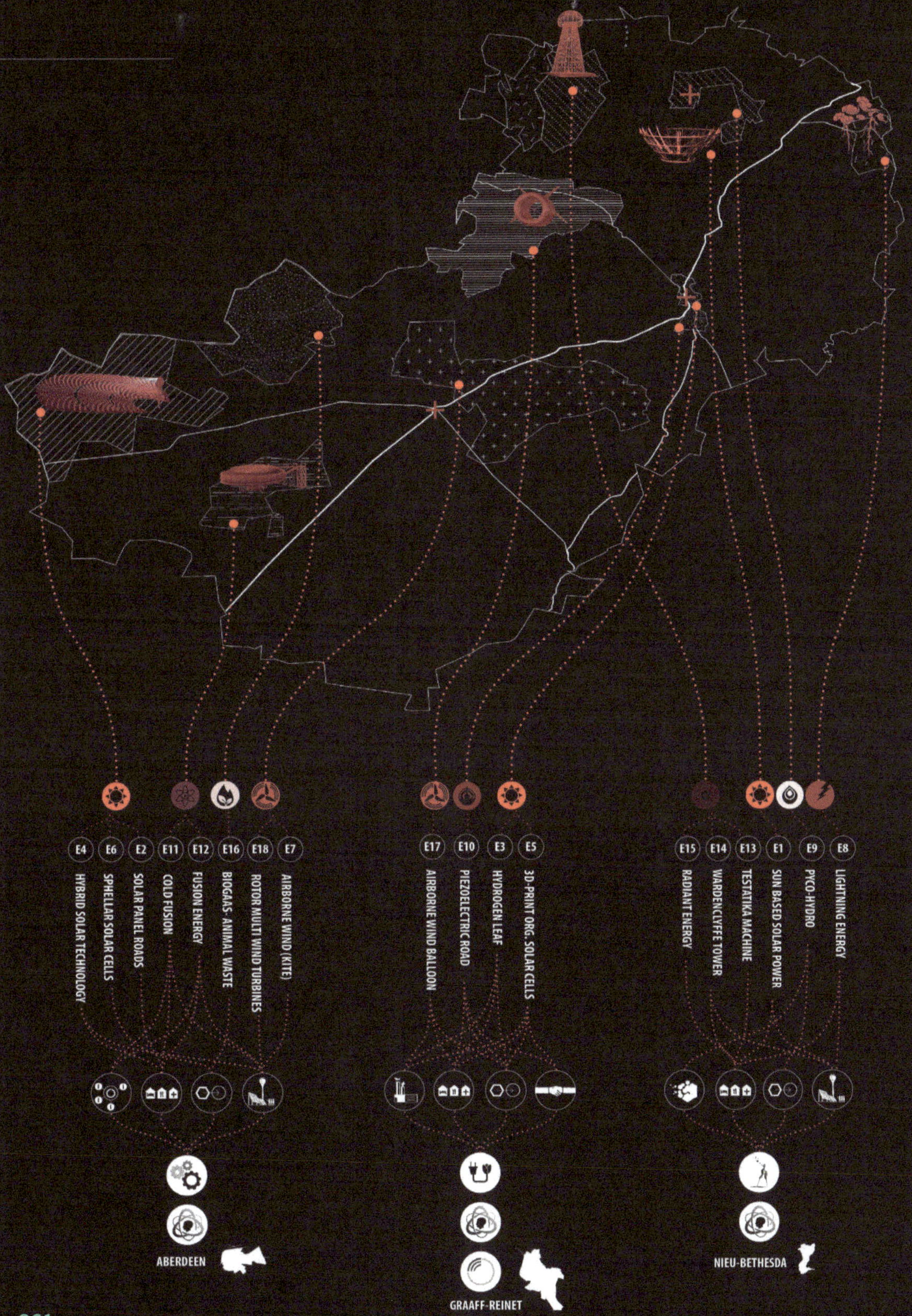

SCIENCITY

CHERYL CHOO

Principles for the Development of a Complete Mind:

Study the science of art; Study the art of science; Develop your senses - especially learn how to see; Realise that everything connects to everything else.

Leonardo Da Vinci (01)

Sciencity is a development that integrates science, art, literature, technology, astronomy, and culture through experimental energy production. This programme begins as a singular notion and expands its grounds through the natural landscape of the Karoo, linking experimental infrastructures as an interconnected network of FREE renewable energy sources such as electrostatic energy, fusion, lightning, and solar energy. Taking clues from the 3-tiered cosmos of San dance rituals to harness cosmic energy, the Sciencity's masterplan is designed in 3 tiers - the Discoverer tier, the Explorer tier and the Producer tier. This transforms Camdeboo into a generative research laboratory of developing senses, an analytical technology-testing ground driven by intellectual curiosity, uncertainty, and the unexpected. This energy network functions as an unrestricted research facility that encourages collaborations and esprit de corps amongst scientists. Freedom of information and data collection allows the network to be an open education institution where a debate assembly comes to life with questions, evidences, arguments, and theories being learnt and challenged. In addition to addressing the future energy demands of South Africa, Sciencity will be the leading innovator of experiments that test and develop Nikola Tesla's wireless electricity transmission theory, involving the movement of people and the energy flow of airborne electricity.

(01) Source: http://www.smarttutorreferrals.com/articles/pespectives-learning/da-vinci

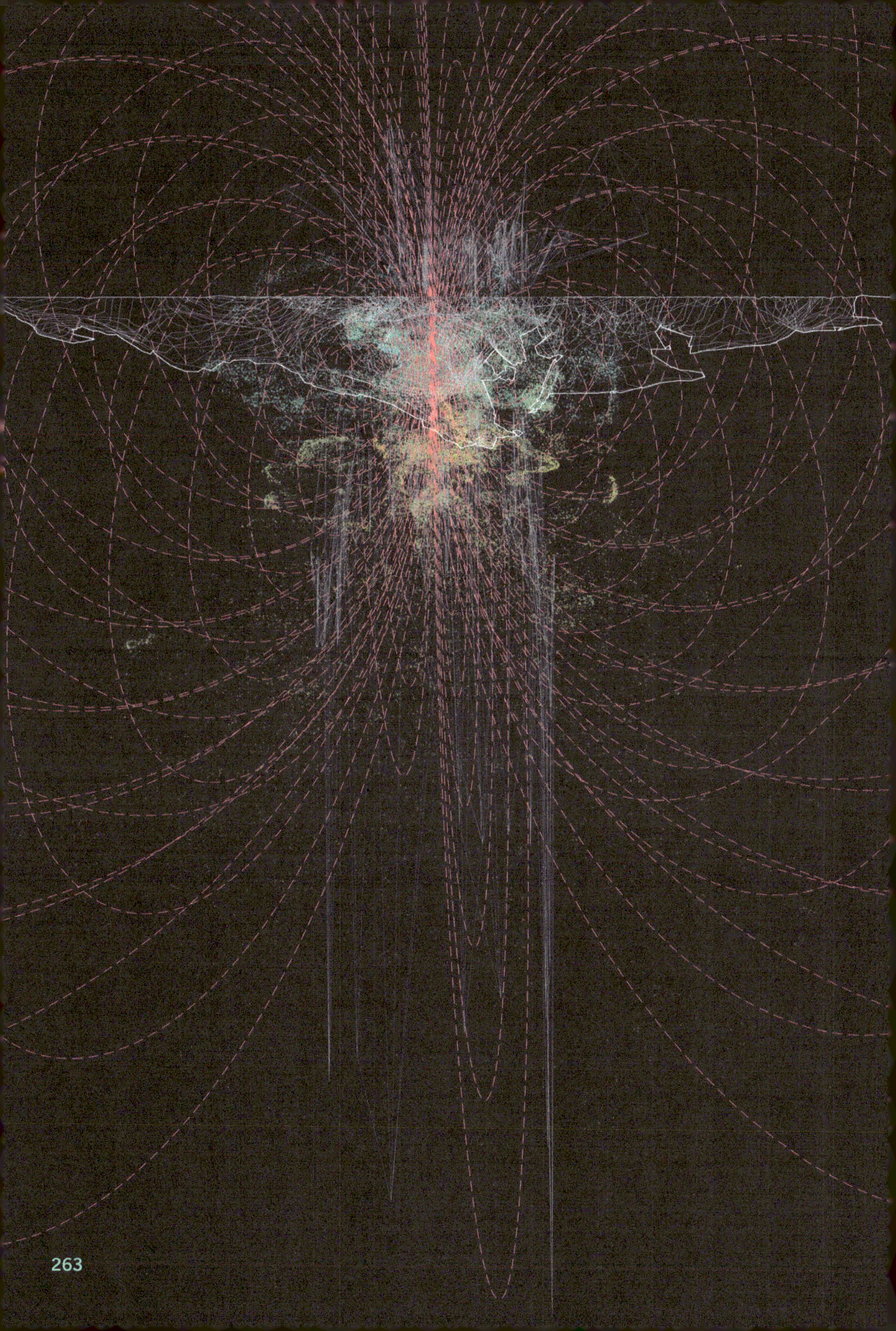

IMAGE CAPTIONS

P263 The Singing Sand
Imaginary section through the earth's distortion during an electromagnetic occurrence.

P264 South Africa Wireless Transmittor Network
[top] Municipality
[middle] District
[bottom] Province

P265 Masterplan of the Camdeboo Local Municipality Wireless Electricity Transmission Network

P266 [top] The Four States of Matter
[middle] Imaginary collection of data of charged particles moving through a scientist's body In 2030
[bottom] Understanding the movement of elctrons through three different environments

P267 Three Tiered Exploded Masterplan showing the Eight Elements of Sciencity

P268 [top] Grid system for the masterplan; layout in accordance with the 24 hr cycle
[middle] Masterplan construction sequence diagram 2020-2030
[bottom] The spatial sorter of theories for the Library of Theories

P269 A list of theories for the Library of Theories
Sources : https://en.wikipedia.org/wiki/Category:Scientific_theories
https://en.wikipedia.org/wiki/Category:Theories

P270 Sciencity at Midnight in 2030

P271 Electrified Physical Site Model

See also drawings on P021 and P039

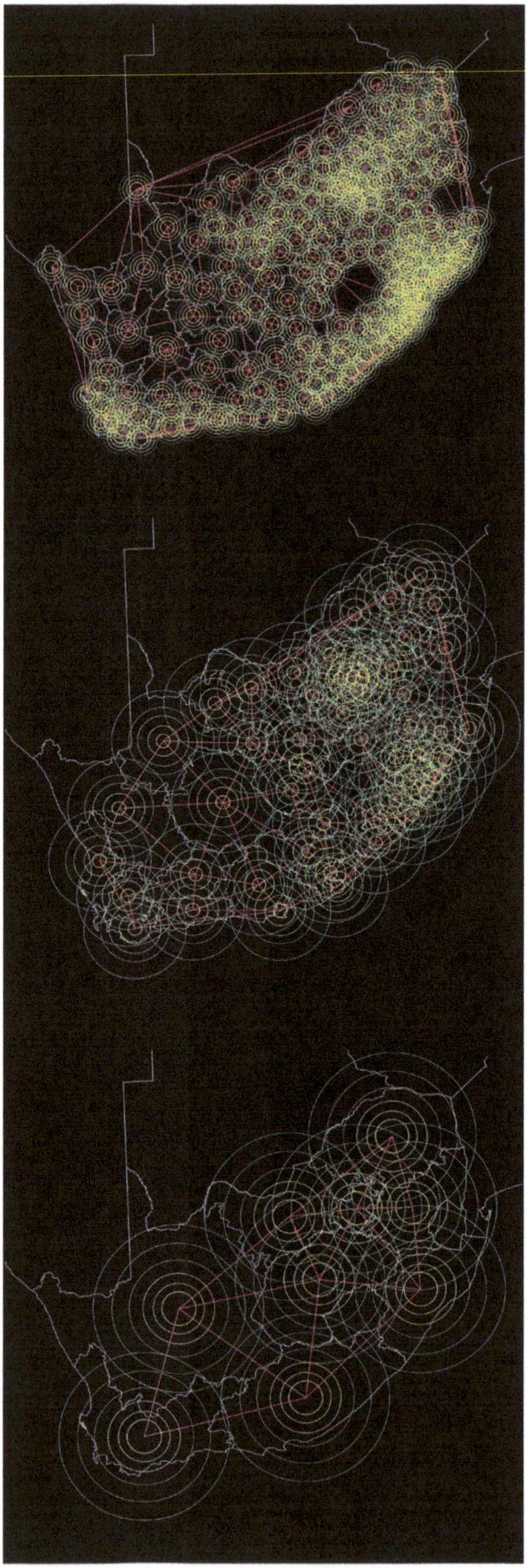

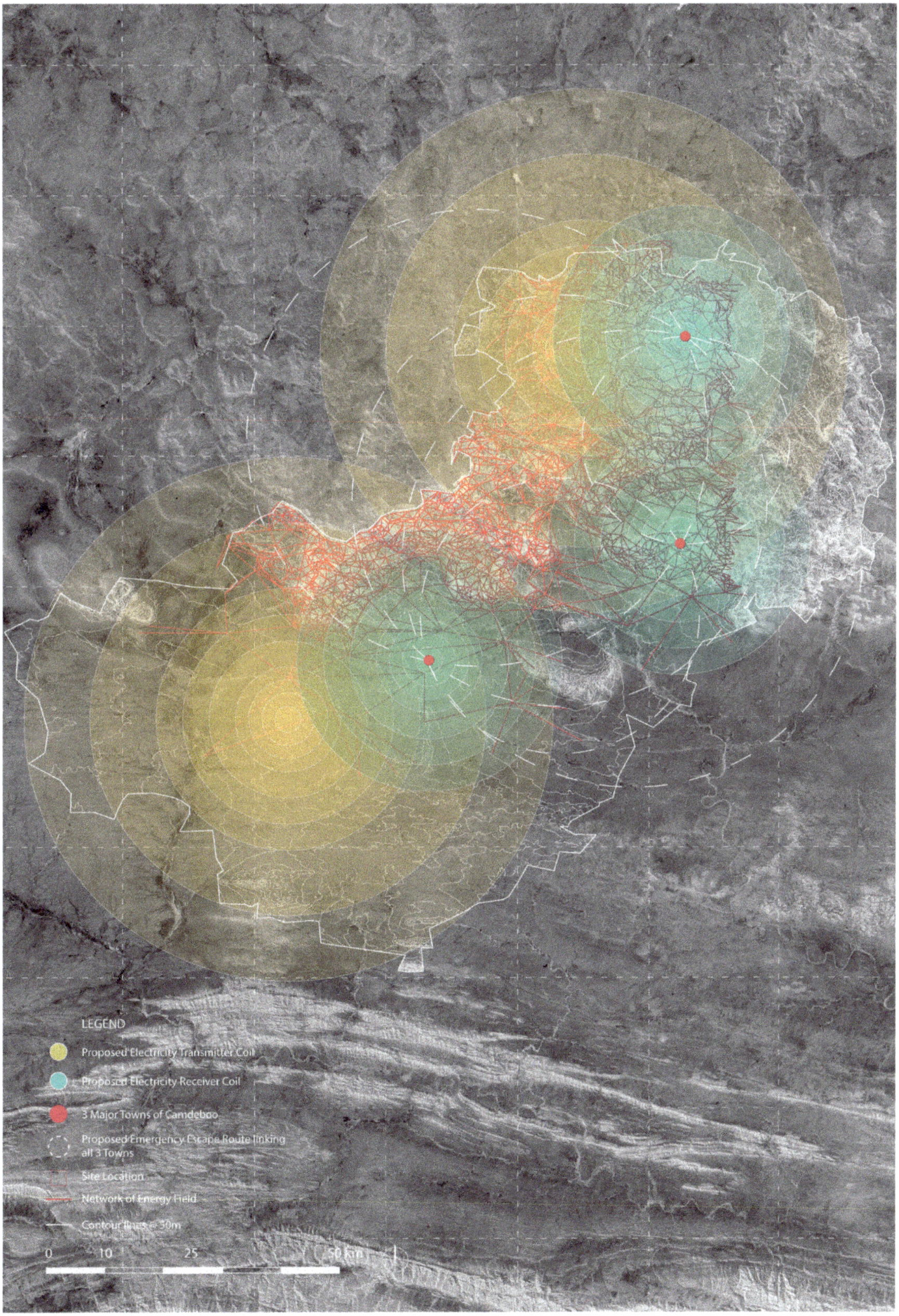
LEGEND
Proposed Electricity Transmitter Coil
Proposed Electricity Receiver Coil
3 Major Towns of Camdeboo
Proposed Emergency Escape Route linking all 3 Towns
Site Location
Network of Energy Field
Contour lines @ 30m
0
10
25
50 km

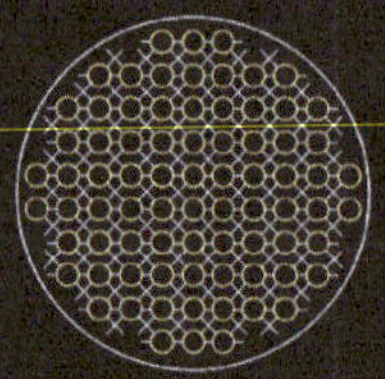

Solid
The molecules that make up a solid are arranged in regular, repeating patterns. They are held firmly in place but can vibrate within limtied area.

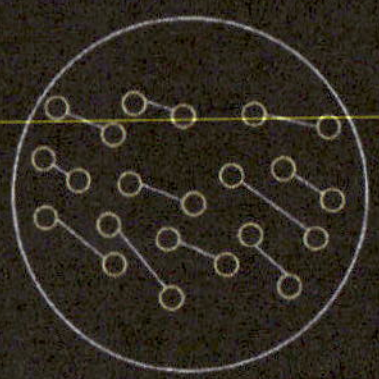

Liquid
The molecules that make up a liquid flow easily around one another. They are kept from flying apart by attractive forces between them.

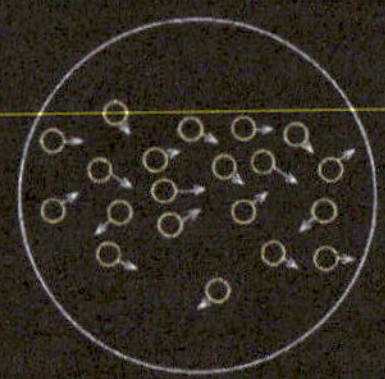

Gas
The molecules that make up a gas moves in all directions at great speed. They are far apart that attractive forces between them are insignificant.

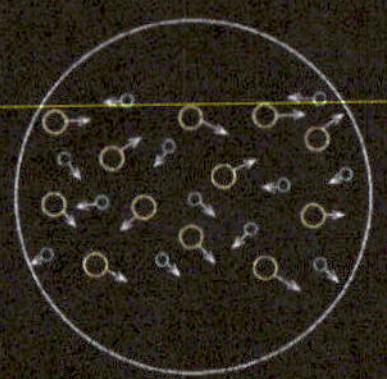

Plasma
At a very high temperature of stars, atoms lose their electrons. The mixture of electrons and nuclei that results is the plasma state of matter.

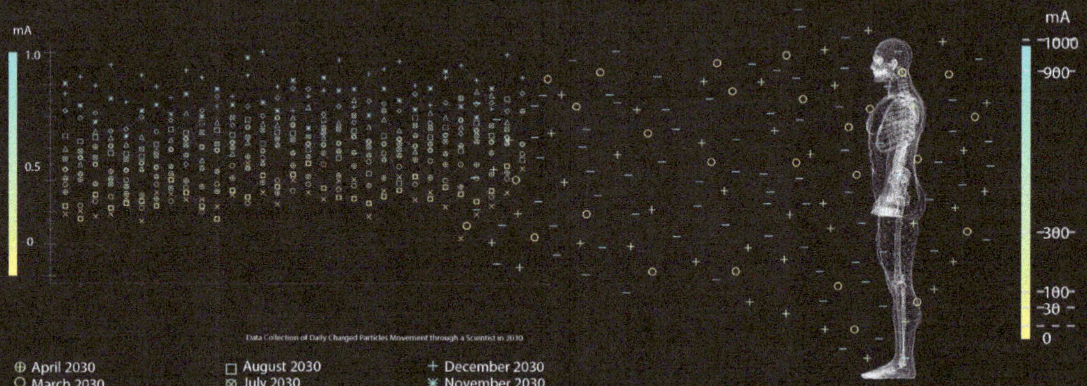

⊕ April 2030
○ March 2030
February 2030
× January 2030
□ August 2030
⊠ July 2030
June 2030
May 2030
+ December 2030
November 2030
◇ October 2030
△ September 2030

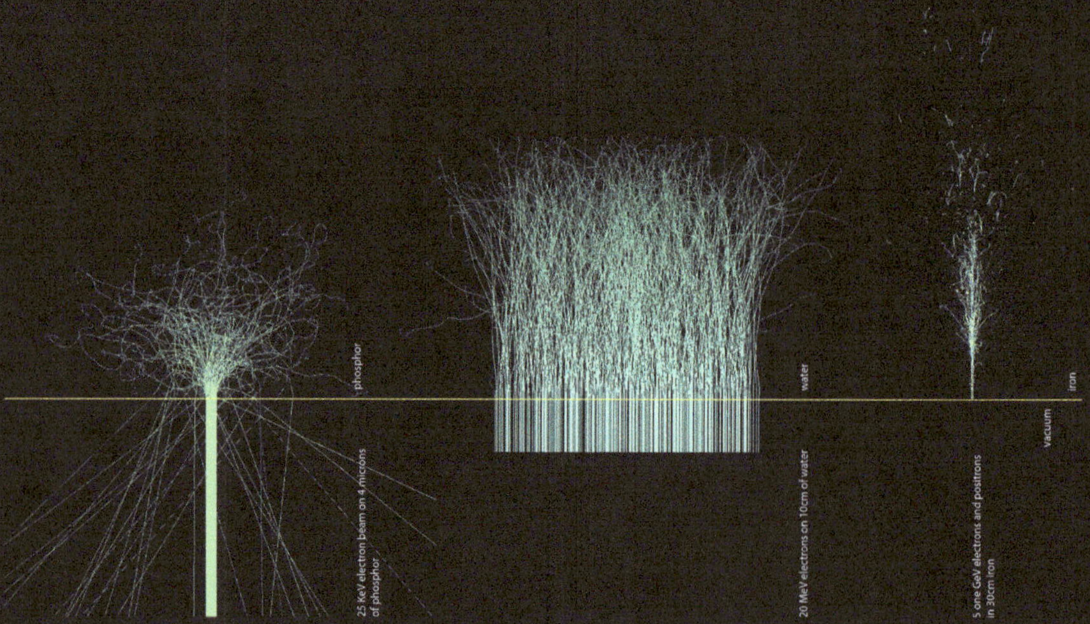

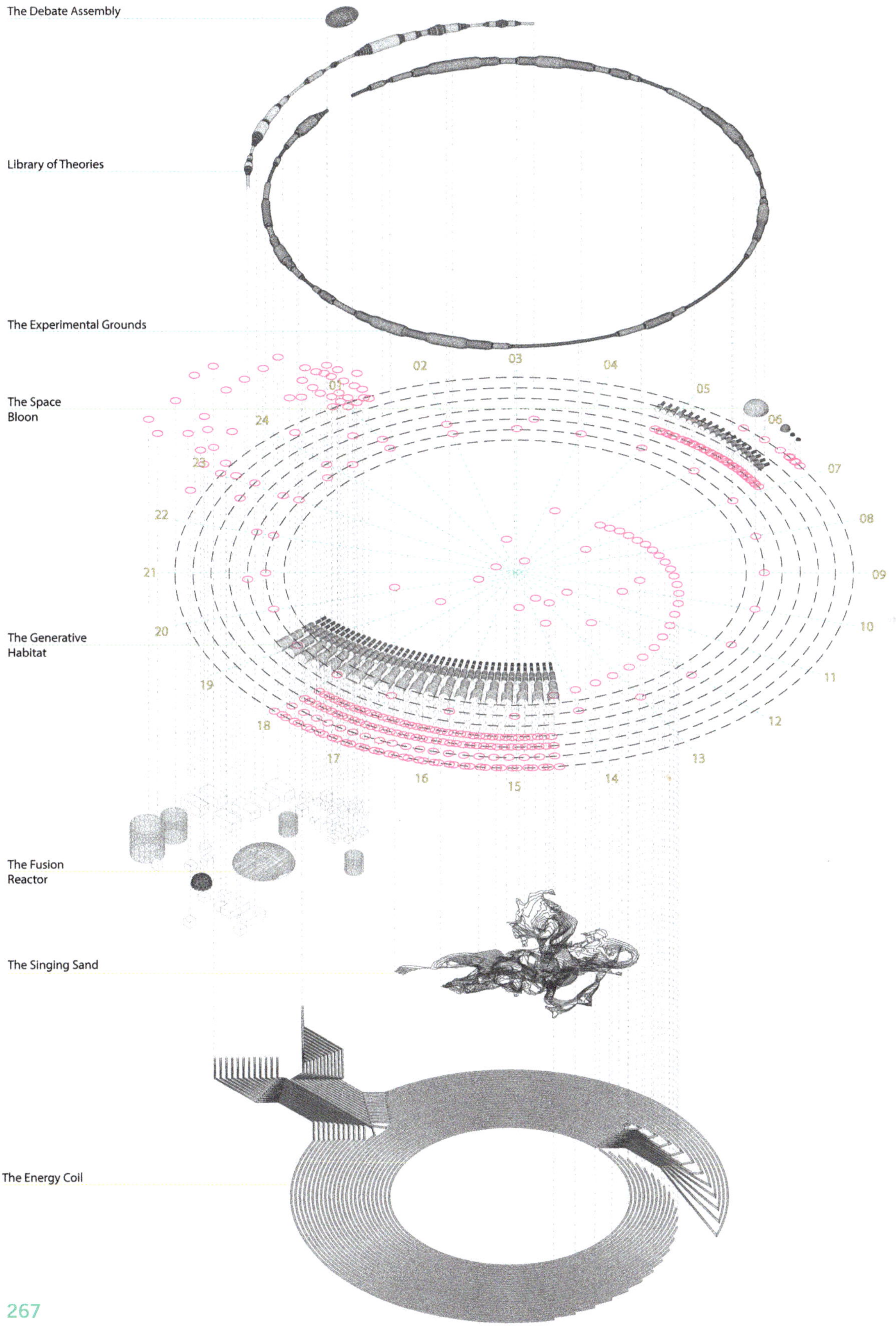
The Debate Assembly
Library of Theories
The Experimental Grounds
The Space
Bloon
The Generative
Habitat
The Fusion
Reactor
The Singing Sand
The Energy Coil
01
02
03
04
05
06
07
08
09
10
11
12
13
14
15
16
17
18
19
20
21
22
23
24

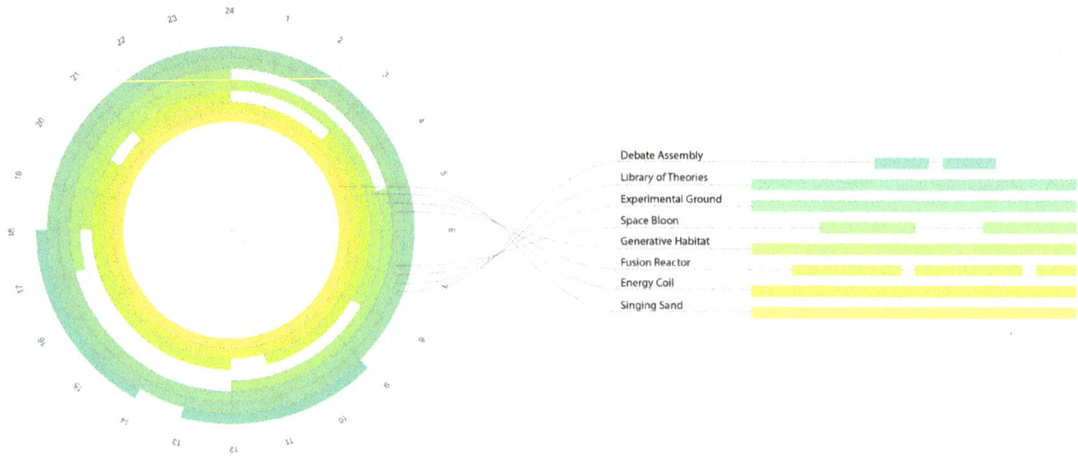

Debate Assembly
Library of Theories
Experimental Ground
Space Bloon
Generative Habitat
Fusion Reactor
Energy Coil
Singing Sand

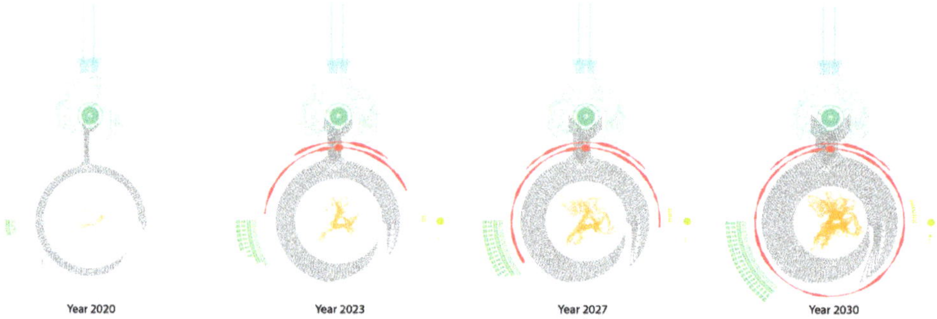

Year 2020
Year 2023
Year 2027
Year 2030

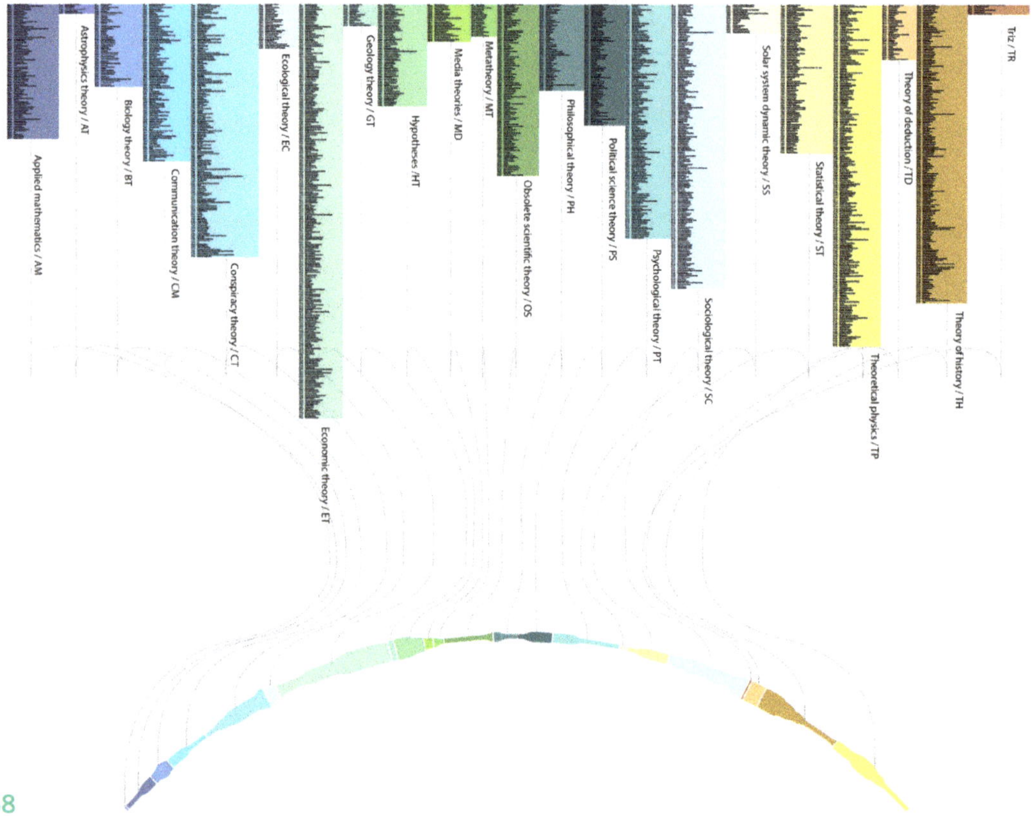

Applied mathematics / AM
Astrophysics theory / AT
Biology theory / BT
Communication theory / CM
Conspiracy theory / CT
Ecological theory / EC
Economic theory / ET
Geology theory / GT
Hypotheses /HT
Media theories / MD
Metatheory / MT
Obsolete scientific theory / OS
Philosophical theory / PH
Political science theory / PS
Psychological theory / PT
Sociological theory / SC
Solar system dynamic theory / SS
Statistical theory / ST
Theoretical physics / TP
Theory of deduction / TD
Theory of history / TH
Triz / TR

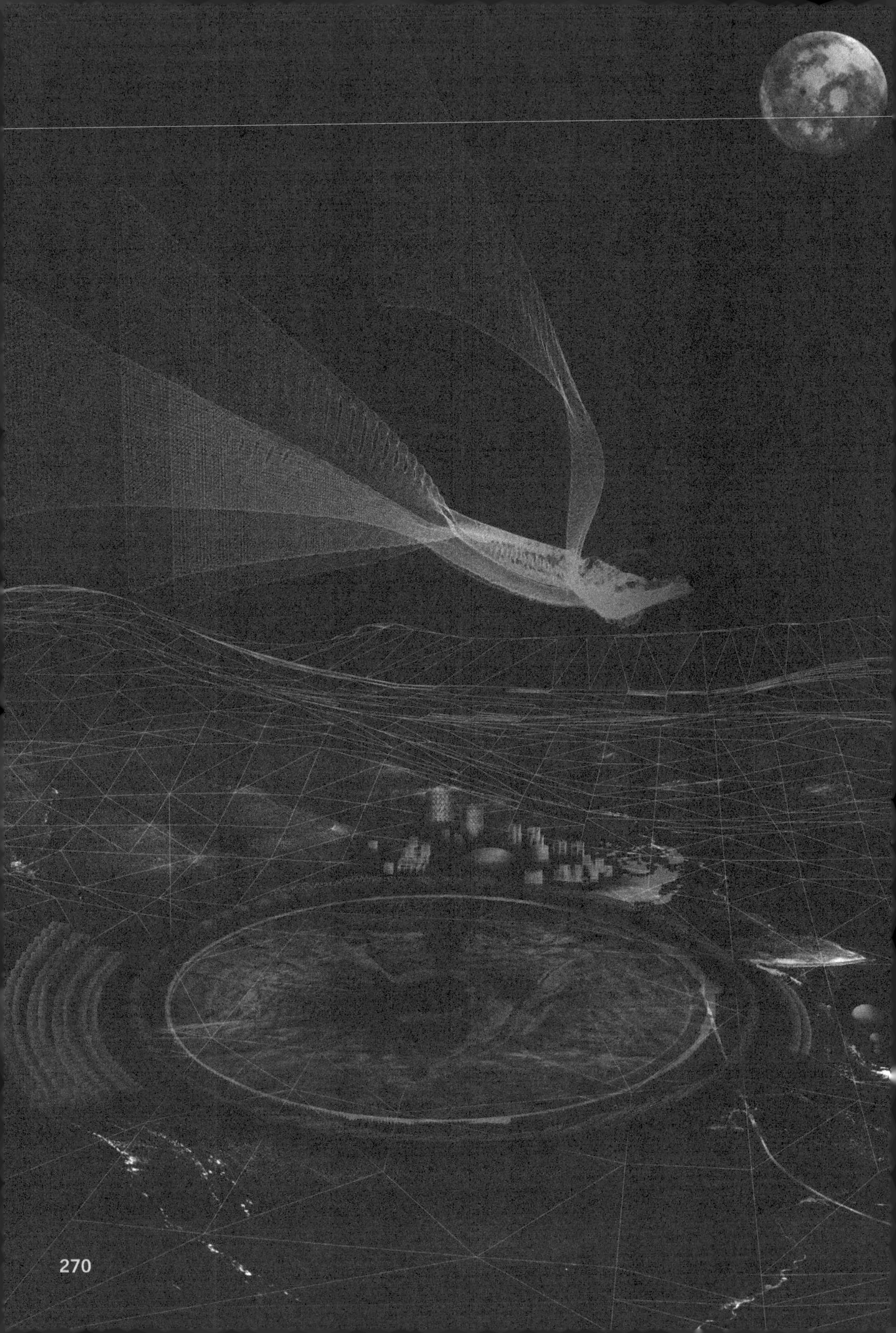

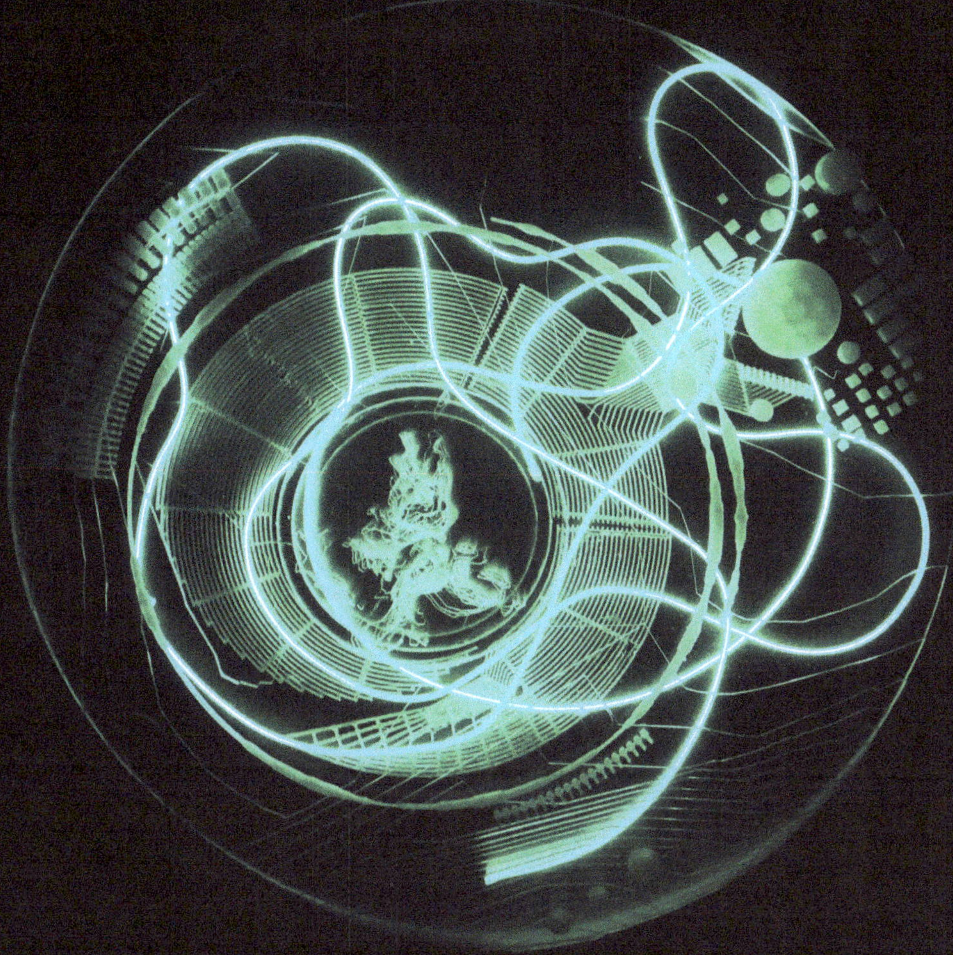

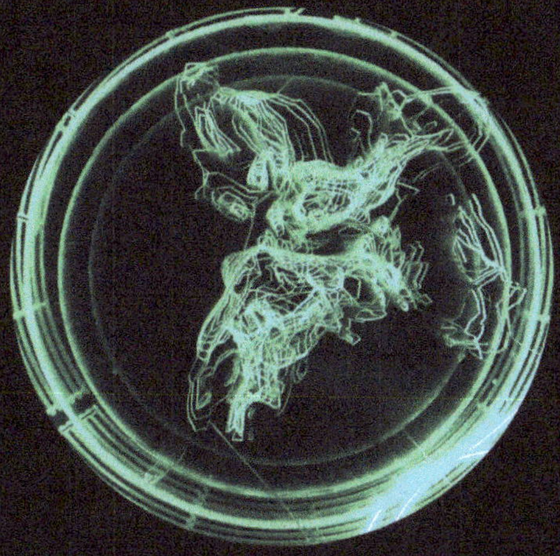

NIEU BETHESDA EXPERIMENTAL WIND COMMUNITY

SHIUE NEE PANG

The Experimental Wind Community proposed for Nieu Bethesda in the Karoo is part of the FREE initiative to pioneer future ways of living in tune with earth energies. Run by Google X, the community consists of an encampment of ten to twelve living units energised by Altaeros's Buoyant Air Turbines (BAT), which capture the energy of air movement 300m above the earth. The community ideology is based on a return to indigenous Khoisan wisdom, particularly beliefs and concepts of life attuned with the phenomenon of wind. From acute observations of the sky and ethnographic knowledge of the wind, economies of pastoralization, animal farming and hunting are cultivated once more.

Familiarity with the sky as an aesthetic object associated with popular forms of weather knowledge (epistemologies of rain, clouds and storms) forms the conceptual ground for integrating culture and science in the design. The proposed community is a composite of nephologists, meteorologists, local families and the energy team. In collectively gathering and sharing atmospheric data using a range of approaches such as a hunters' intuition of the wind, age old radiosonde, photo apps, and BAT infrastructure, knowledge of the air will be reclaimed for future energy and environmental developments.

Also considered in the project is that air tourism generates enthusiasts, both scientists and hobbyists, seeking total immersion in the activities of observing and recording the atmosphere. Geared with their own personal instruments such as charts of the mountains, polariscopes and theodolites they seek refuge in a community that will provide a place of solidarity between humans and the sky.

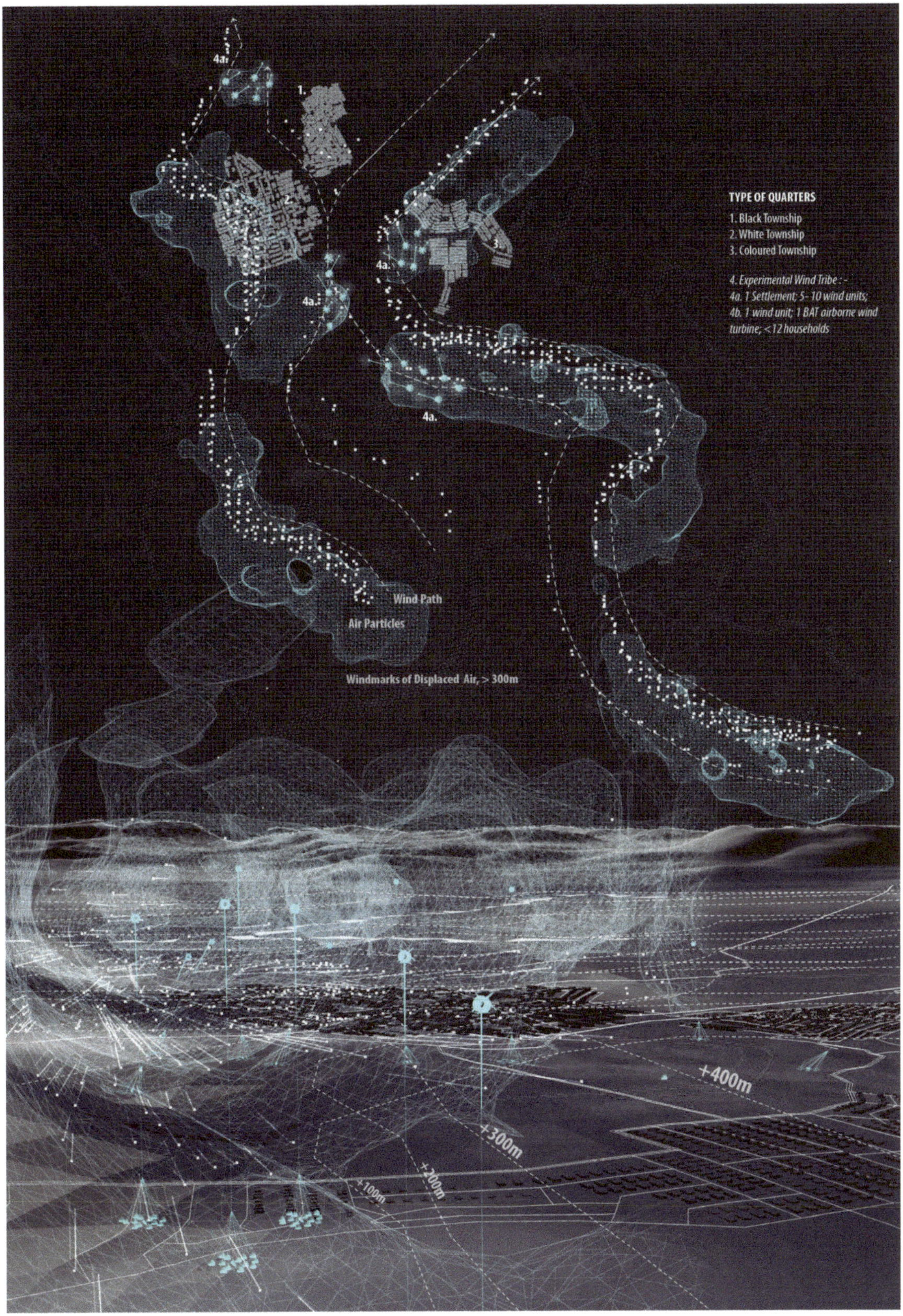
4a.
1.
2.
3.
4a.
4a.
4a.
TYPE OF QUARTERS
1. Black Township
2. White Township
3. Coloured Township
4. Experimental Wind Tribe : -
4a. 1 Settlement; 5- 10 wind units;
4b. 1 wind unit; 1 BAT airborne wind turbine; <12 households
Wind Path
Air Particles
Windmarks of Displaced Air, > 300m
+400m
+300m
+200m
+100m

P273 Experimental Wind Tribe

P274 Farmer's Almanac: A Guide to Reading the Sky

[top] Almanac 1: "The higher the clouds, the better the weather."

a. Cirrus: These filamentous calligraphy are exclusively made of ice crystals, and a sure sign of a good day.

b. Stratocumulus: These clouds are seen often and everywhere. They are individual rounded clouds that have been pushed together to form a layer. They bring little precipitation.

c. Cumulus: Small, flat-bottomed puffs, with rounded tops that drift across blue skies and bring good weather.

d. Stratus: Uniform, flat clouds, with little definition. They bring cover from the sun but little precipitation.

[middle] Almanac 2

e. Cirrocumulus: High, ice crystal clouds with visible rows or ripples. They exist briefly, with no immediate weather consequence.

f. Cirrostratus: These high icy clouds form a featureless, hazy layer. Sometimes they produce characteristic halos as sun or moonlight refracts through their crystals. Folklore states: "Halo around the sun or moon, rain or snow is coming soon."

g. Altocumulus: Middle-level, heap-like clouds, often appearing in rows with blue sky between reminiscent of the sand ripples in a tidal pool. They produce the sailor's fabled mackerel sky: "Mare's tails and mackerel scales make tall ships take in their sails."

[bottom] Almanac 3: Stormy Weather

h. Cumulonimbus: Very dark, flat bottomed clouds with immense tops that frequently build into an anvil shape. Folklore says: "when clouds appear like rocks and towers, the earth will be washed by frequent showers."

i. Nimbostratus: Dark gray to deep blue, layered clouds, parts of which are blurry because of visible precipitation. In areas with distant horizons, one can see these clouds as an approaching storm.

j. Altostratus: Flat, uniform sheets of gray or blue-gray flannel, layered across the sky without feature or differentiation. While they are cloud light, they weigh on the mind.

Sources: L. Howard, On the modifications of clouds, London: J. Taylor, 1804; http://www.metoffice.gov.uk/learning/clouds/cloud-names-classifications; http://nenes.eas.gatech.edu/Cloud/Clouds.pdf; http://www.almanac.com/weather.

P275 B.A.T (Buoyant Air Turbine) Masterplan: Altaeros's B.A.T wind turbines are located according to geographical factors and the local economy within a 3km x 3km grid of Nieu Bethesda

P276 Reading the Sky: Assimilating scientific and indigenous knowledge of the sky

P277 Tracking Wind Movement above Nieu Bethesda

P278 Inhabiting the Air: In trying to understand the form of the air, a gridded structure manifests from the assembly of loops along a spline

P279 [top left] Structural Model

[top right] The architecture relies on the environment to both generate energy and provide passive ventilation. Much like the burrowing of air moving through space, the building itself becomes a burrowed dwelling

[bottom] Section

P280 Ground Floor Plan

P281 The Cloud House

a.
b.
c.
d.

"If you see these clouds, you are likely to have a full and fine day ahead."

e.
f.
g.

"Okay Today, Not so Much Tomorrow"

h.
i.
j.

"When clouds appear like rocks and towers, The earth will be washed by frequent showers."

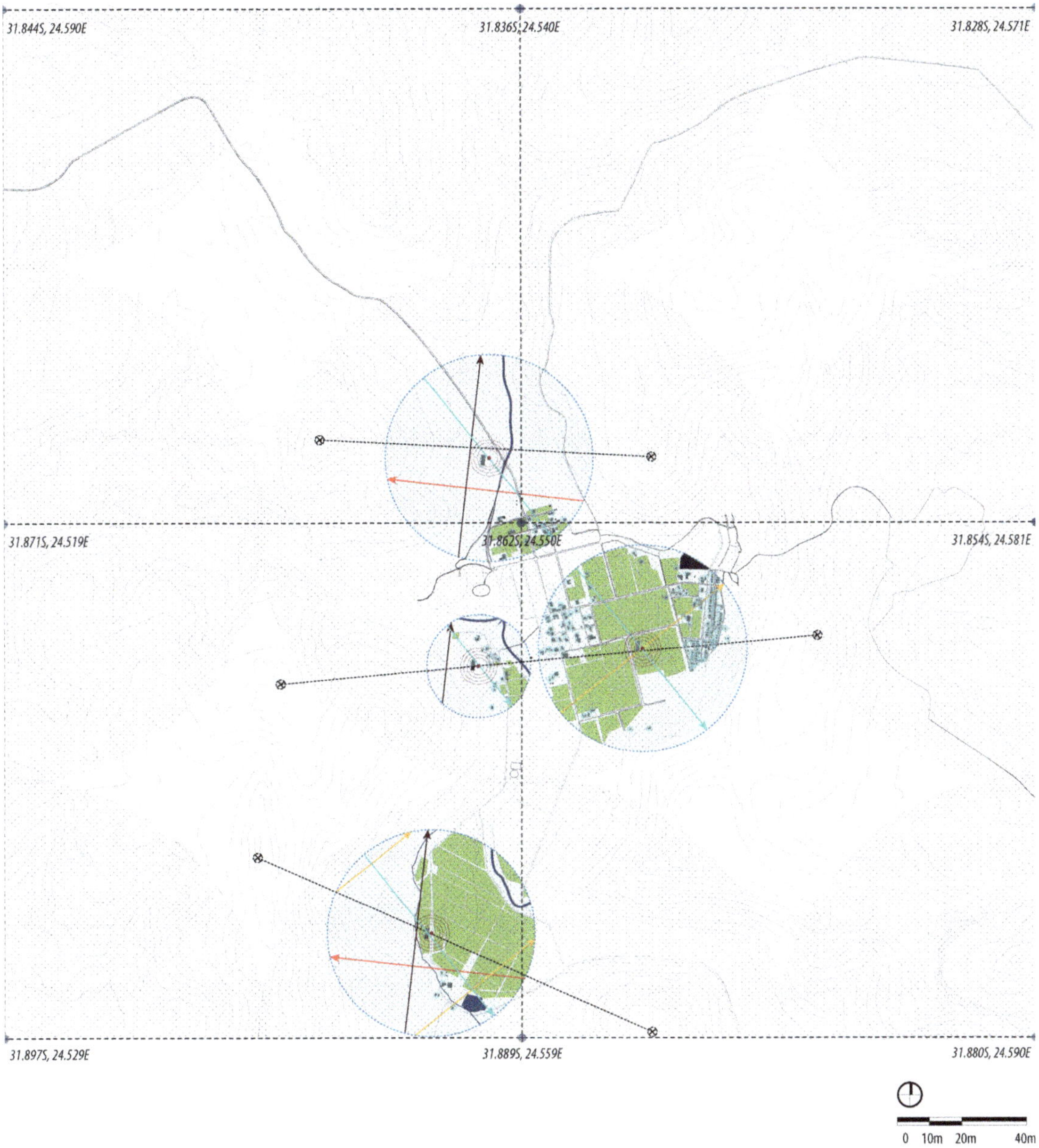

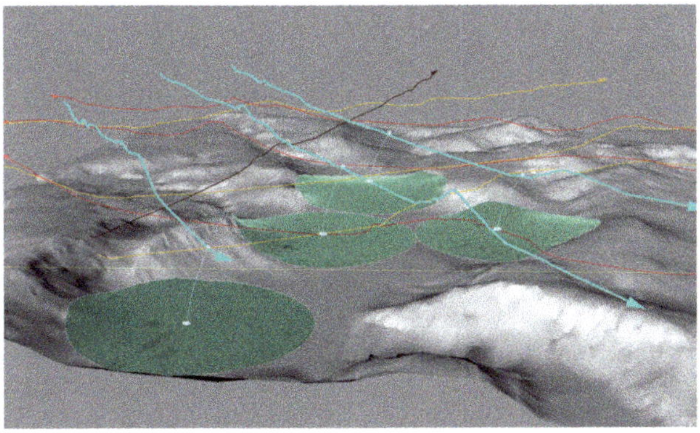

LEGEND

North-Eastern Wind (Prevailing Wind)
South-Eastern Wind
Southern Wind
Eastern Wind

Valleys- create naturally controlled wind paths which optimises wind energy for BAT infrastructures.

Agricultural Land

Wind Data Areas:
Radiosonde Release Points

BAT infrastructure - Data Collection + Wifi Station

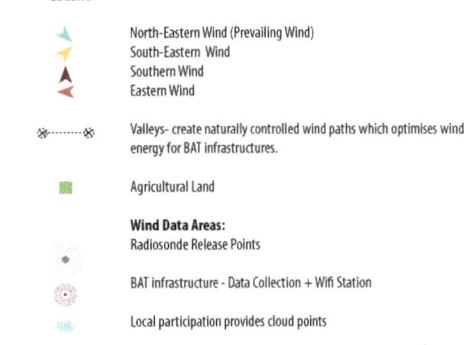

Local participation provides cloud points

AUTUMN / STORMY SEASON
NORTH-EASTERN WIND/ PREVAILING WIND

WINTER / HARVESTING SEASON
SOUTH-EASTERN WIND

SPRING / HUNTING SEASON
SOUTHERN WIND

SUMMER /ASTRONOMICAL VIEWING
EASTERN WIND

North-Eastern Wind (Prevailing Wind)
South-Eastern Wind
Southern Wind
Eastern Wind

Building Site — *Nieu-Bethesda Radiosonde Release Station*
B.A.T radius — *Google X Floating Wind Energy Infrastucture*
31.862S, 24.550E — *World Atlas of South Africa Wind Grid*

Integrating Wind Data Analysis

0 300m 600m

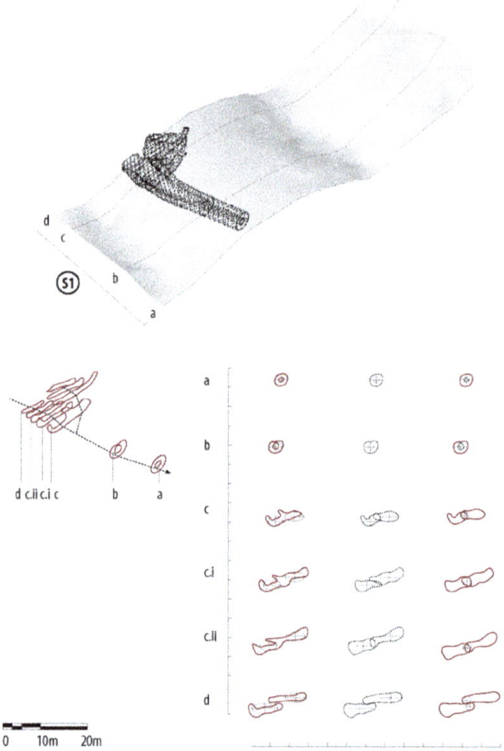
S1
d
c
b
a
d c.ii c.i c b a
a
b
c
c.i
c.ii
d
0 10m 20m

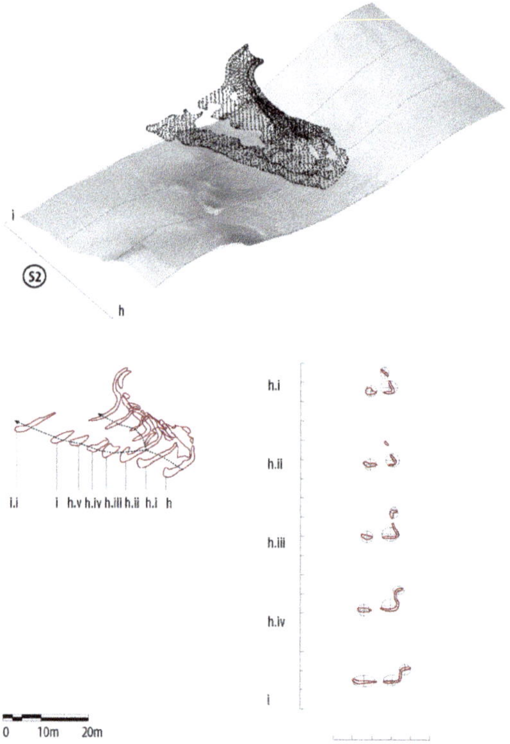
S2
i
h
i.i i h.v h.iv h.iii h.ii h.i h
h.i
h.ii
h.iii
h.iv
i
0 10m 20m

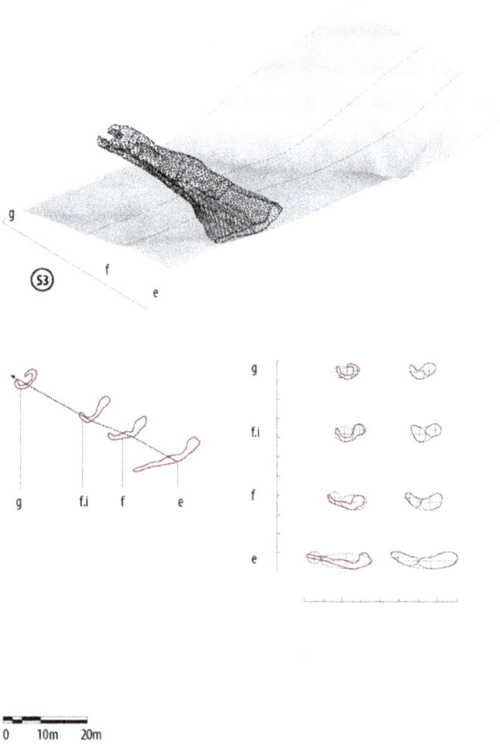
S3
g
f
e
g f.i f e
g
f.i
f
e
0 10m 20m

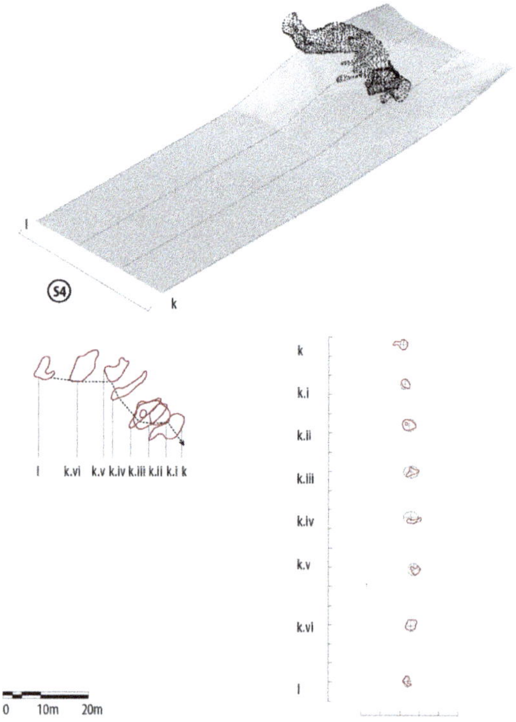
S4
l
k
l k.vi k.v k.iv k.iii k.ii k.i k
k
k.i
k.ii
k.iii
k.iv
k.v
k.vi
l
0 10m 20m

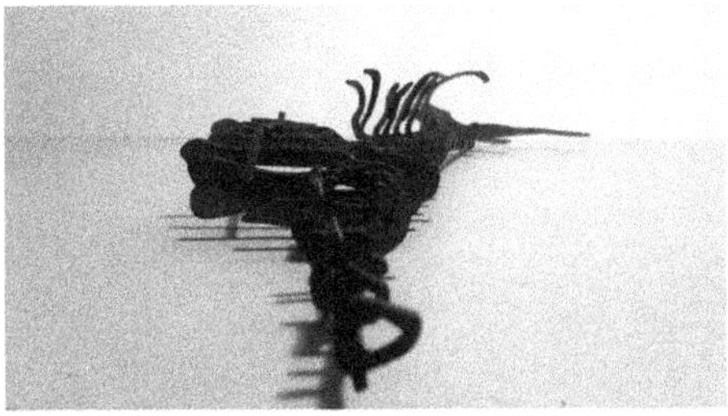

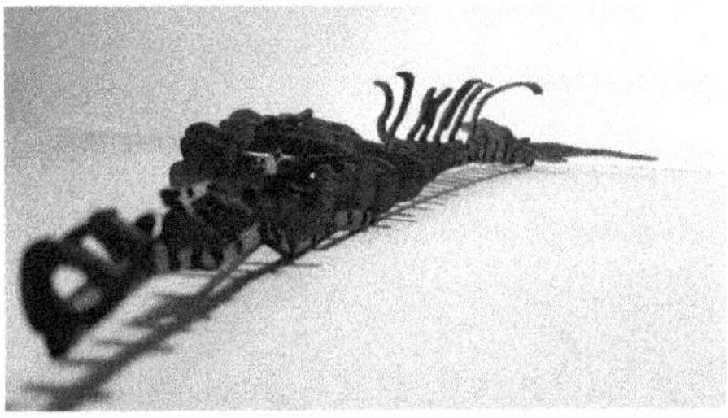

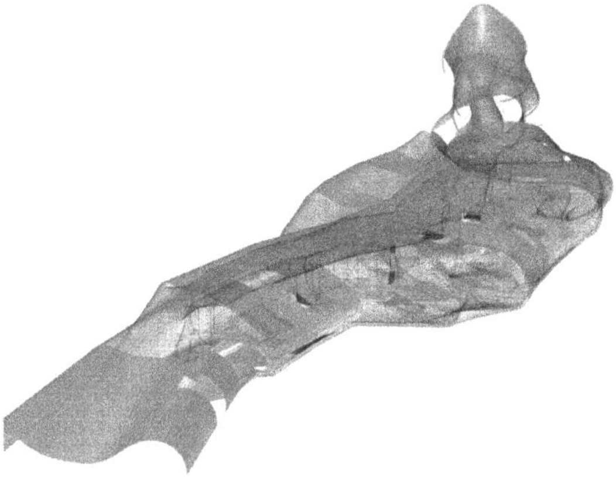

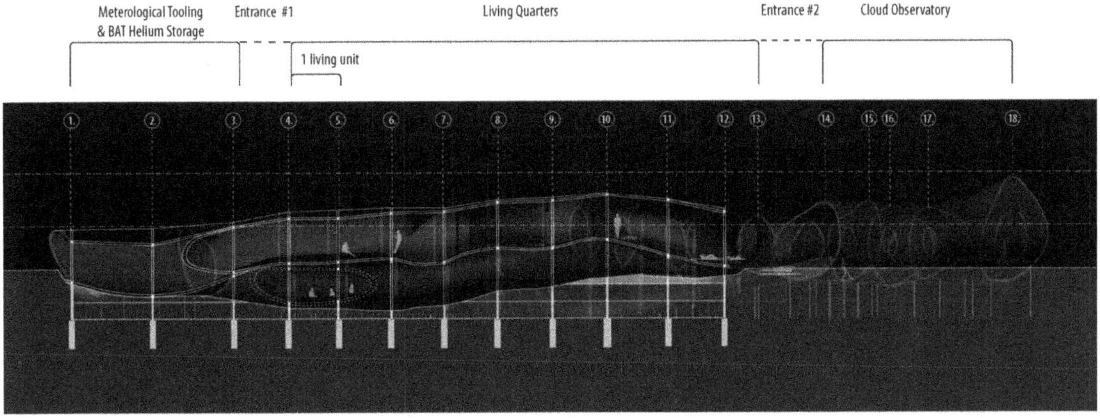
Meterological Tooling
& BAT Helium Storage
Entrance #1
Living Quarters
Entrance #2
Cloud Observatory
1 living unit

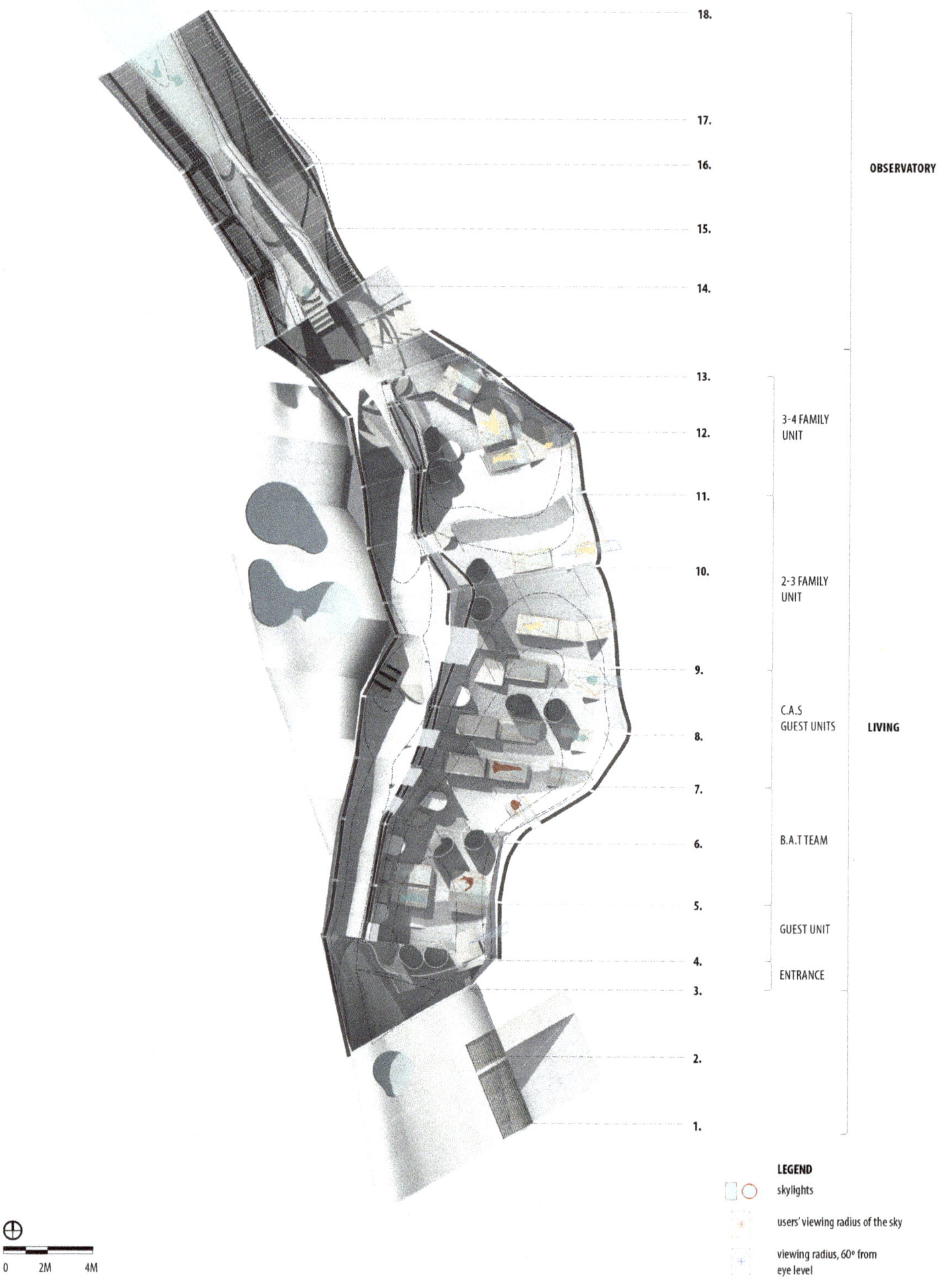
18.
17.
16.
15.
14.
13.
12.
11.
10.
9.
8.
7.
6.
5.
4.
3.
2.
1.
OBSERVATORY
3-4 FAMILY UNIT
2-3 FAMILY UNIT
C.A.S GUEST UNITS
LIVING
B.A.T TEAM
GUEST UNIT
ENTRANCE
LEGEND
skylights
users' viewing radius of the sky
viewing radius, 60° from eye level
0
2M
4M

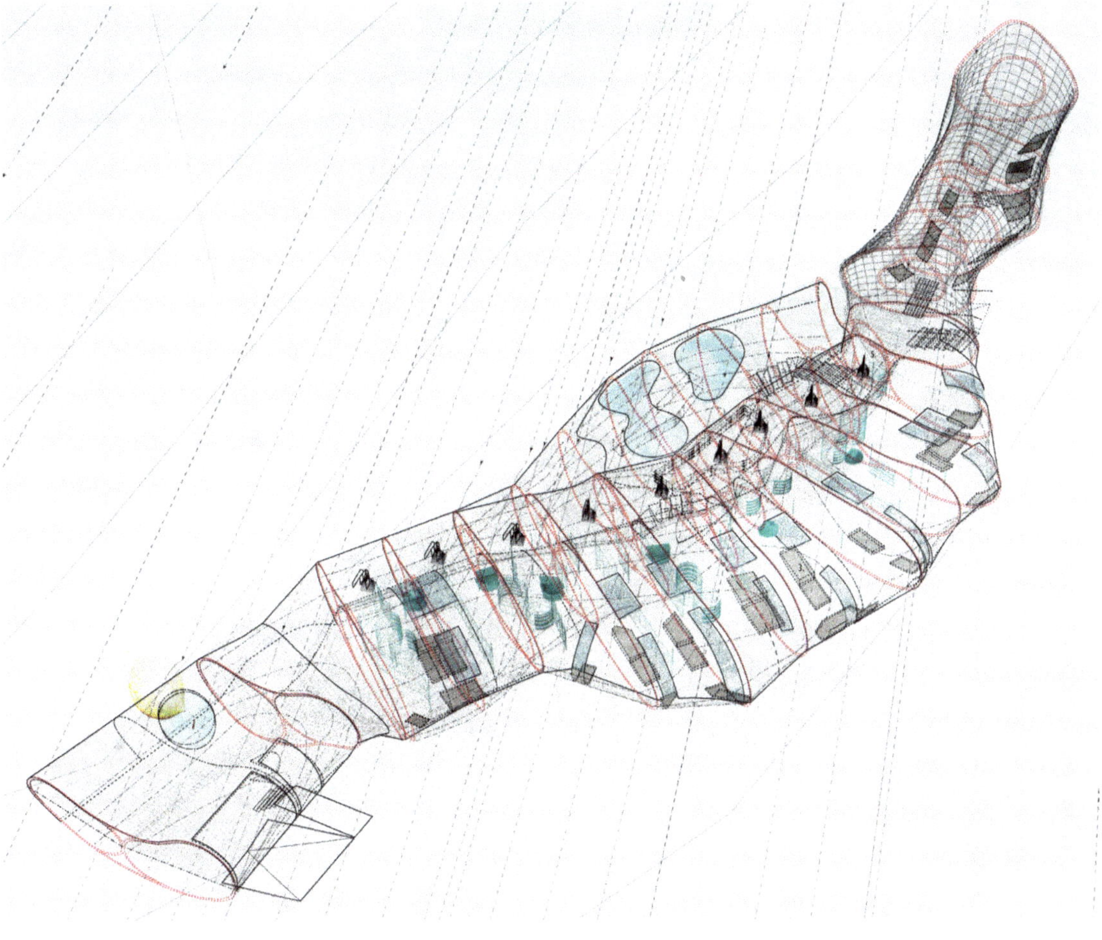

WIND SEED 01

OSCAR MCDONALD

The objective of Wind Seed 01 is to address low education and unemployment in the town of Graaff Reinet in the Karoo, South Africa. The basic Seed is an easily transportable kit containing five one-kilowatt wind turbines and a single simple workshop. This workshop once assembled facilitates the production of further turbines and goods and, if successful, the next generation of units. Members of the project are encouraged to create a multidisciplinary workforce suited to small-scale entrepreneurship. A single 'Seed' can be fully sourced and constructed for around £7500 but is envisioned to contain the potential to reprogram the spatial and economic properties of entire settlements for the better.

Drawing on ideas from Von Neumann's cellular automatons and Conway's Game of Life, the seeded elements subtract and evolve with the life of the project. The buildings and turbines respond in an adaptive manner to the fluctuations in the wind and are flexible in their design, with density, height, orientation and even detailed construction parametrically flexible within the rules of the game.
The existence of objects is an emergent property of the social, economic and natural conditions in which they are produced. The goal of this evolutionary approach is to allow successes to multiply and failures to recede in a way that avoids the white elephants and constrained successes common in community based projects.

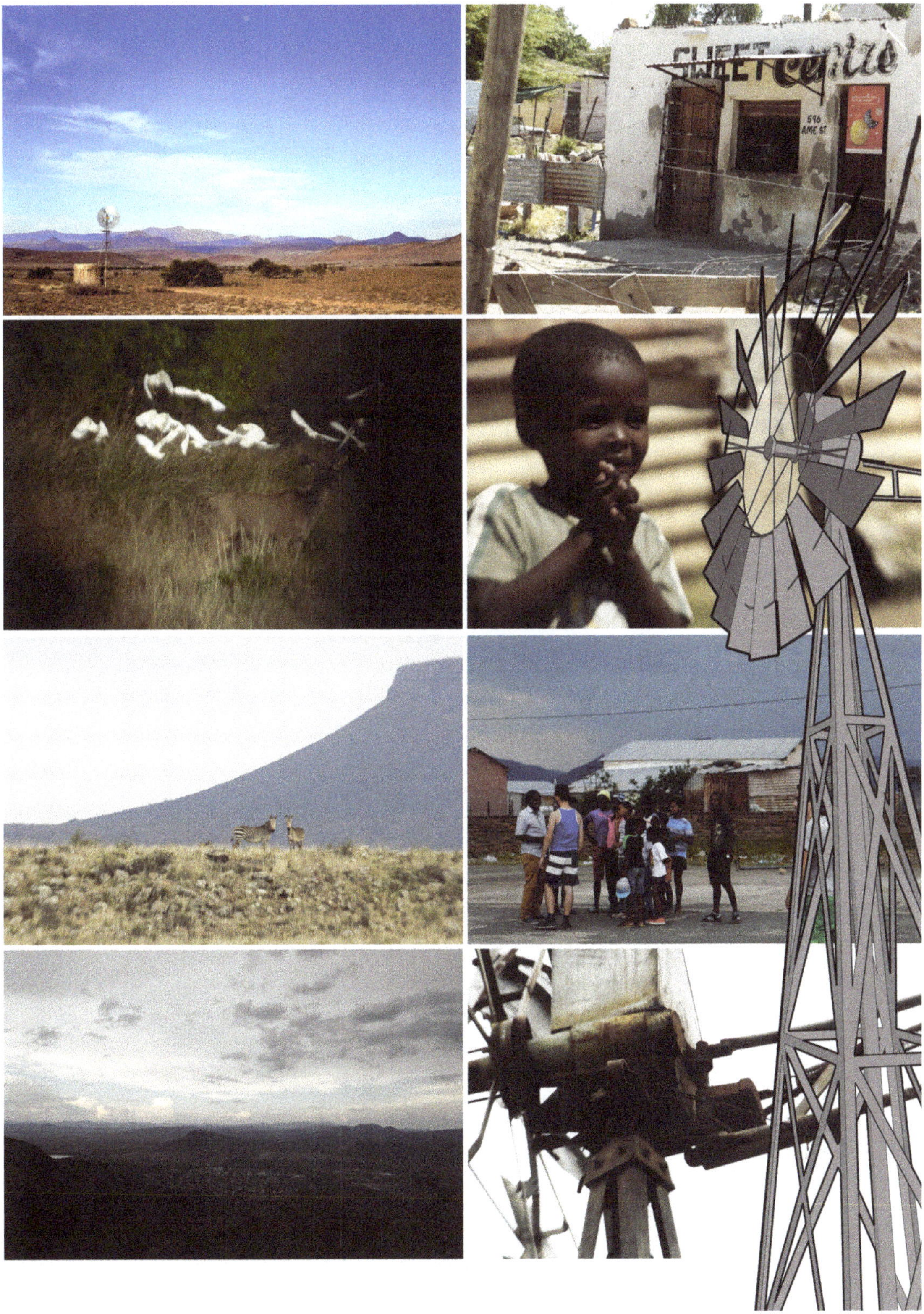

P283 The Karoo: Beauty, Poverty, Segregation
Images show the stark contrast between the beauty of the natural environment and the difficult realities of life for the poor in the Karoo. There is an unemployment rate of 45% with HIV prevalence estimated at 25%. 65% of the population have no secondary school education.
Photographs: Oscar McDonald except second from bottom on the right by Martino Gasparrini

P285 Workshop 01 : Initial 'Seed' Workshop and Five Turbines

P286 Growth Concept: An Adaptable, Self-generating System
This diagram shows the business plan for the Wind Seed project beginning with the purchasing of the initial 'Seed' and showing the growth of the system in perpetuity. The project functions as a community collective, which simultaneously educates and employs the local population. The first circle shows the erection of the 'Seed' workshop, the training of staff and the progressively fast production and sale of wind turbines. Turbine sales allow a certain amount of profit to be accrued, which eventually allow the purchase of new turbines and workshops for on site expansion. If successful these new workshops or commercial units then begin the process of production and expansion themselves.

P287 Value and Density: Developing a Universal Grid and Data Driven Language
This drawing describes the system by which the grid and spatial elements of the project are determined. Computational techniques are used to analyse the site according to universal factors and a number is assigned which relates to 'value' to the Wind Seed project. This value allows a density to be acquired; the number of units or centres in a certain area, this density rises in tandem with value meaning a centre always has a consistent unit of value in its vicinity. It also gives rise to a triangulated grid, which can easily adapt to the vast majority of site constraints.

P288 Site Selection: The Intermediate Zone
This process determines areas of high wind speed and human capital potential that are neither rural nor fully urban; both factors are crucial to the success of the project. Numbers relate to value as determined by average wind speed, urban adjacency amongst other considerations. Importantly, use of intermediate areas also begin to exploit the non-urban environment in a more intensive but crucially small scale way. This fills the role of low intensity farming in more fertile landscapes

P289 Site 01 Scenario 01
The project at around +300 weeks following successful growth of the system at a mid-high rate. This shows the growth of the units within the variable grid along with earlier turbines. Adjacent wind turbines determine workshop heights and locations. New units may require the height to be reduced to allow better wind flow or the movement of local turbines.

P290 Game Piece 01: Workshop / Commercial Unit
The unit itself is designed to be easily transportable and highly functional. The unit is fully parametric and angles away from the intense sun using corrugated plastic to allow in light on the northern sides. The building also hovers above the ground to dissipate heat and allow the better passage of wind around the building. The walls are constructed using the corrugated sheet metal common to the area in conjunction with the more vernacular use of adobe, which is sandwiched between two layers of sheet metal. The units maximise access and interaction with communal space; a vital component of this project.

P291 The Semi-Urban Windscape: Public Square and Informal Restaurant

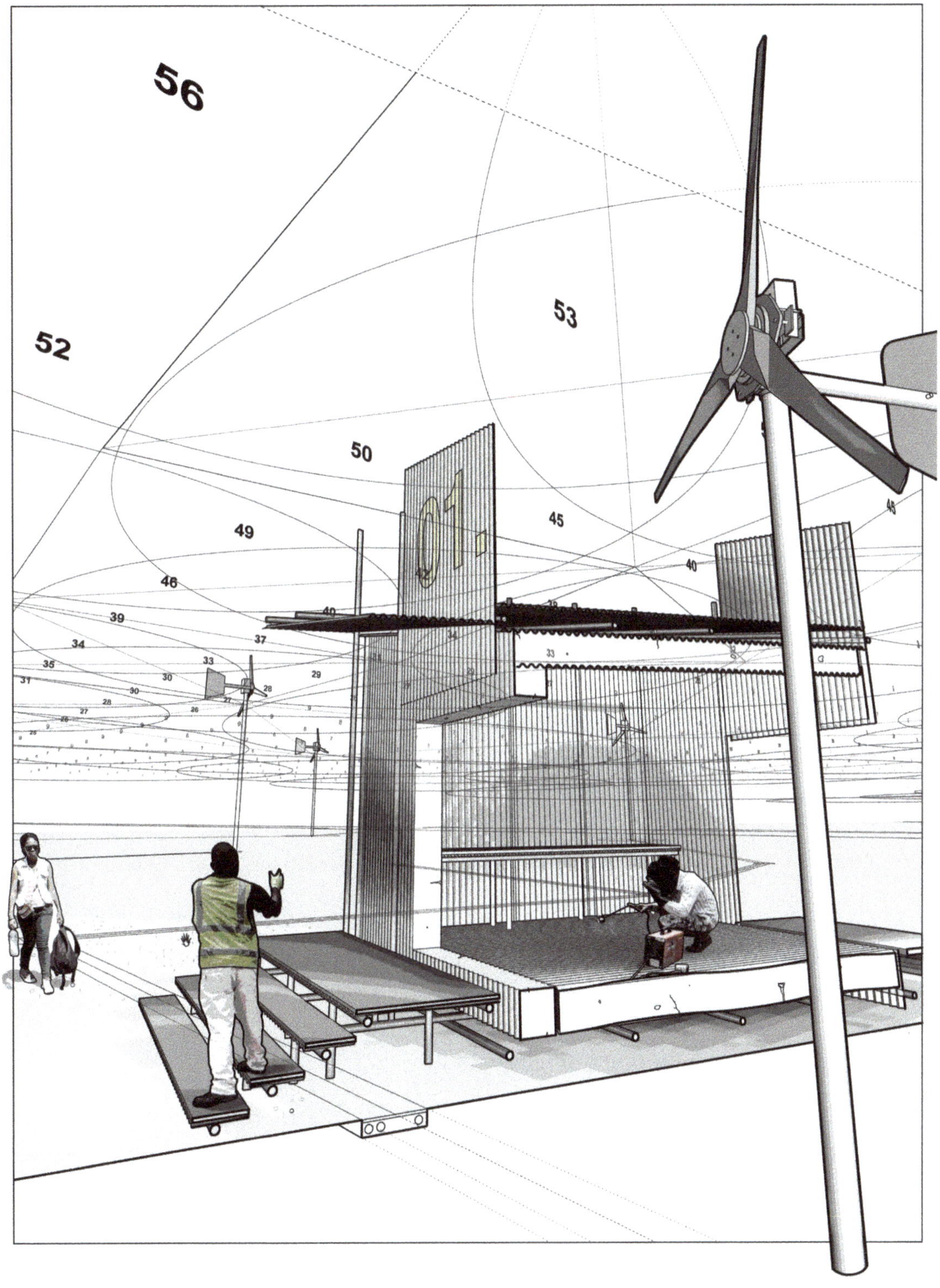

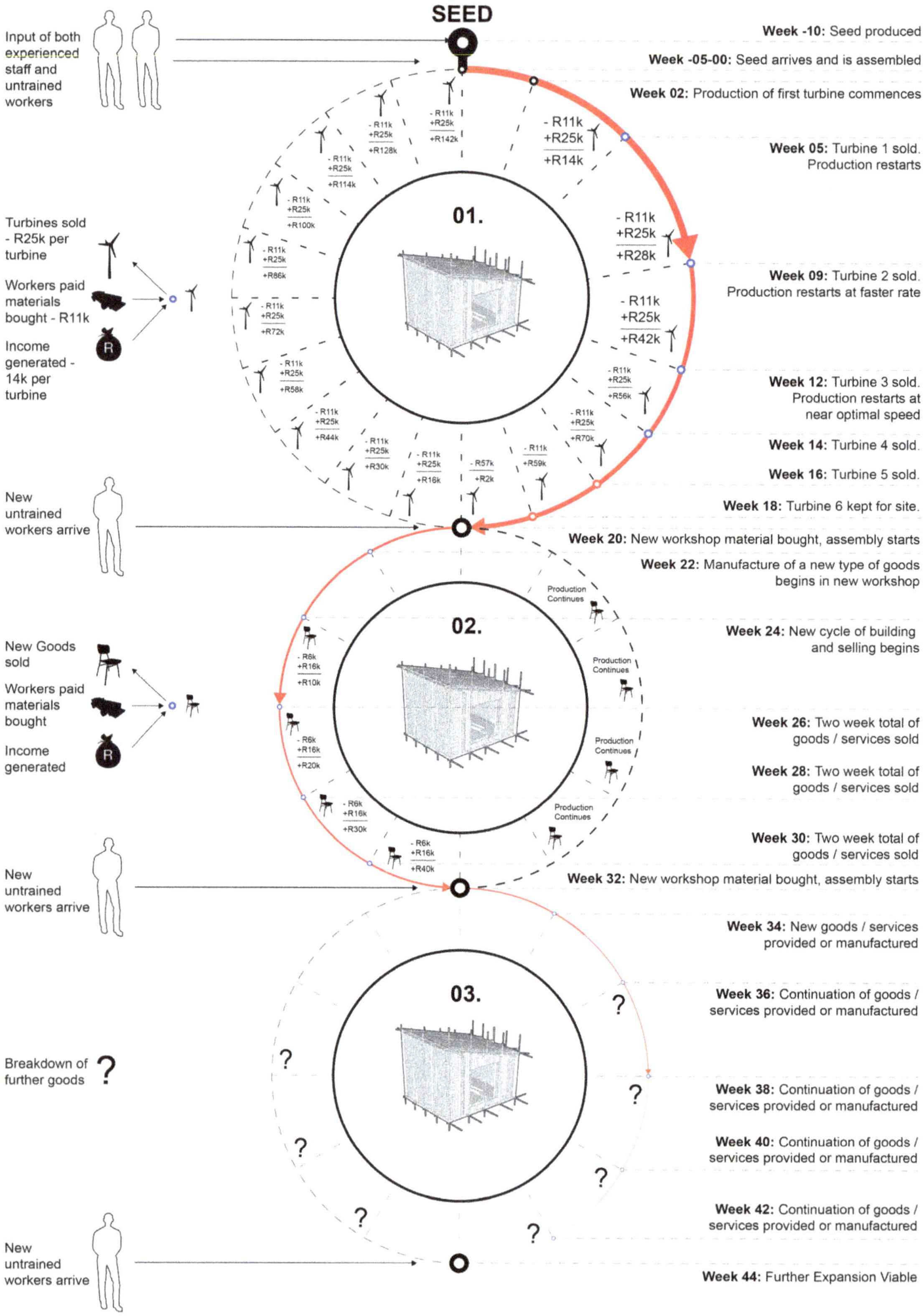

SEED
Input of both experienced staff and untrained workers
Week -10: Seed produced
Week -05-00: Seed arrives and is assembled
Week 02: Production of first turbine commences
Week 05: Turbine 1 sold. Production restarts
Week 09: Turbine 2 sold. Production restarts at faster rate
Week 12: Turbine 3 sold. Production restarts at near optimal speed
Week 14: Turbine 4 sold.
Week 16: Turbine 5 sold.
Week 18: Turbine 6 kept for site.
01.
- R11k +R25k +R14k
- R11k +R25k +R28k
- R11k +R25k +R42k
- R11k +R25k +R56k
- R11k +R25k +R70k
- R11k +R59k
- R57k +R2k
- R11k +R25k +R16k
- R11k +R25k +R30k
- R11k +R25k +R44k
- R11k +R25k +R58k
- R11k +R25k +R72k
- R11k +R25k +R86k
- R11k +R25k +R100k
- R11k +R25k +R114k
- R11k +R25k +R128k
- R11k +R25k +R142k
Turbines sold - R25k per turbine
Workers paid materials bought - R11k
Income generated - 14k per turbine
New untrained workers arrive
Week 20: New workshop material bought, assembly starts
Week 22: Manufacture of a new type of goods begins in new workshop
Week 24: New cycle of building and selling begins
Week 26: Two week total of goods / services sold
Week 28: Two week total of goods / services sold
Week 30: Two week total of goods / services sold
02.
Production Continues
Production Continues
Production Continues
Production Continues
- R6k +R16k +R10k
- R6k +R16k +R20k
- R6k +R16k +R30k
- R6k +R16k +R40k
New Goods sold
Workers paid materials bought
Income generated
New untrained workers arrive
Week 32: New workshop material bought, assembly starts
Week 34: New goods / services provided or manufactured
Week 36: Continuation of goods / services provided or manufactured
Week 38: Continuation of goods / services provided or manufactured
Week 40: Continuation of goods / services provided or manufactured
Week 42: Continuation of goods / services provided or manufactured
03.
?
Breakdown of further goods
New untrained workers arrive
Week 44: Further Expansion Viable

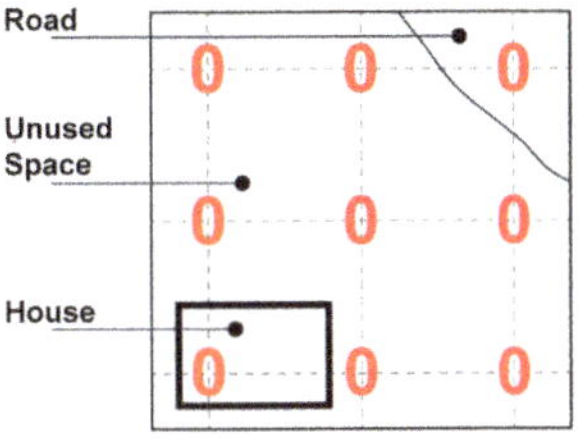

01.
THE GRID

The grid is laid out over the plan of the existing site. A number is assigned to each point which will eventually represent the 'value' of that point.

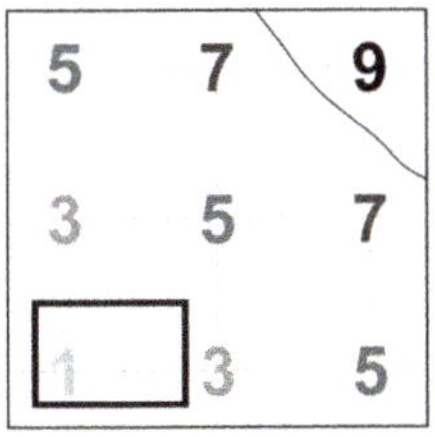

02.
HUMAN CAPITAL 01. CONECTIVITY

These parameters determine value primarily associated with ease of access, conectivity and visibility. This example shows a simple distance to road caluculation.

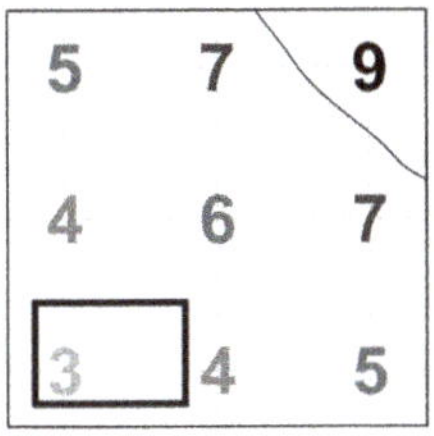

03.
HUMAN CAPITAL 02. ADJACENCY

Value due proximity to certain buildings. Proximiy to homes for example is less positive then to shops but it still has some value.

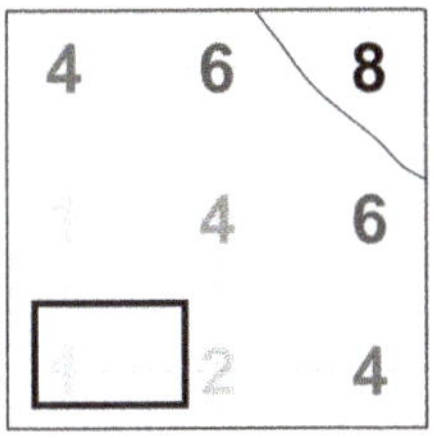

04.
WIND SHADOWS

Adjacency to obstructive objects causes reduction in the usefull wind in that point. This effect is increaced in the direction of the prevailing wind.

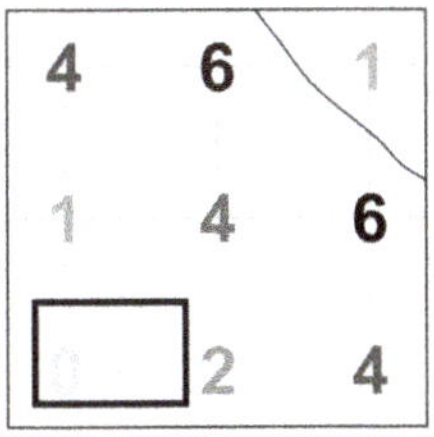

05.
USABILITY

Existing usage of the point reduces its value from the point of view of the project. Roads can be diverted or crossed and buildings demolished in extreme cases (the '1' value on the road shows some tendancy for crossing / other use at this point).

VALUE

This process determines the value to the project of each point on the master plan / plan. This value can then be used to establish the density and use of the space available.

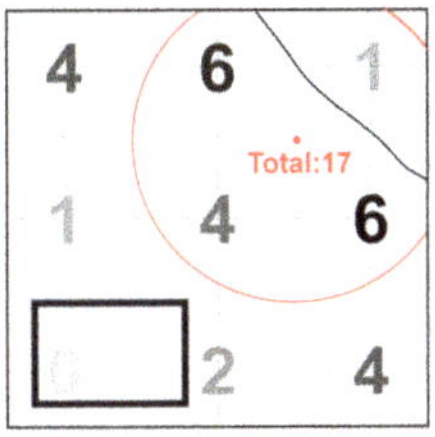

06.
DENSITIES FROM VALUES

Density is established as a concept proportional to value (higher values mean higher densities). Groups of values are ringed and totaled and the circle expands until reaching a certain total.

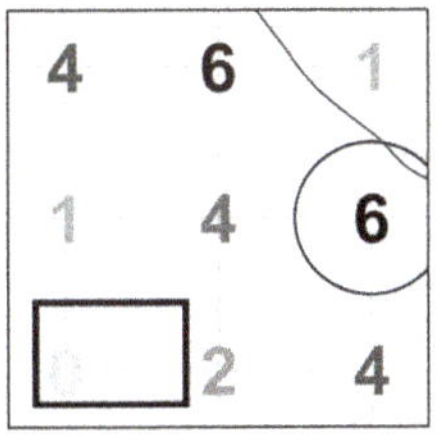

07.
MINIMUM DENSITY AND 'TARGET VALUE'

This finds the lowest usefully allowable distance between the two highest value areas. For example if we want a single grid cell distance between each centre, this makes the 'target value' equal to the highest value, in this case 6.

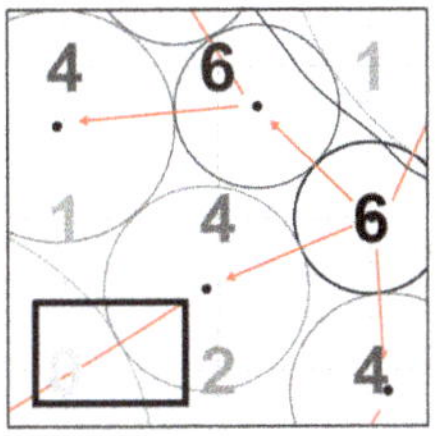

08.
GROWTH FROM HIGHEST POINT

The density calculation then grows out from the highest point in the direction of the next highest unused point and so on. The circles are adjusted to allow for the best teselation.

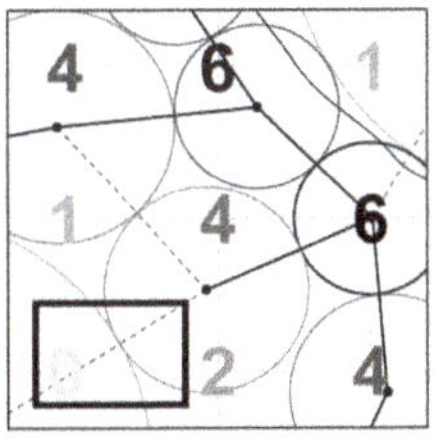

09.
CRYSTALISATION OF GRID

With centre points aquired a grid can be drawn and from which buildings and turbinescan be set out. Each centre froms an infrastructural link to the closest adjacent centre. Where routes are obstructed links are harder to form.

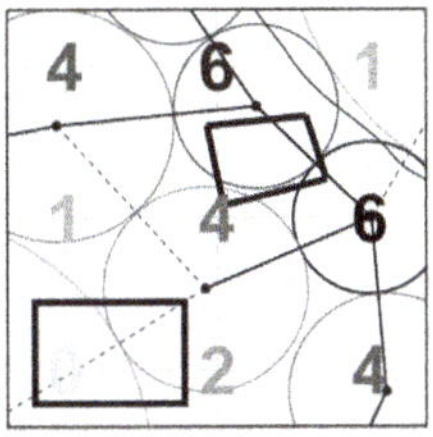

10.
BUILT FORM FORMATION AROUND GRID

The game elements (buildings, turbines...etc) can now start to be set out within this grid according to rules described later in the game.

DENSITY

This process gives an adaptable density that can respond to any multitude of constraints aswell as ensure the most appropriate use of space.

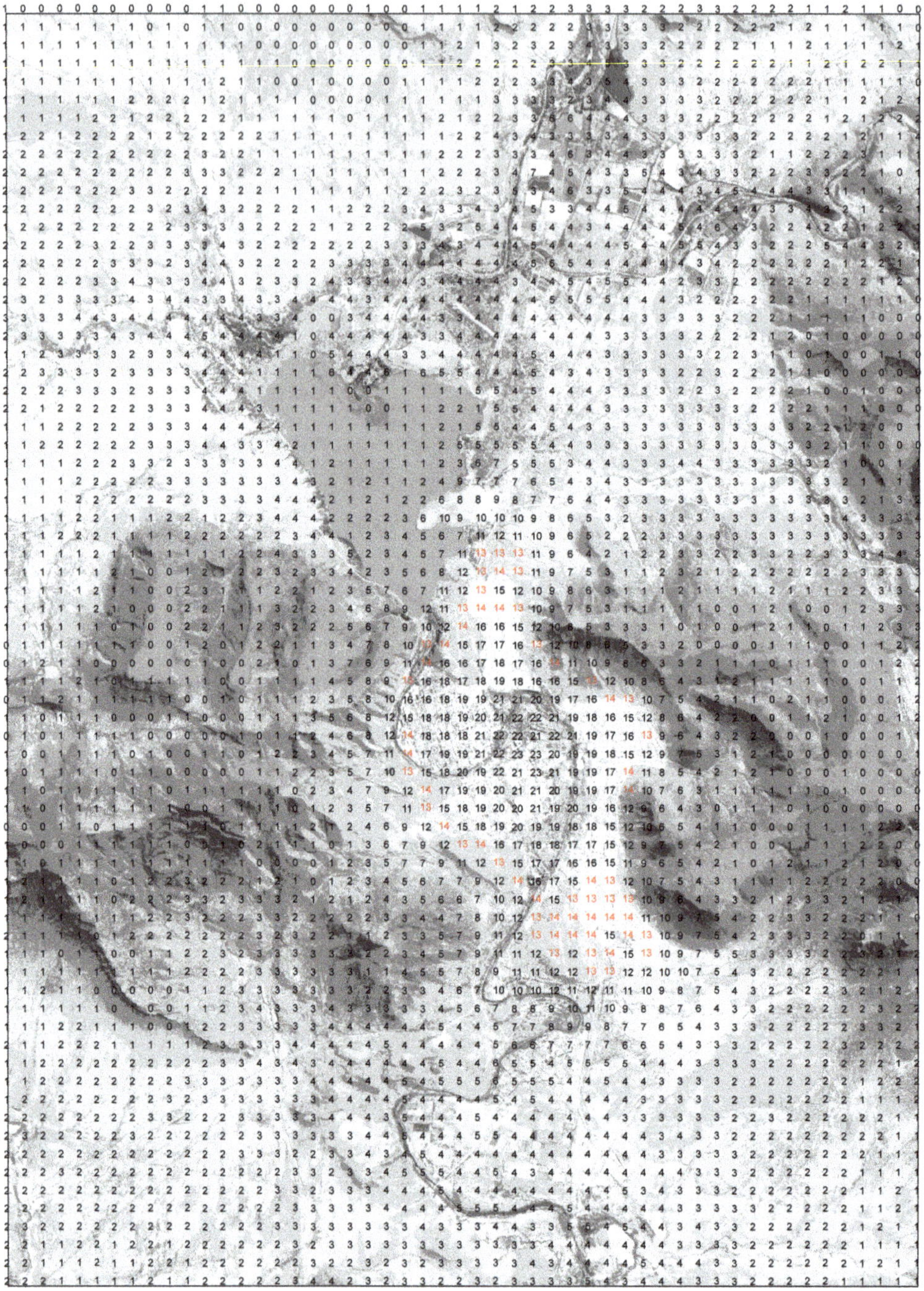

0km 1km 2km 3km 4km 5km

1:50000

N

Unit defined by total of Human Capital, Agricultural and Wind Potential (Arbitrary Scale) **1**

Unit defining intermediate value (suited to intermediate density) **13**

Windrose

Site

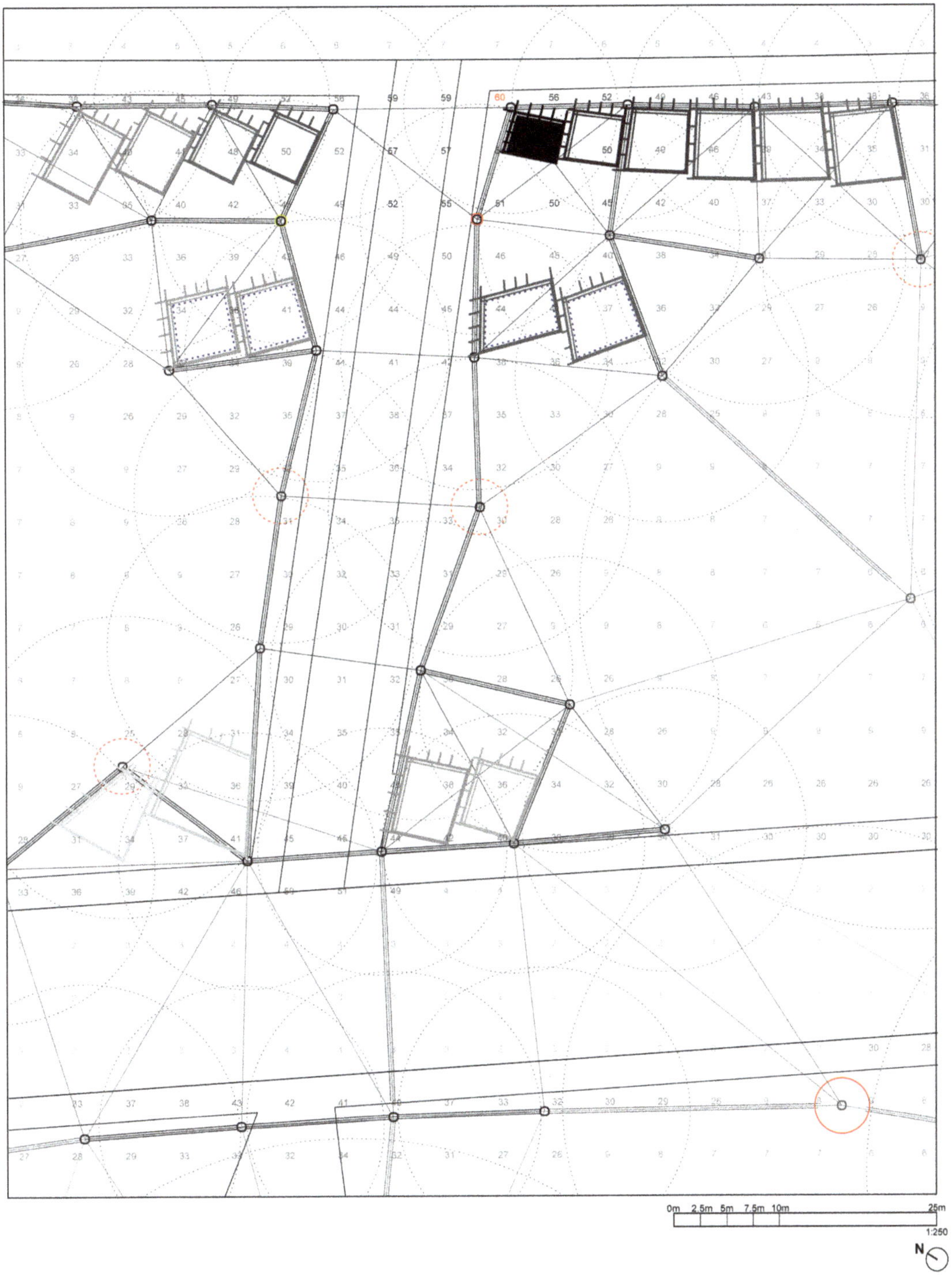

Total Power Windicator
Power Available Windicator
Service route linking each centre to its nearest centre
Crystallisation line linking each centre to its nearest 5 centres

Centre point of value totalling radius
Arbitrary value figure 45
Highest value, defining adjacent Workshop 01. 60
Turbine radius at centre, dotted line denotes relocation to accommodate growth
Public buildings for use as toilets, teaching blocks, water storage and batteries

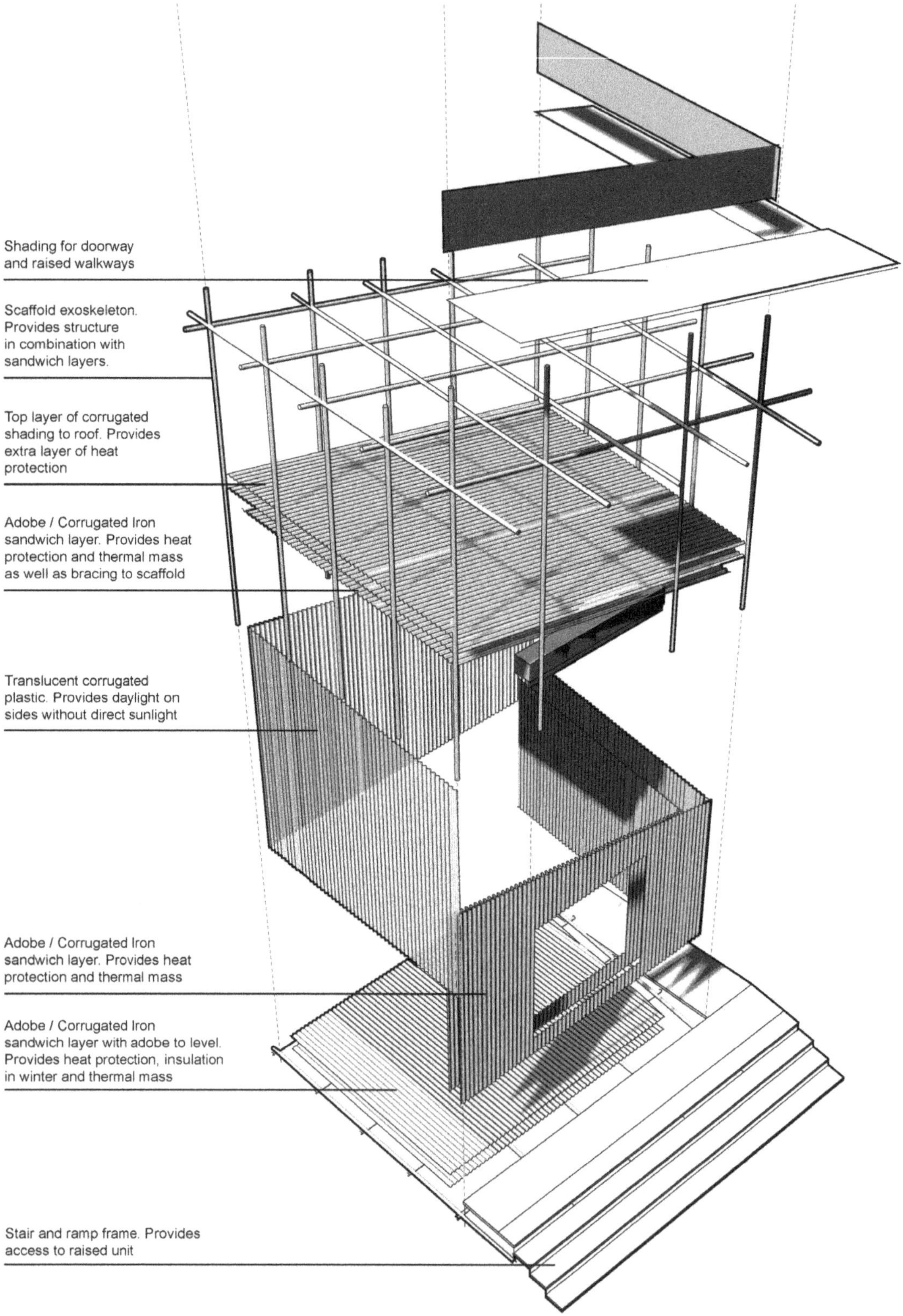
Shading for doorway and raised walkways
Scaffold exoskeleton. Provides structure in combination with sandwich layers.
Top layer of corrugated shading to roof. Provides extra layer of heat protection
Adobe / Corrugated Iron sandwich layer. Provides heat protection and thermal mass as well as bracing to scaffold
Translucent corrugated plastic. Provides daylight on sides without direct sunlight
Adobe / Corrugated Iron sandwich layer. Provides heat protection and thermal mass
Adobe / Corrugated Iron sandwich layer with adobe to level. Provides heat protection, insulation in winter and thermal mass
Stair and ramp frame. Provides access to raised unit

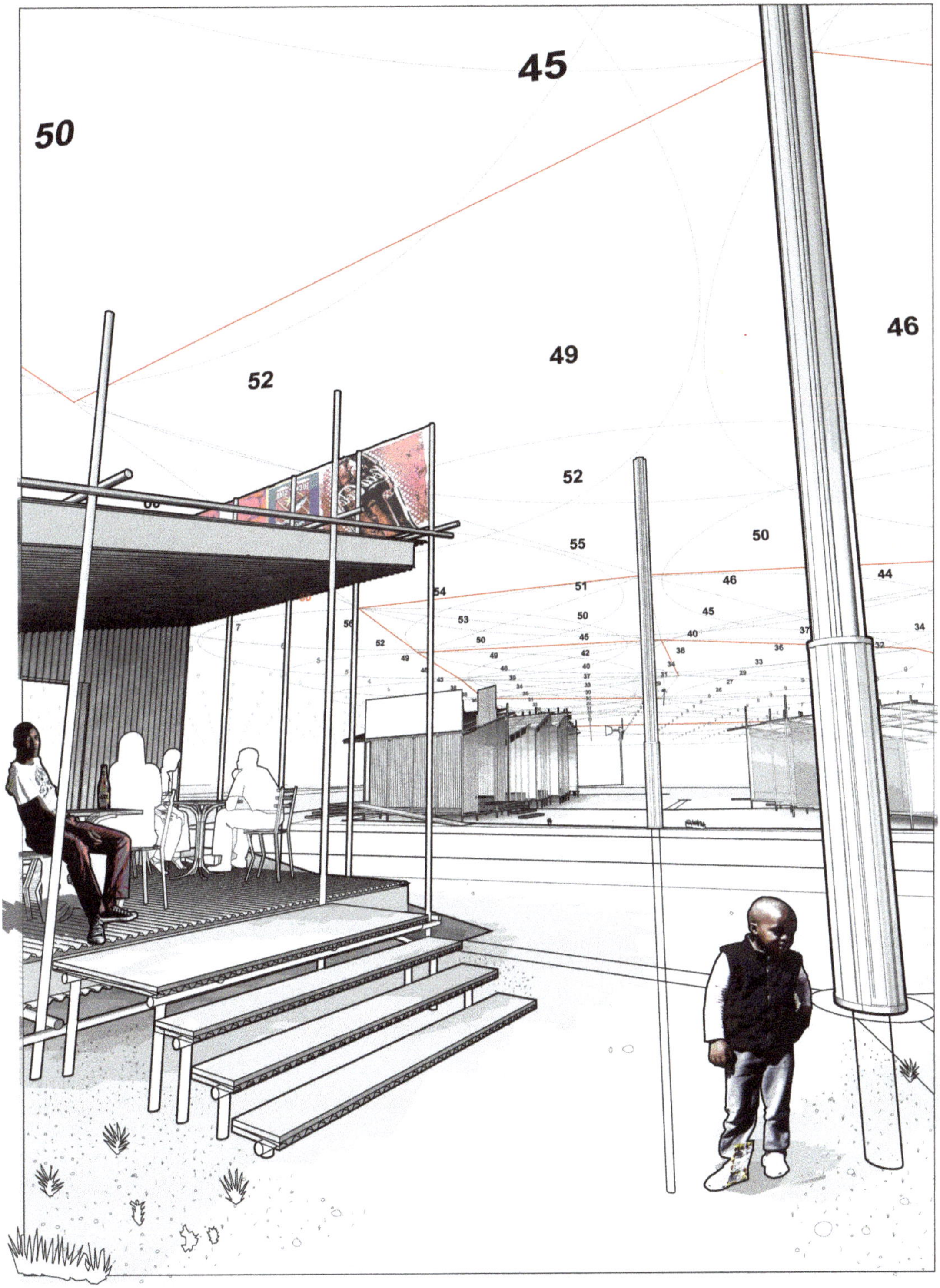

50
45
46
52
49
52
50
55
46
44
51
54
45
53
50
34

URBAN REGENERATOR

JACK THOMPSON

This project confronts the reality of malnutrition in South Africa, a country where about 35% of the population suffers from food poverty. It is designed for Umasizakhe, the apartheid era township located adjacent Graaff Reinet in the Karoo, South Africa. It implements an agricultural system called a Continuous Productive Urban Landscape (CPUl) a concept devised by Andre Viljoen and Katrin Bohn [(01)]. to bring agricultural production into urban settings. In Umasizakhe, crops such as maize, cabbage and carrots are combined with poultry pens and spirulina tanks to help provide a balanced diet for residents. This strategy is combined with a central MicroPublicPlace called the Urban Generator that serves as an educational facility for the community. It helps inform community members about nutrition through a series of classrooms and allows people to trade and share ideas in an energy producing market space.

The urban generator is a part prefabricated, part on site build. A steel exoskeleton frame is transported to site and is infilled through various building components which are recycled, sourced locally or made on site. These include sandbags filled with spoil from the foundations, recycled glass bottles, and a hydroponic facade made from plastic bottles found around Umasizakhe. Concrete 'c' channels are cast on site to form structural joists, locally sourced bricks make up new ground floor extension and corrugated metal panels form the roofs. Solar panels are built into the market to provide energy, and stormwater is channeled to two large underground tanks which feed the facade and external classrooms. The architecture and spaces combine to make a building supported by the government and made by the community, for the community.

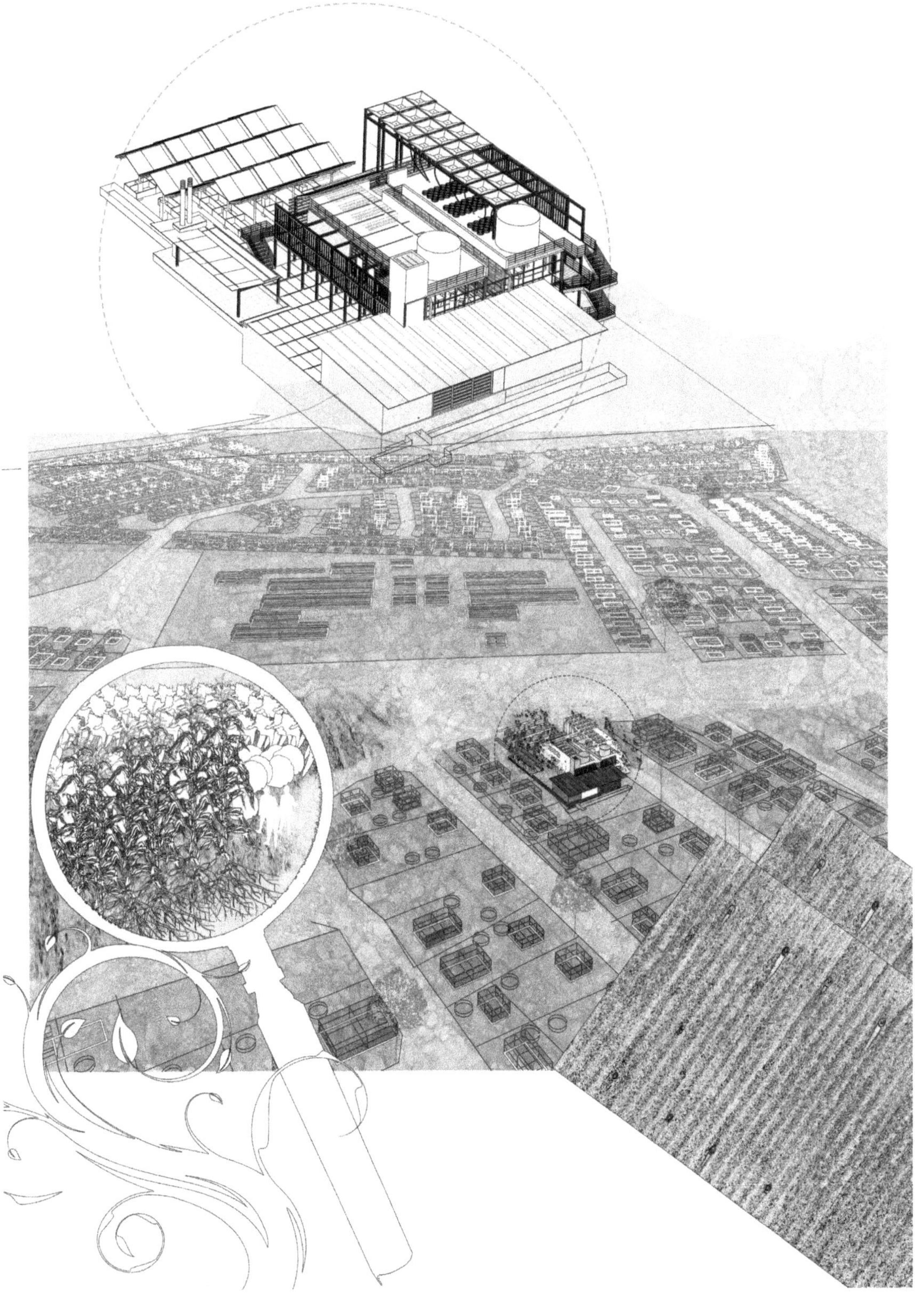

P293 The Urban Generator

The Urban Generator is a MicroPubicPlace at the heart of the Continuous Productive Urban Landcape (CPUL) threading through Umasizakhe. People come here to trade food in a market space and to learn about urban agriculture, water preservation and nutrition in the various classrooms and outdoor spaces tailored to experimenting with food production techniques in the semi arid environment. The centre also provides computers for communal use and spaces to socialise and relax.

P295 Umasizakhe Township

[top left] A local spaza shop

[top right] A small freshly watered vegetable garden

[middle left] A shack sits behind a brick post apartheid Reconstruction and Development (RPD) house

[middle right] The barren ground around RDP houses

[bottom] A view from the east of Umasizakhe looking across the township.

Photographs: Jack Thompson

P296 Urban Generator Plans

The Urban Generator's different programmes overlap to create free flowing vibrant spaces. The market is closely entwined with open community spaces which wrap around the exterior of the building. These spaces are all overlooked by classrooms screened by a hydroponic facade. A larger hall to the rear of the building provides a cooler area for open meeting and seminars. A centrally located kitchen is used all year round for community events. A succulent garden sits between the market, classrooms and community spaces.

P297 Exploded axonometric of the Urban Generator

P298-299 Long section showing [left to right]: market, circulation, classroom and community spaces

P300-301 Members of the community set up for an event at the Urban Generator

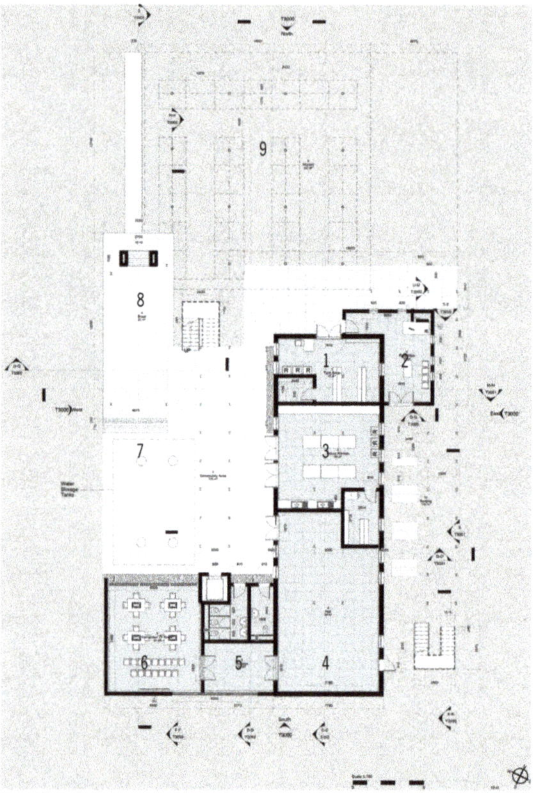

Ground Floor Plan

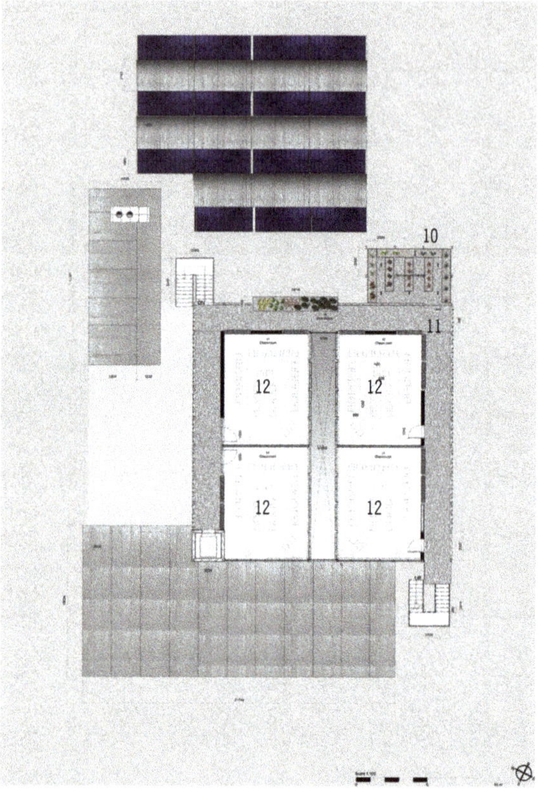

First Floor Plan

1 Tuck Shop
2 Reception
3 Kitchen
4 Hall
5 Storage
6 Computer Room
7 External Community Space
8 Braai Area
9 Market

10 Succulent Garden
11 Circulation
12 Classroom

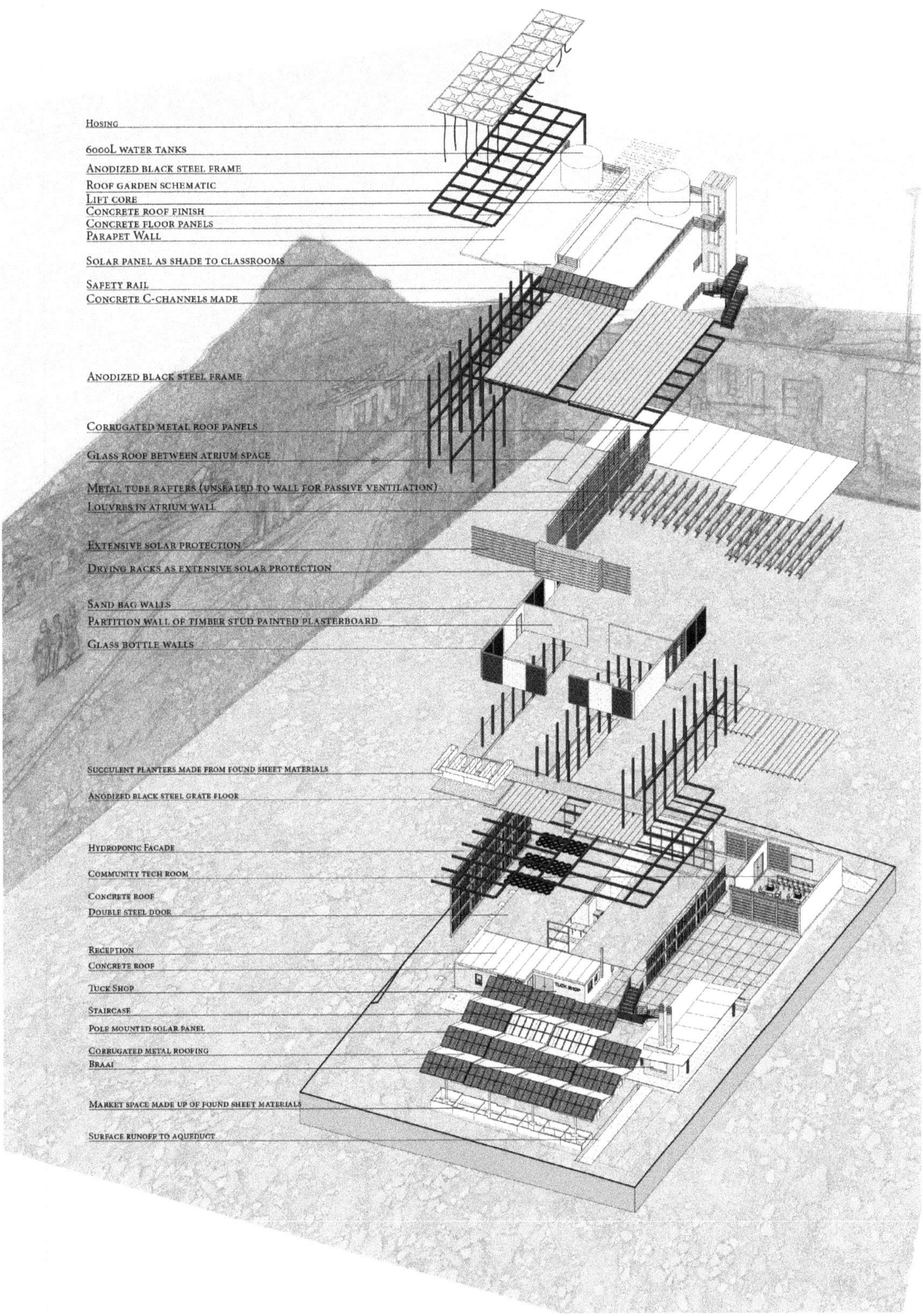
Hosing
6000L water tanks
Anodized black steel frame
Roof garden schematic
Lift core
Concrete roof finish
Concrete floor panels
Parapet Wall
Solar panel as shade to classrooms
Safety rail
Concrete C-channels made
Anodized black steel frame
Corrugated metal roof panels
Glass roof between atrium space
Metal tube rafters (unsealed to wall for passive ventilation)
Louvres in atrium wall
Extensive solar protection
Drying racks as extensive solar protection
Sand bag walls
Partition wall of timber stud painted plasterboard
Glass bottle walls
Succulent planters made from found sheet materials
Anodized black steel grate floor
Hydroponic Facade
Community tech room
Concrete roof
Double steel door
Reception
Concrete roof
Tuck Shop
Staircase
Pole mounted solar panel
Corrugated metal roofing
Braai
Market space made up of found sheet materials
Surface runoff to aqueduct
TUCK SHOP

Energy Market

Site Entrance

Circulation

Roof Garden

Classroom

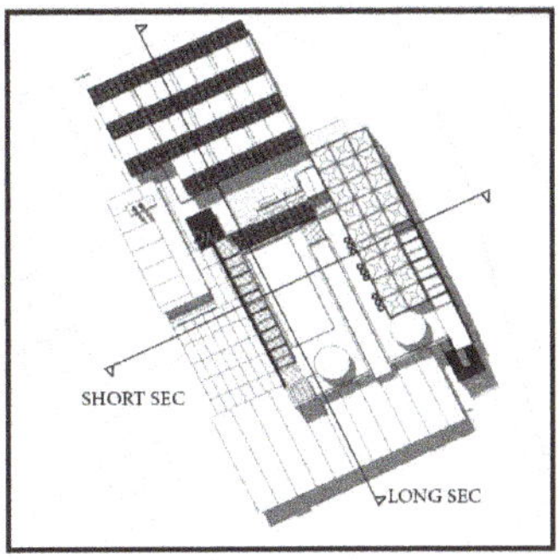

Section Key

APPENDICES

DATA MINING

EDITED BY LAWRENCE CARLOS, IMOGEN WEBB AND JOHN COOK

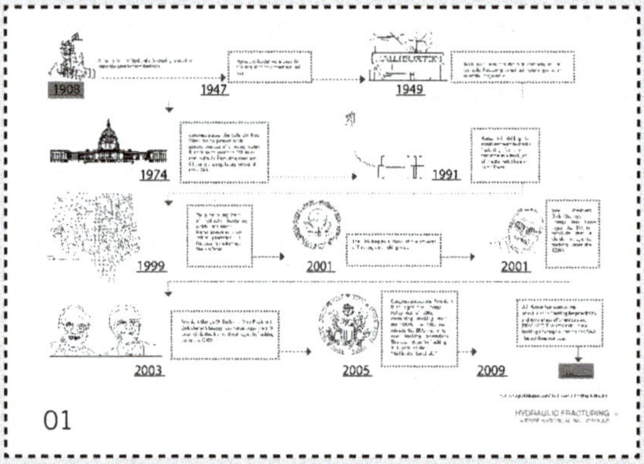

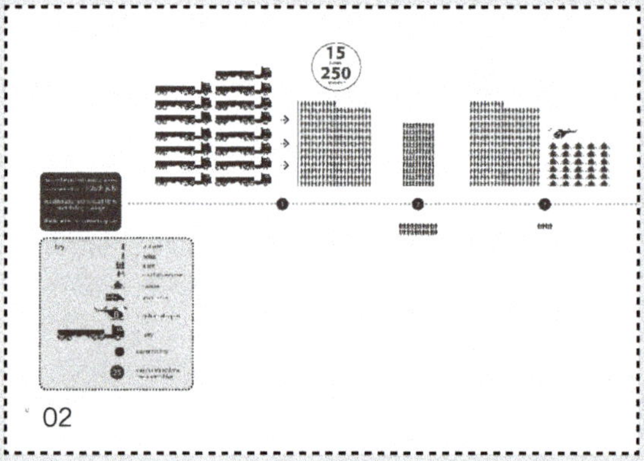

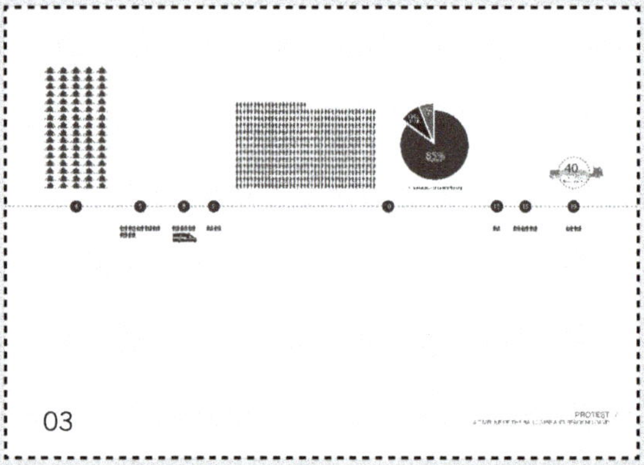

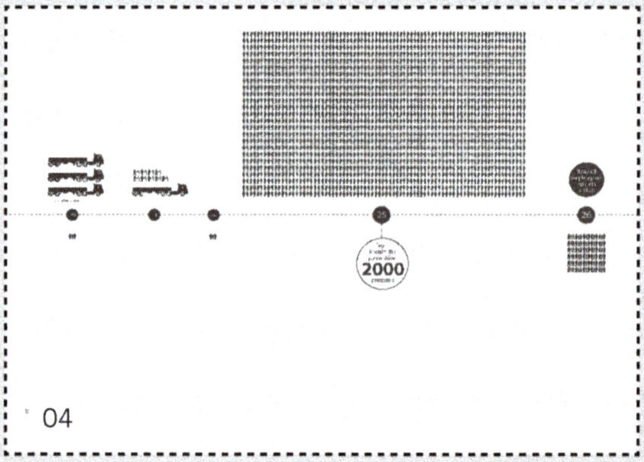

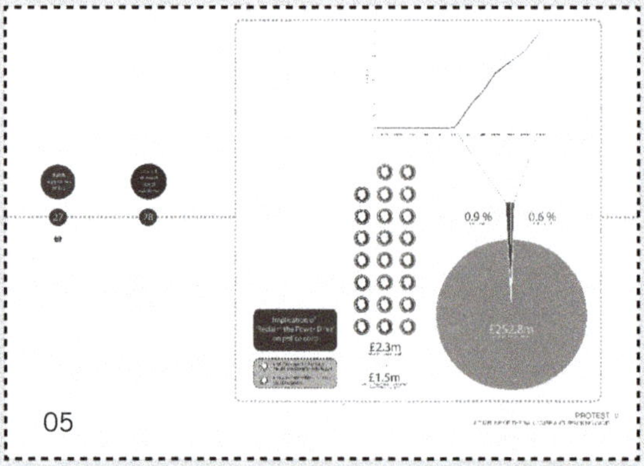

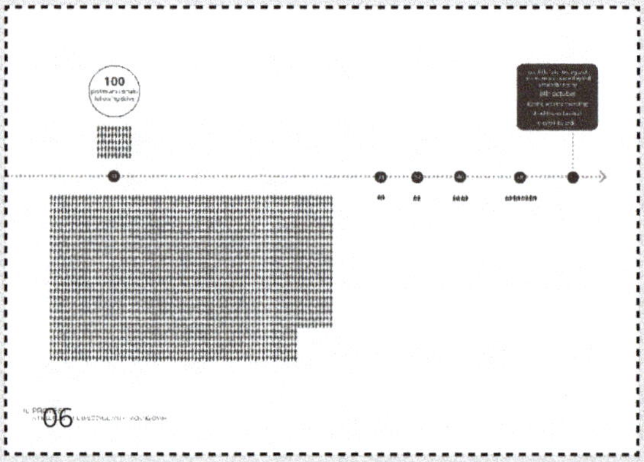

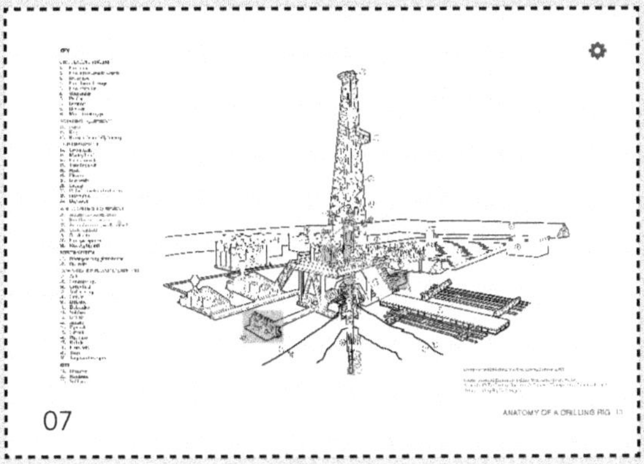

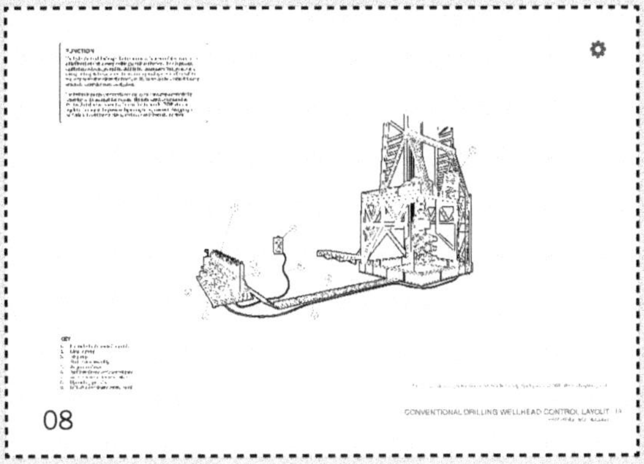

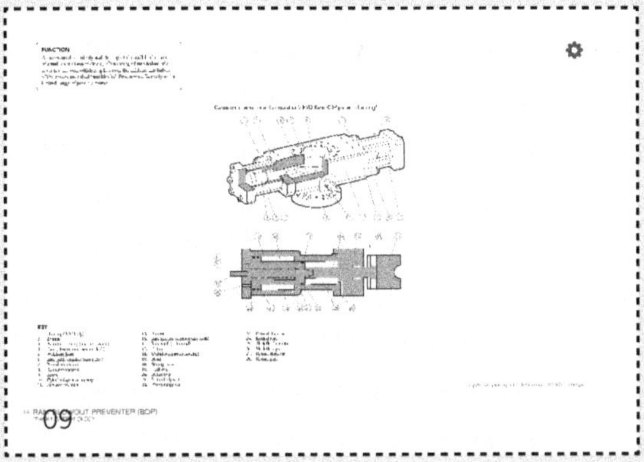

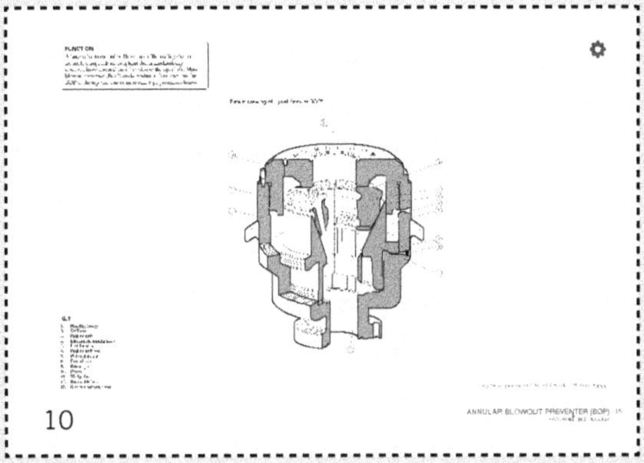

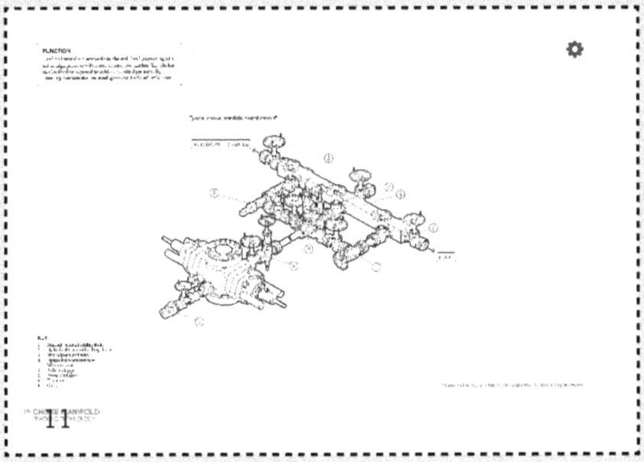

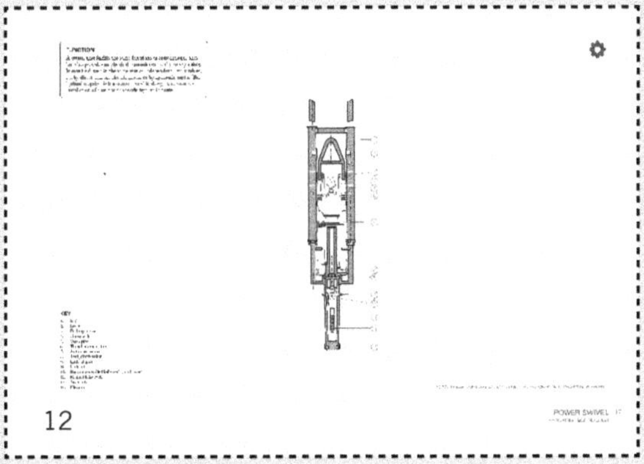

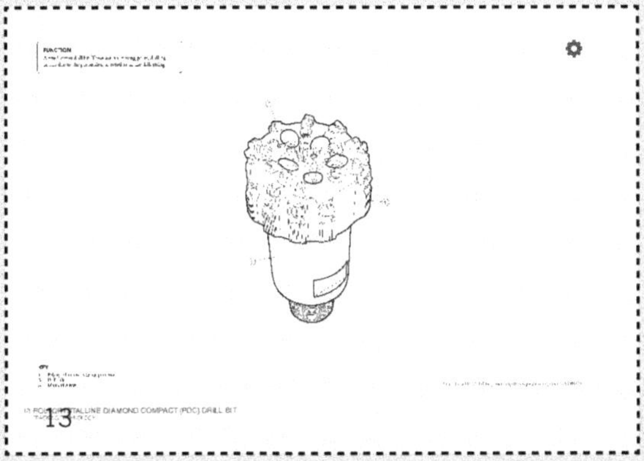

[01] Hydraulic Fracturing: A Historical Background
[02-06] Time-line of a Fracking Protest Balcombe, Sussex, 2013
[07] Anatomy of a Drilling Rig
[08] Conventional Drilling Well-head Control Layout
[09] Ram Blow-out Preventer (BOP)
[10] Annular Blow-out Preventer (BOP)
[11] Choke Manifold
[12] Power Swivel
[13] Polycrystalline Diamond Compact (PDC) Drill-bit

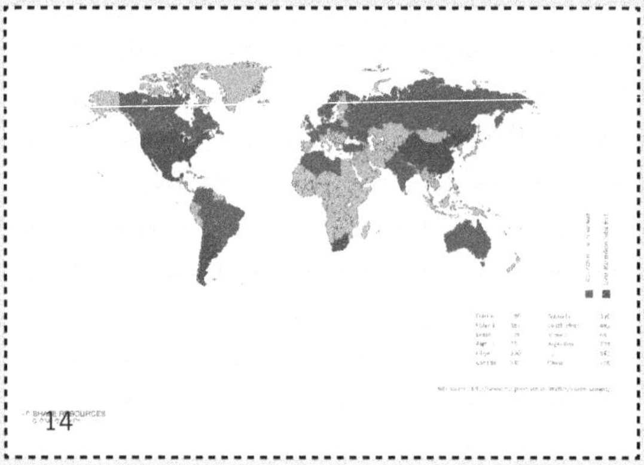
SHALE RESOURCES
14

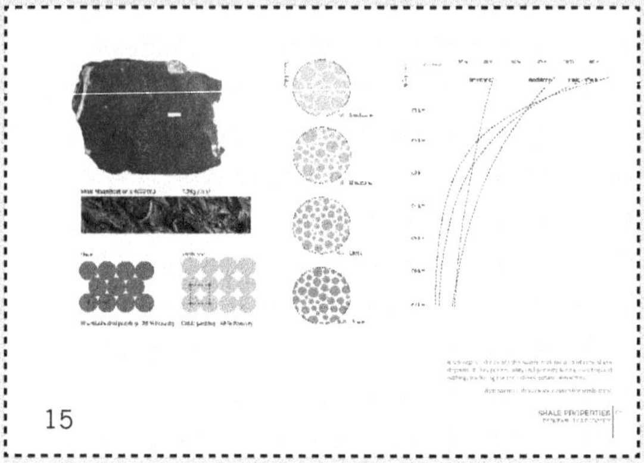
15
SHALE PROPERTIES

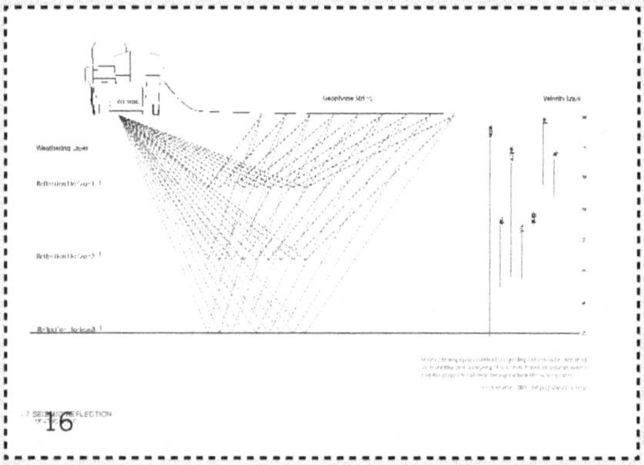
Geophone String
Velocity Layer
Weathering Layer
SEISMIC REFLECTION
16

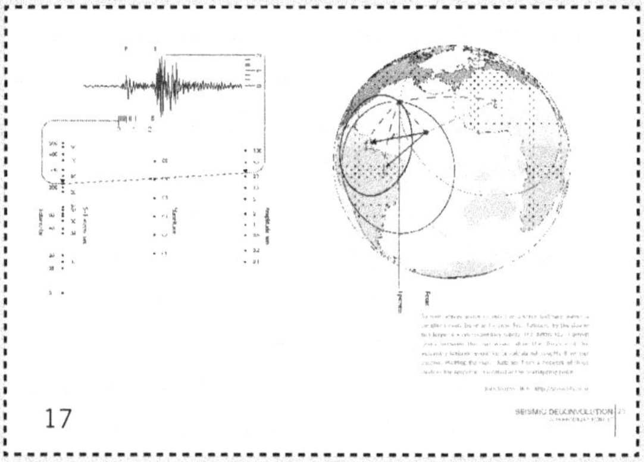
17
SEISMIC DECONVOLUTION

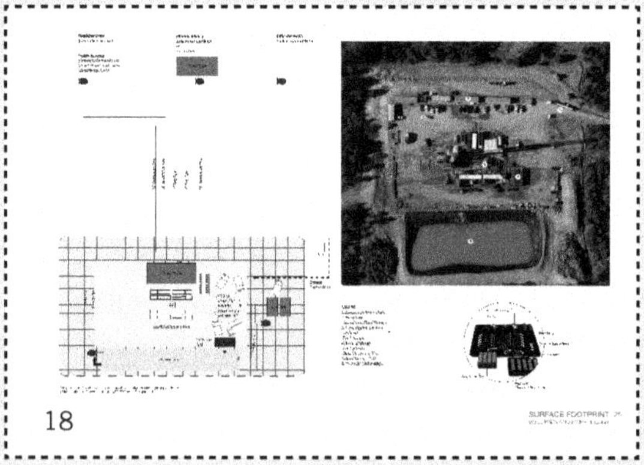
18
SURFACE FOOTPRINT

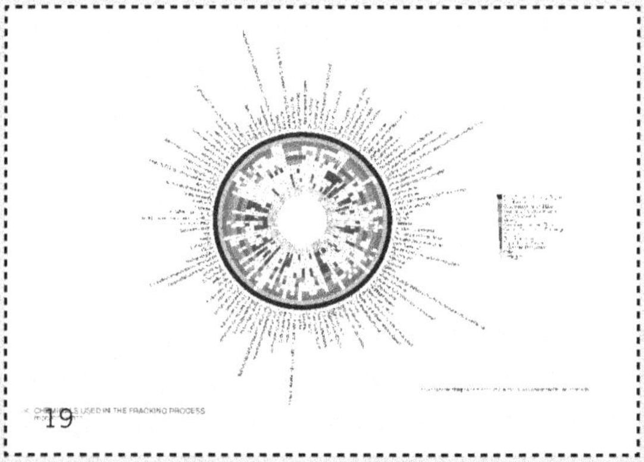
CHEMICALS USED IN THE FRACKING PROCESS
19

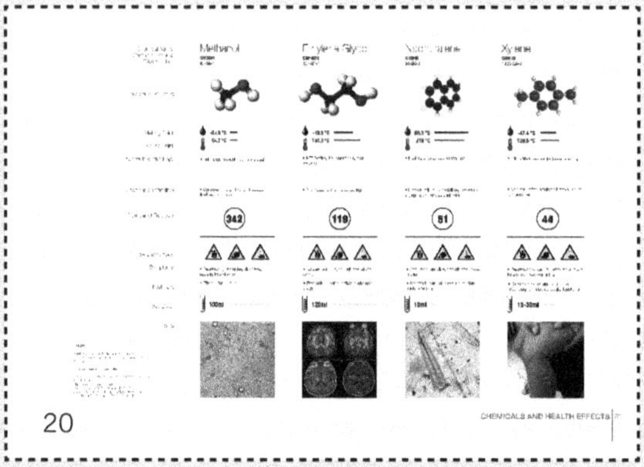
Methanol
Naphthalene
Xylene
342
119
51
48
20
CHEMICALS AND HEALTH EFFECTS

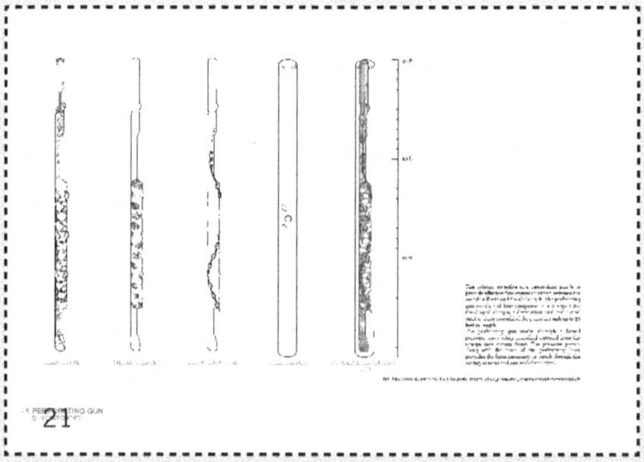
PERFORATING GUN
21

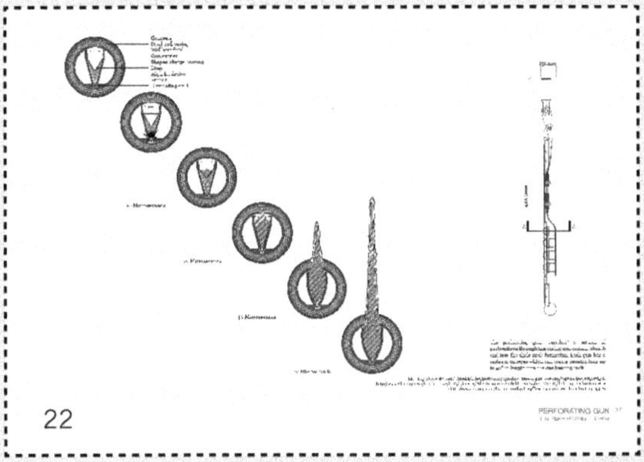

22

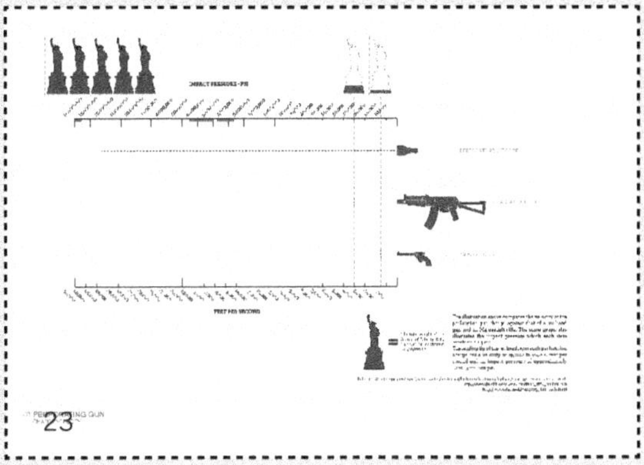

23

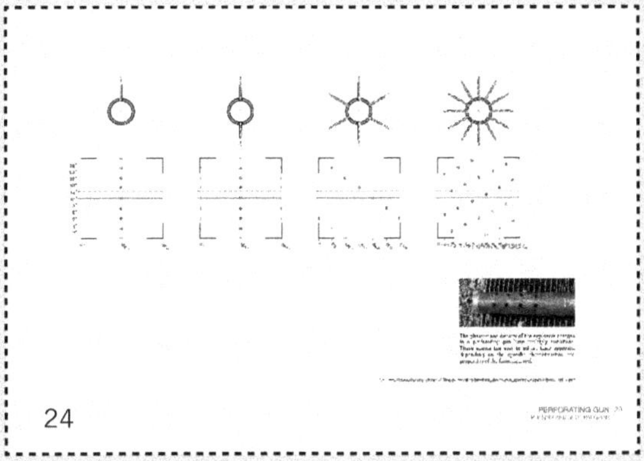

24

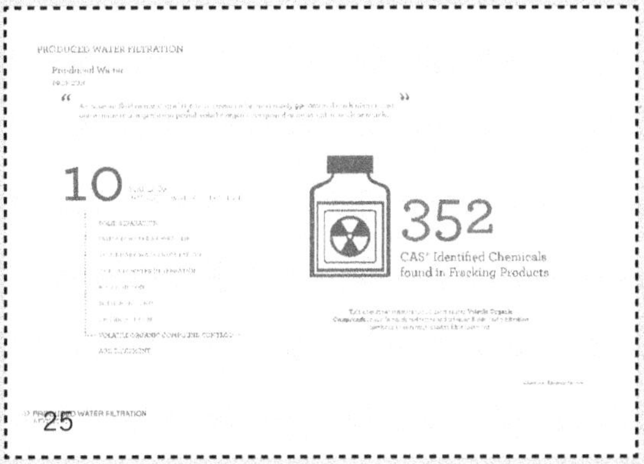

25

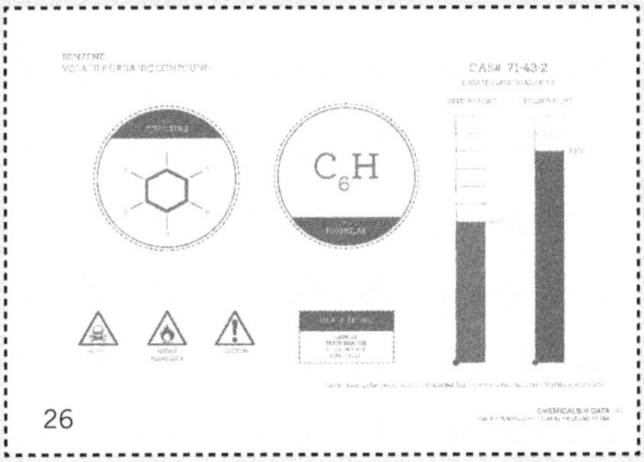

26

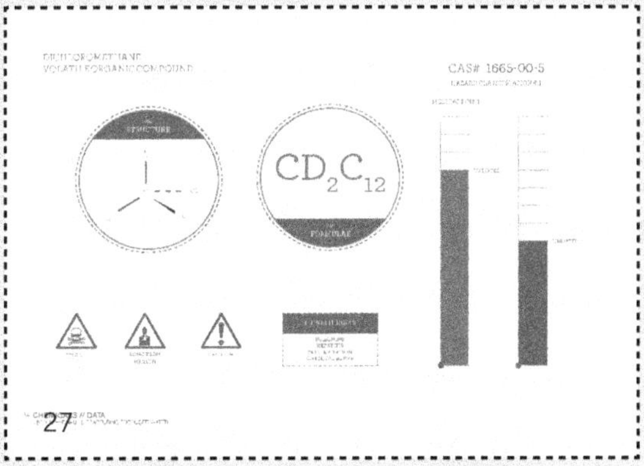

27

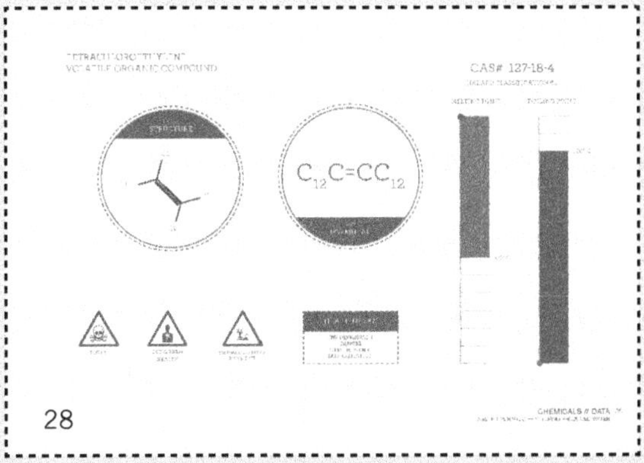

28

[14] Global Shale Resources
[15] Shale Properties
[16] Seismic Reflection
[17] Seismic Deconvolution
[18] Surface Footprint of a fracking well
[19] Chemicals used in the Fracking Process
[20] Chemicals and Health Effects
[21] Perforating Gun
[22] Perforating Charge
[23] Charge Velocity
[24] Phasing and Shot Patterns
[25] Produced Water Filtration
Chemicals used in the Fracking Process:
[26] Benzene
[27] Dichloromethane
[28] Tetrachloroethylene

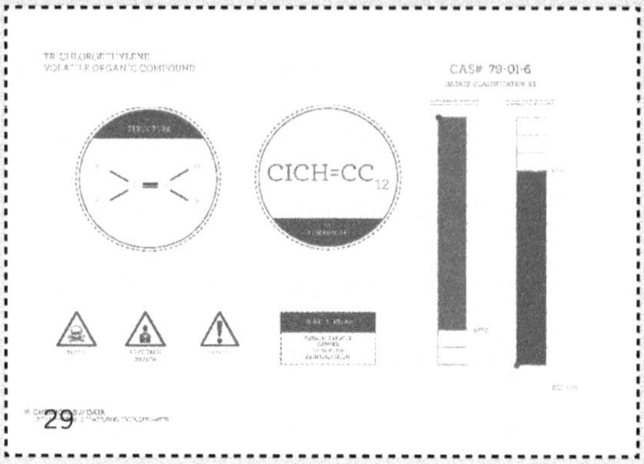
CAS# 79-01-6
CICH=CC12
29

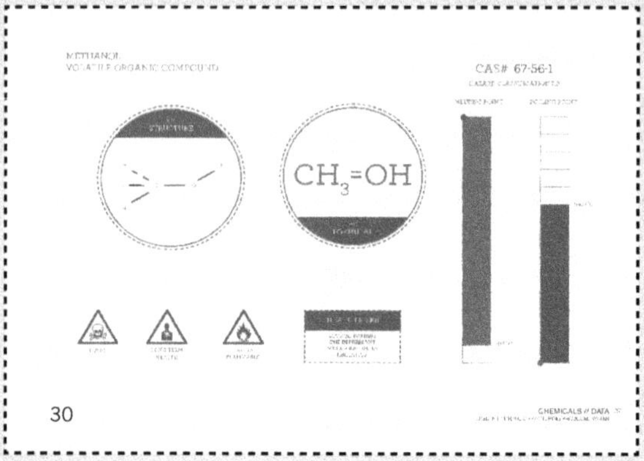
CAS# 67-56-1
CH3=OH
30
CHEMICALS // DATA

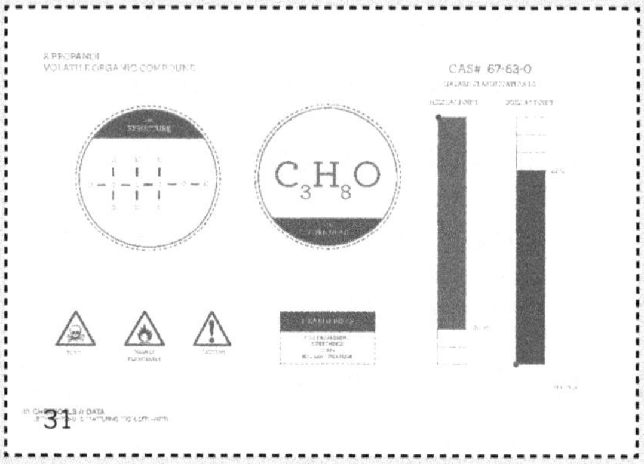
CAS# 67-63-0
C3H8O
31

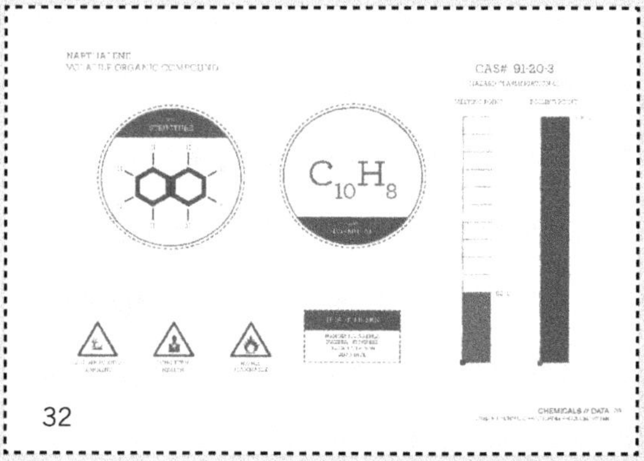
CAS# 91-20-3
C10H8
32
CHEMICALS // DATA

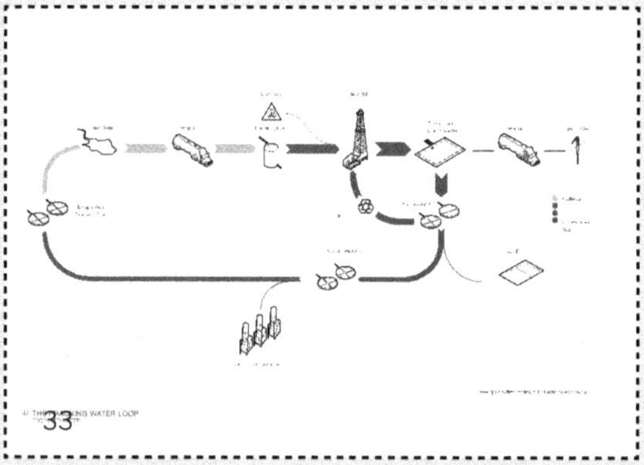
33

34
FRACKING FLUID & PRODUCED WATER

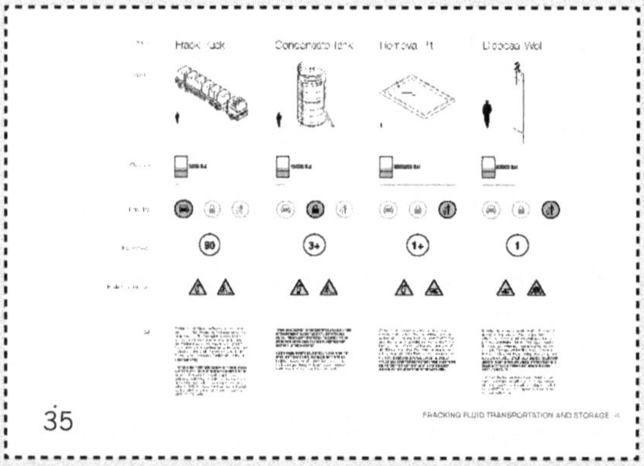
90
3+
1+
1
35
FRACKING FLUID TRANSPORTATION AND STORAGE

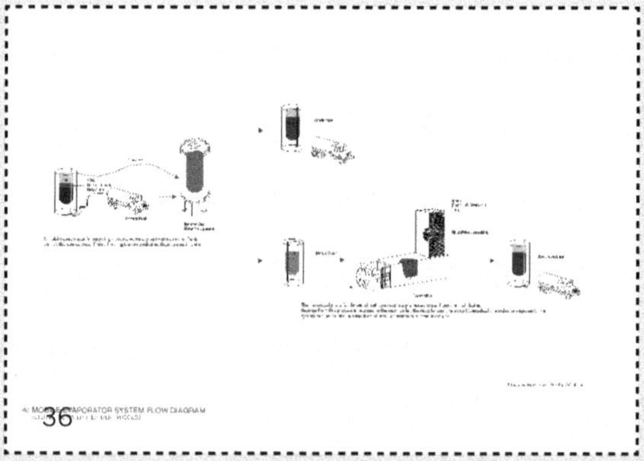
36

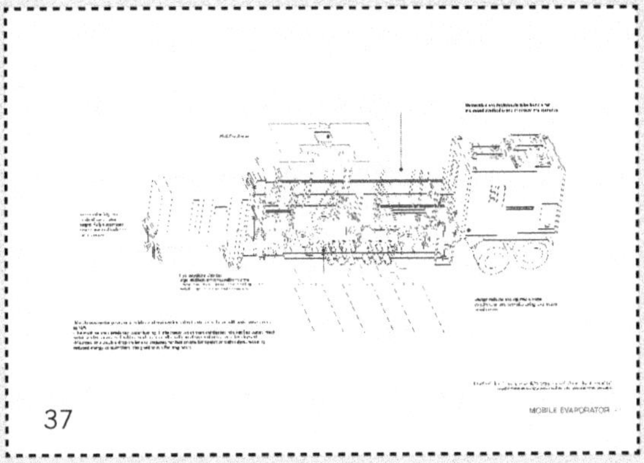

37

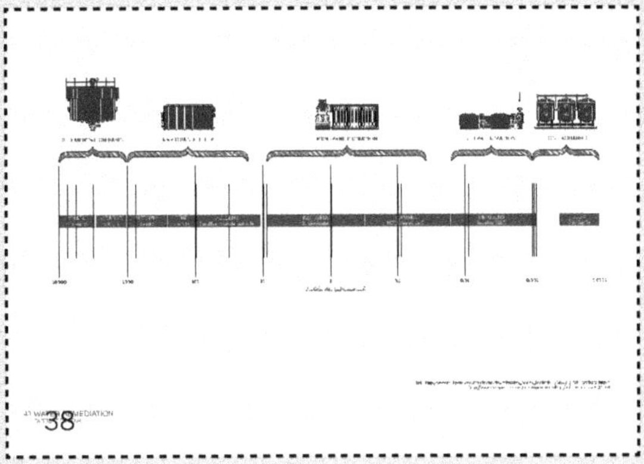

38

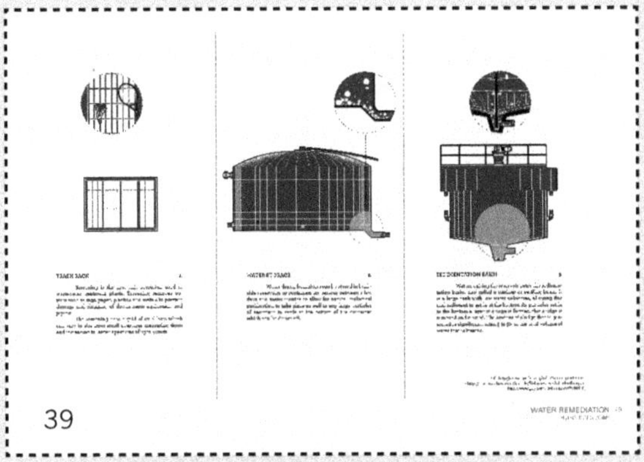

39

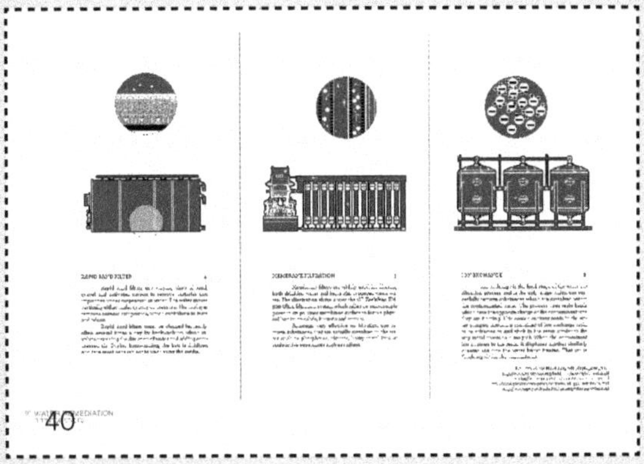

40

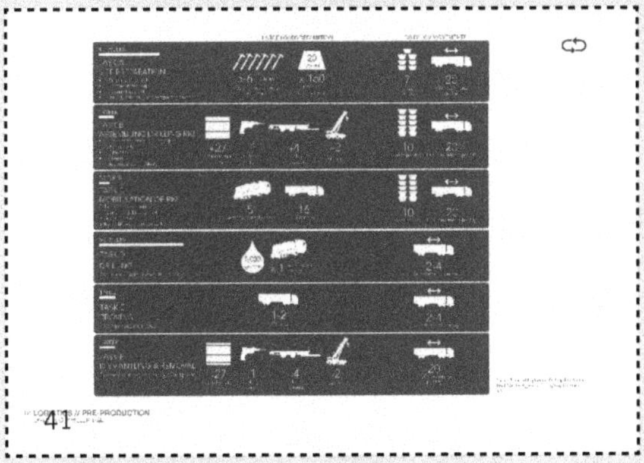

41

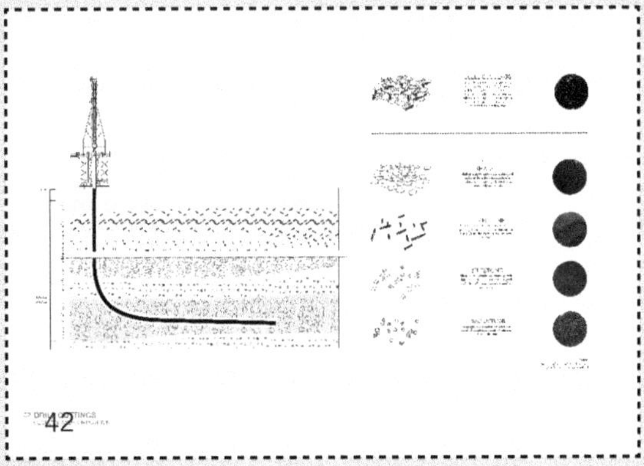

42

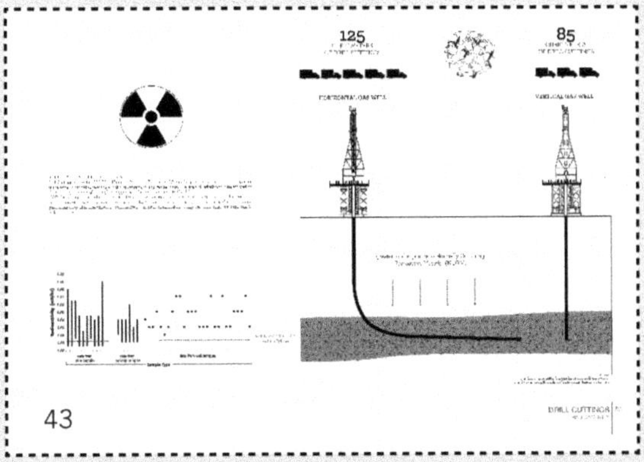

43

[29] Trichloroethylene
[30] Methanol
[31] Propanol
[32] Napthalene
[33] The Fracking Water Loop
[34] Fracking Fluid and Produced Water
[35] Fracking Fluid Transportation and Storage
[36] Mobile Evaporator System Flow Diagram
[37] Mobile Evaporator
[38] Water Remediation - Particle Removal
[39-40] Filtration Stages
[41] Well-head Pre-production
[42] Drill Cuttings - Geology and Composition
[43] Drill Cuttings - Radioactivity

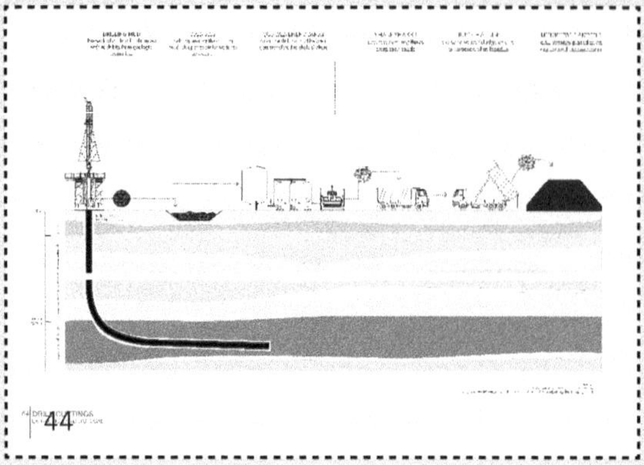
DRILL CUTTINGS
44

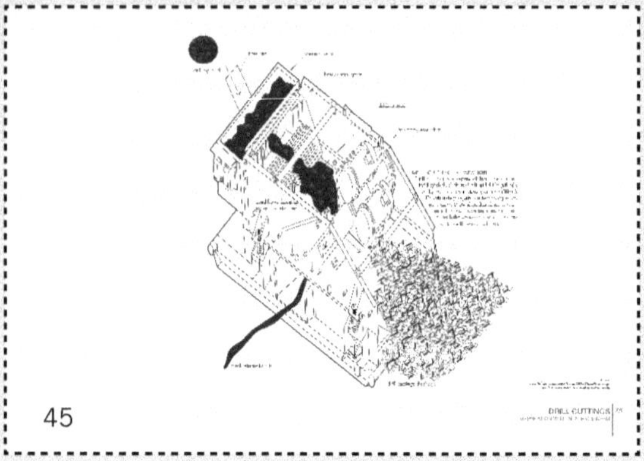
45
DRILL CUTTINGS

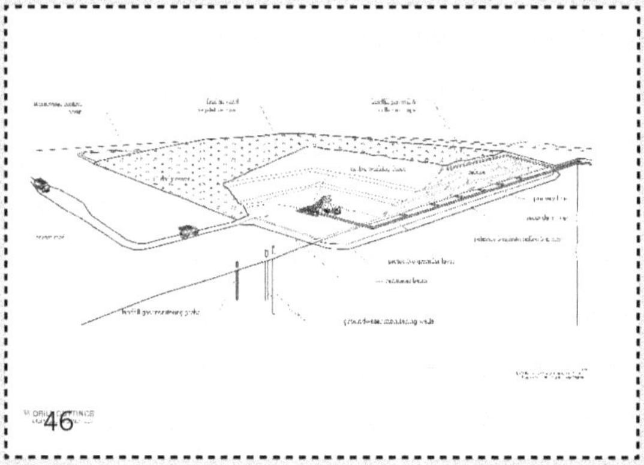
46

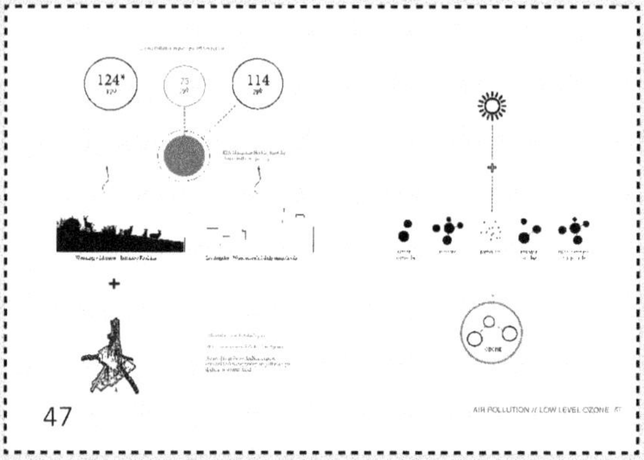
124
114
47
AIR POLLUTION // LOW LEVEL OZONE

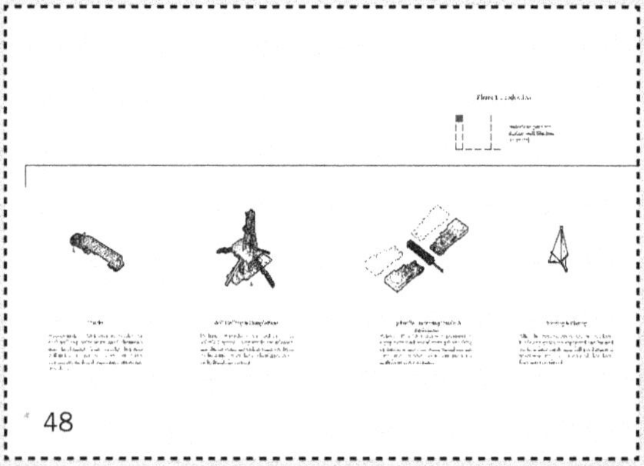
48

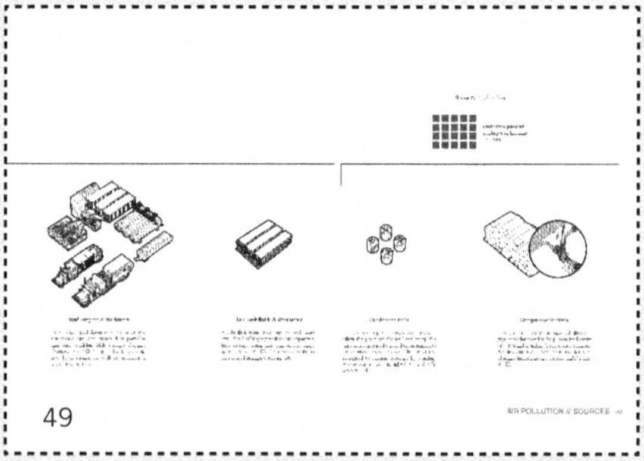
49
AIR POLLUTION // SOURCES

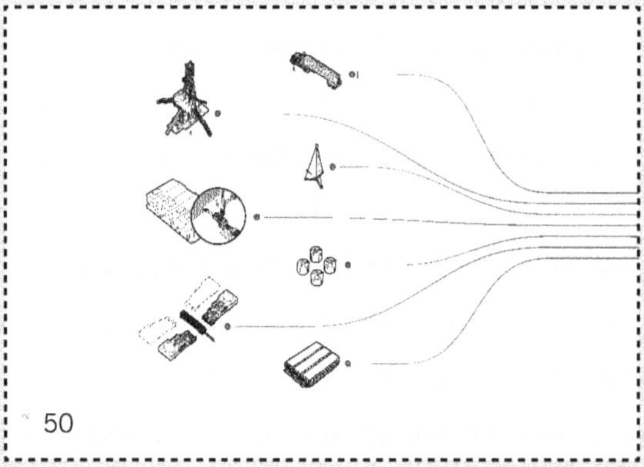
50

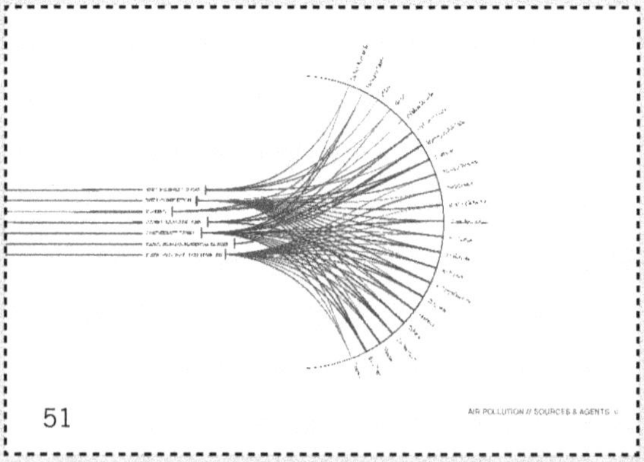
51
AIR POLLUTION // SOURCES & AGENTS

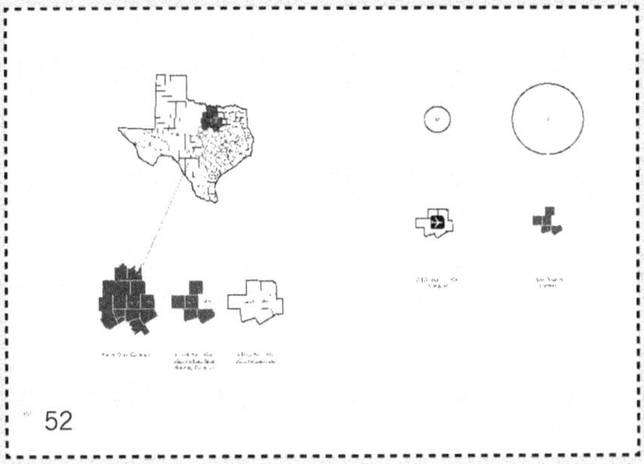

52

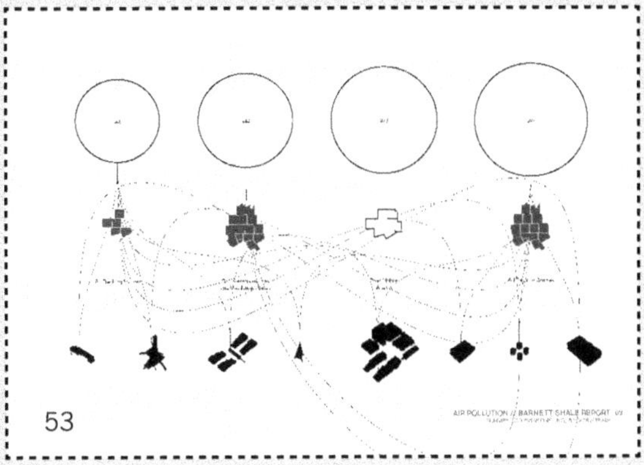

53

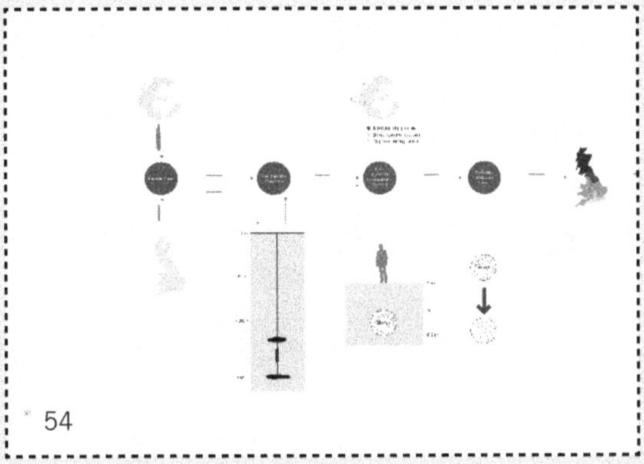

54

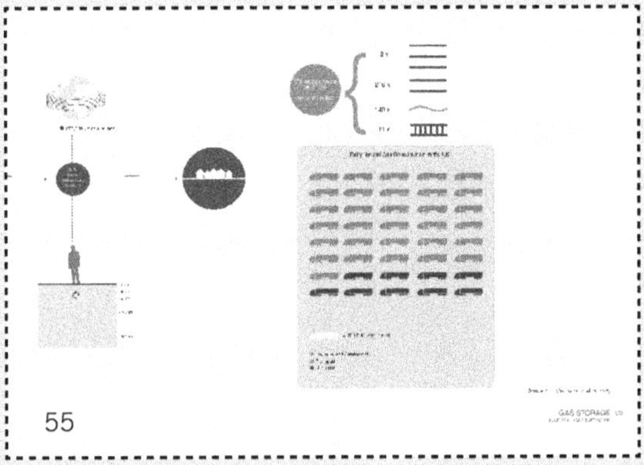

55

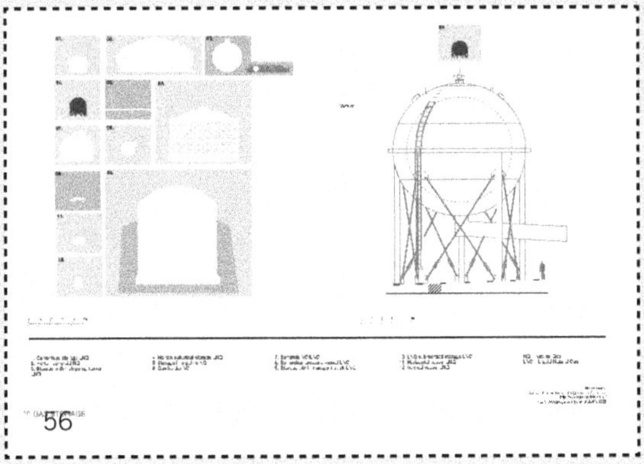

56

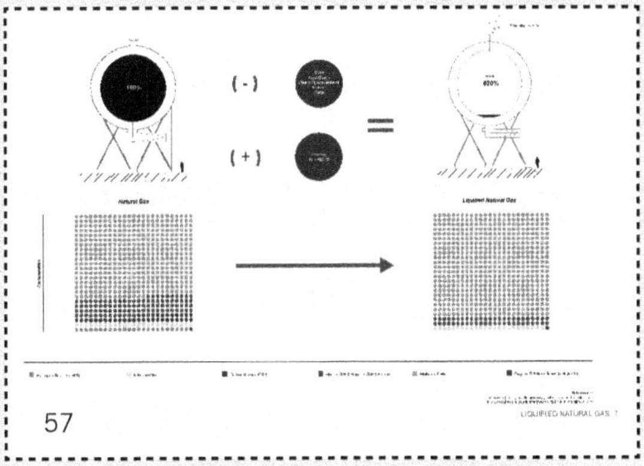

57

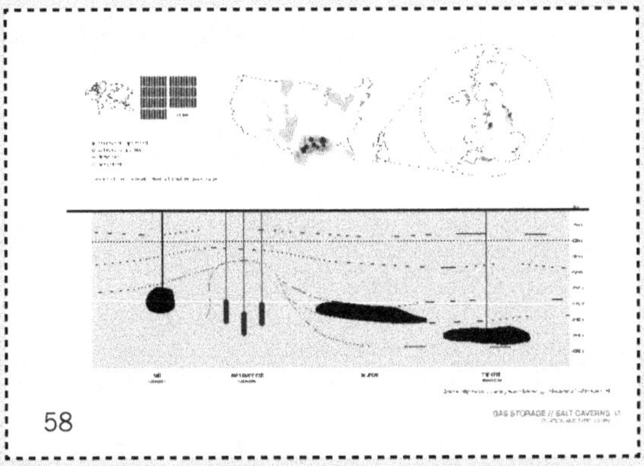

58

[44] Drill Cuttings: Extraction
[45] Drill Cuttings: Separation
[46] Drill Cuttings: Disposal
[47] Air Pollution: Low Level Ozone
[48-49] Air Pollution: Sources
[50-51] Air Pollution: Sources + Agents
[52-53] Air Pollution: Barnett Shale Report
[54-55] Gas Storage: Natural Gas Network
[56-57] Gas Storage: Liquified Natural Gas
[58] Gas Storage: Salt Caverns

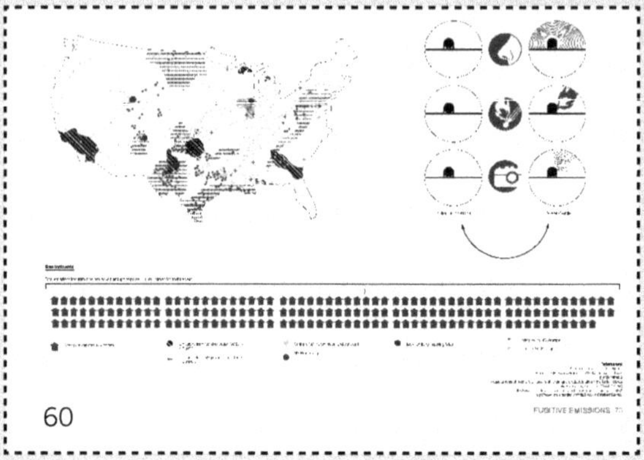

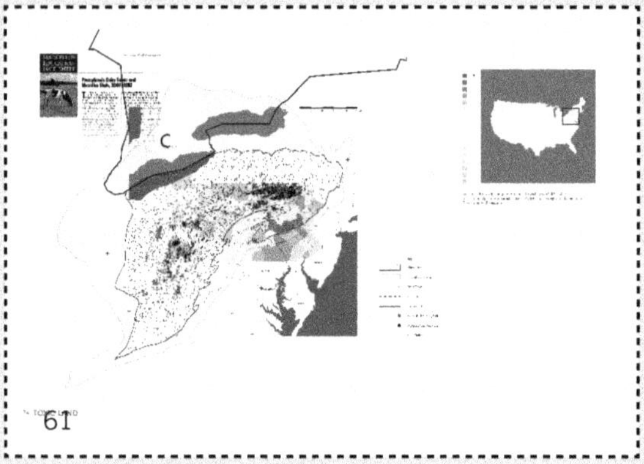

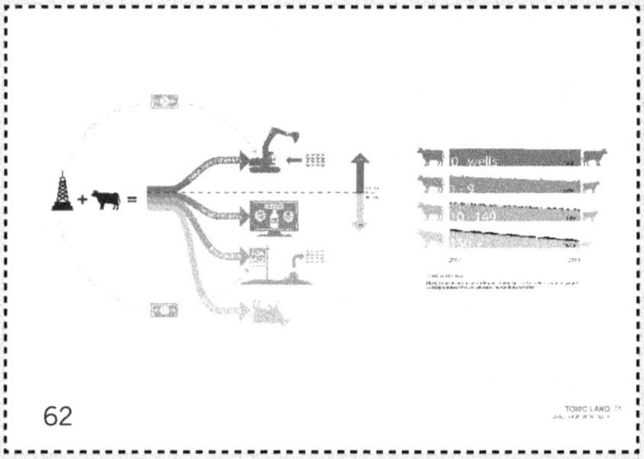

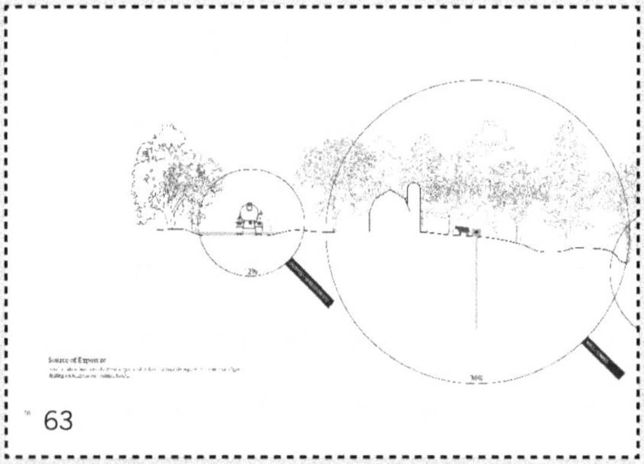

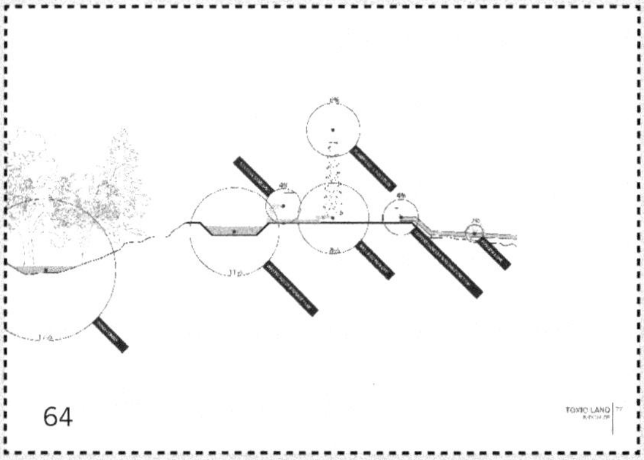

[59] Gas Blasts
[60] Fugitive Emissions
[61] Toxic Land
[62] Toxic Land: Livestock Mortality
[63-64] Toxic Land: Exposure

BIOGRAPHIES

NICK AXEL

Nick Axel is Managing Editor of Volume Magazine and Interactive Specialist at Forensic Architecture. Originally trained as an architect in the United States, he received an MA with distinction from the Centre for Research Architecture at Goldsmiths, University of London. His work spans across the practices of writing, criticism, speculation, research, cartography and design, and has been published widely. Nick has taught at Strelka Institute, Design Academy Eindhoven, Royal Academy of Art Den Haag, Bauhaus-Universität Weimar, and The Bartlett UCL.

LINDSAY BREMNER

Lindsay Bremner is Director of Architectural Research at the University of Westminster where she teaches DS18 with Roberto Bottazzi. She was previously Professor of Architecture in the Tyler School of Art at Temple University in Philadelphia (2006 – 2011) and Chair of Architecture at the University of the Witwatersrand in Johannesburg (1998 – 2004). She is author of Writing the City in Being: Essays on Johannesburg 1998 - 2008 (Fourthwall Books, 2010) and her writings have appeared in Bracket, Cities, Domus, Public Culture, Social Identities, and The Journal of the Indian Ocean Region, amongst others. She was recently awarded a European Research Council grant for a five-year research project titled Monsoon Assemblages that will investigate changing monsoon climates in three south Asian cities - Chennai, Delhi and Dhaka.

ROBERTO BOTTAZZI

Roberto Bottazzi is an architect, researcher, and educator based in London. He runs the RC14 in the Urban Design Programme at the Bartlett, UCL and is a Lecturer at University of Westminster where co-runs DS18. Previously, Roberto was Research Coordinator and Master tutor at the Royal College of Art in London (2005-15). He studied architecture in Italy and Canada before moving to London. His research has been exhibited internationally including: FuturePlace, Porto, 35 Degrees in Chattanooga, USA and FACT, Liverpool, UK. His writings have appeared in Architectural Review, Abitare, DomusWeb, Opere, Critical Cities vol.4 amongst others. He is currently completing a book on the history of computation due to be published in 2017.

DAVID CHANDLER

David Chandler is Professor of International Relations and Director of the Centre for the Study of Democracy, Department of Politics and International Relations, University of Westminster, London. He is the founding editor of the journal Resilience: International Policies, Practices and Discourses. His most recent books are Resilience: The Governance of Complexity (Routledge, 2014) and Freedom vs Necessity in International Relations: Human-Centred Approaches to Security and Development (Zed Books, 2013).

LESLEY GREEN

Lesley Green is an Associate Professor of Anthropology and the director of Environmental Humanities South, a research and graduate teaching initiative at the University of Cape Town. Her current research explores the challenges of decoloniality in tandem with the critique of modernist thought for the sciences and social sciences in Southern Africa, along with questions of race and the making of an environmental public in a time of climate change. She is the editor of Contested Ecologies: Dialogues in the South on Nature and Knowledge (HSRC Press, 2013) and co-author of Knowing the Day, Knowing the World: Engaging Amerindian Thought in Public Archaeology (Arizona University Press, 2013).

TOMAS HOLDERNESS

Tomas Holderness is a Geomatics Research Fellow at the SMART Infrastructure Facility, University of Wollongong, where his research focuses on the use of geospatial analysis, Earth observation, and network modeling techniques applied to urban infrastructure resilience and Earth systems engineering. He is director of the SMART OSGeo Lab, and co-director and co-principal investigator of PetaJakarta.org, an international research collaboration studying infrastructure resilience through the novel collection of social media information and urban data. Before joining the SMART Infrastructure Facility, Dr. Holderness worked as a spatial modeller for the Geospatial Engineering Research Group at Newcastle University, where he was responsible for the development of an open source integrated modelling environment for urban systems research. His PhD thesis analyzed long time series thermal Earth observation data to quantify intra-urban spatio-temporal temperature dynamics in Greater London, UK.

KIEL MOE

Kiel Moe is a registered practicing architect and Associate Professor of Architecture & Energy in the Department of Architecture at Harvard University Graduate School of Design where he is the co-director of the MDES design research program and the Energy, Environments, & Design research lab. He is author of Insulating Modernism: Isolated and Nonisolated Thermodynamics in Architecture (Birkhauser, 2014) Convergence: An Architectural Agenda for Energy (Routledge, 2013) and Thermally Active Surfaces (Princeton Architectural Press, 2010) amongst other books.

ETIENNE TURPIN

Etienne Turpin is a philosopher researching, designing, curating, and writing about complex urban systems, the political economies of data and infrastructure, visual culture and aesthetics, and Southeast Asian colonial-scientific history. In Jakarta, Indonesia, Etienne is the director of anexact office and the co-director and co-principal investigator of PetaJakarta.org. He is the editor of Architecture in the Anthropocene (Open Humanities Press, 2013) and co-editor of Art in the Anthropocene (Open Humanities Press, 2015), Fantasies of the Library (K. Verlag, 2015), Land & Animal & Nonanimal (K. Verlag, 2015), and Jakarta: Architecture + Adaptation (Universitas Indonesia Press, 2013).

STUDENTS

2013/2014

1ST YEARS
Andrew Baker-Falkner
Jared Baron
Sam Cady
John Cook
Jimi Deji-Tijani
Dylan Main
Shiue Nee Pang
Michael O'Hanlon

2ND YEARS
Rashad Al-Karooni
Hoki Au
Agustina Briano
Lawrence Carlos
Claire Holton
Philip Hurrell
Alex Jaggs
William Liu
Nzinga B. Mboup
Rupert Rathbone
Aishah Suhaimi
Alexander Watt
Imogen Webb

2014/2015

1ST YEARS
Jessica Hillam
Oscar McDonald
Ben Pollock
Alice Thompson
Iulia Stefan

2ND YEARS
Andy Baker-Falkner
Jared Baron
Rupert Calvert
Cheryl Choo
John Cook
Sophie Fuller
Martino Gasparrini
Niall Green
Matthew Hedges
Natasha Khambhaita
Shiue Nee Pang
Michael O'Hanlon
Anna-Maria Papasotiriou
Emma Swarbrick
Jack Thompson

2015/2016

Work by the following students appears in this book:

1ST YEARS
Calvin Sin

2ND YEARS
Jessica Hillam
Ben Pollock
Iulia Stefan
Alice Thompson

ACKNOWLEDGEMENTS

TEACHING ASSISTANTS

Christos Antonopoulos
Jeg Dudley

GUEST CRITICS

2013/2014
Nabil Ahmed, Laura Allen, Nick Axel, David Dernie, Tom Fox, Jon Goodbun, Kate Heron, Janike Kampevold Larsen, Adrian Lahoud, Constance Lau, Lilit Mnatsakanyan, Douglas Murphy, John Palmesino, Ann-Sofi Ronnskog, Rosa Schiano-Phan, Fransesco Sebregondi, Ronald Wall, Liam Young.

2014/2015
Christos Antanopoulos, Jennifer Beningfield, Harry Charrington, Richard Difford, Jeg Dudley, Neil Dusheiko, Kostas Grigoriadis, Jon Goodbun, Constance Lau, Lorenzo Pezzani, Dimitar Pouchnikov, Etienne Turpin, Maria Veltcheva, Filip Visnjic, Alex Watt

SPONSORS

2013 - 2015
Student Licenses for Realflow by Next Limit Technologies

2014
Field Trip: The Reinet Foundation

OTHER

2013
Field Trip: Alan Holiday

2014
Field Trip: Jan Glazewski, Stefan Kramer, Derek Light
Guest Lectures: Jennifer Beningfield, Ian Fraser, John Sadar

P320-323 Frames from stop-frame animation of construction of DS18 Open Exhibition Model, June 2014
Animation: Michael O'Hanlon

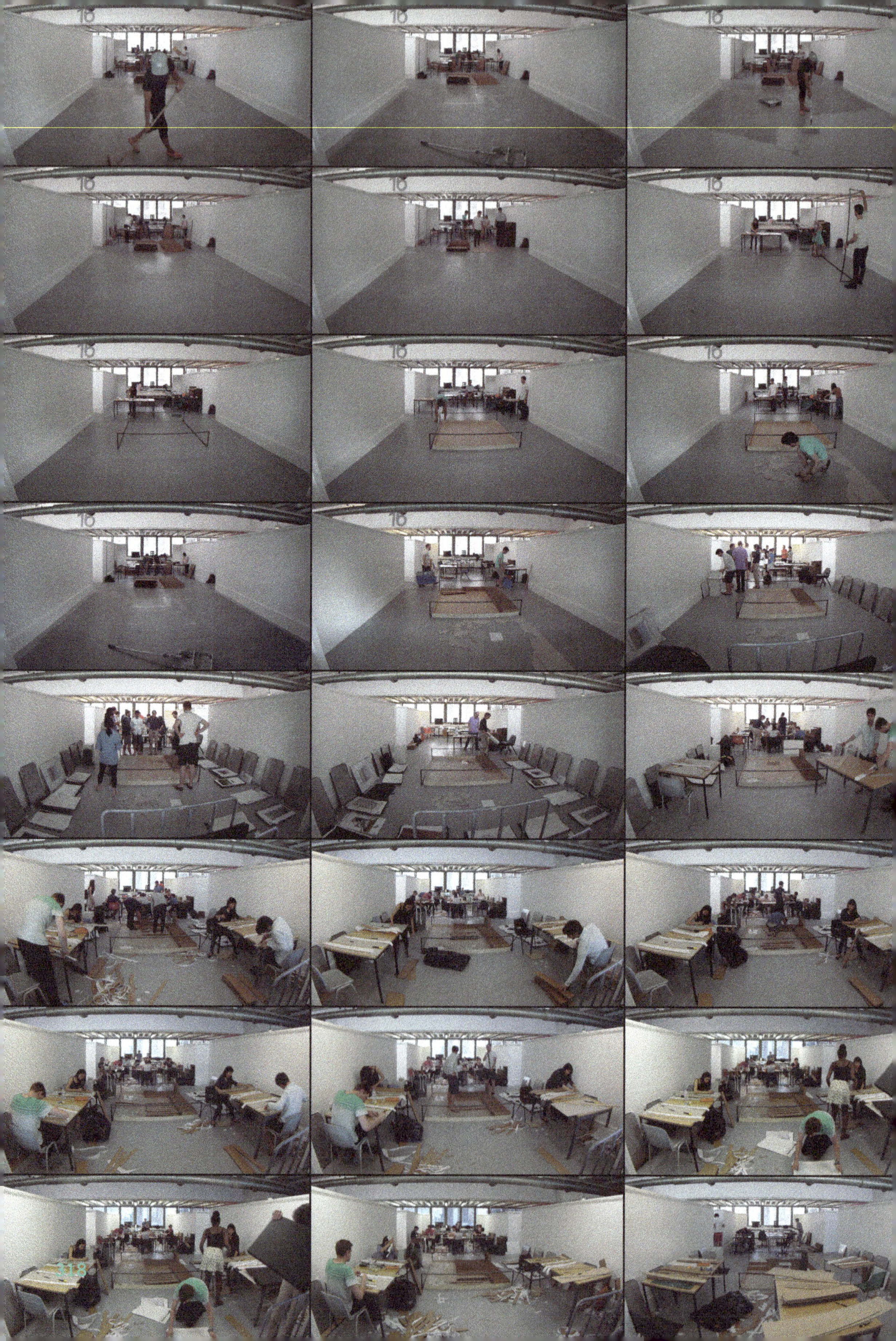

18

Architecture, Energy, Matter

DS18: 2013-2015

Edited by Lindsay Bremner and Roberto Bottazzi
Assisted by John Cook and Andrew Baker-Falkner

A University of Westminster
Department of Architecture Publication
Designed by Mark Boyce
Printed by Lightning Source

Cover image: Michael O'Hanlon

ISBN 978-0993398612

The Studio as Book series are available to purchase through www.studioasbook.org and other online stores.

The editors have attempted to acknowledge all sources of images used and apologise for any errors or omissions.

Department of Architecture
University of Westminster
35 Marylebone Road
London
NW1 5LS

www.ingramcontent.com/pod-product-compliance
Lightning Source LLC
Chambersburg PA
CBHW041240240426
43668CB00023B/2444
* 9 7 8 0 9 9 3 3 9 8 6 1 2 *